The Drift of Sea Ice

Matti Lepparanta

The Drift of Sea Ice

Springer

Published in association with
Praxis Publishing
Chichester, UK

Professor Matti Leppäranta
Division of Geophysics
University of Helsinki
Helsinki
Finland

SPRINGER–PRAXIS BOOKS IN GEOPHYSICAL SCIENCES
SUBJECT *ADVISORY EDITOR*: Dr Philippe Blondel, C.Geol., F.G.S., Ph.D., M.Sc., Senior Scientist, Department of Physics, University of Bath, Bath, UK

ISBN 3-540-40881-9 Springer-Verlag Berlin Heidelberg New York

Springer is part of Springer-Science + Business Media (springeronline.com)

Bibliographic information published by Die Deutsche Bibliothek

Die Deutsche Bibliothek lists this publication in the Deutsche Nationalbibliografie; detailed bibliographic data are available from the Internet at http://dnb.ddb.de

Library of Congress Control Number: 2004110445

Apart from any fair dealing for the purposes of research or private study, or criticism or review, as permitted under the Copyright, Designs and Patents Act 1988, this publication may only be reproduced, stored or transmitted, in any form or by any means, with the prior permission in writing of the publishers, or in the case of reprographic reproduction in accordance with the terms of licences issued by the Copyright Licensing Agency. Enquiries concerning reproduction outside those terms should be sent to the publishers.

© Praxis Publishing Ltd, Chichester, UK, 2005
Printed in Germany

The use of general descriptive names, registered names, trademarks, etc. in this publication does not imply, even in the absence of a specific statement, that such names are exempt from the relevant protective laws and regulations and therefore free for general use.

Cover design: Jim Wilkie
Project Management: Originator Publishing Services, Gt Yarmouth, Norfolk, UK

Printed on acid-free paper

Contents

Preface ix

List of figures xiii

List of tables xvii

List of symbols xix

List of abbreviations xxiii

1   Introduction 1

2   Drift ice material 9
    2.1   Sea ice cover 9
        2.1.1   Sea ice landscape 10
        2.1.2   Sea ice zones 15
        2.1.3   Sea ice charting 17
    2.2   Ice floes to drift ice particles 19
        2.2.1   Scales 19
        2.2.2   Size and shape of ice floes 22
    2.3   Thickness of drift ice 27
        2.3.1   Basic characteristics 27
        2.3.2   Measurement methods 30
        2.3.3   Ice thickness distribution 34
    2.4   Sea ice ridges 35
        2.4.1   Structure of ridges 35
        2.4.2   Statistical distributions of ridge size and occurrence 38

		2.4.3	Ridging measures	40
		2.4.4	Hummocked ice	41
		2.4.5	Total thickness of deformed ice	42
	2.5	Drift ice state		42

3 Ice kinematics — 45

	3.1	Description of ice velocity field		45
		3.1.1	Motion of a single floe	46
		3.1.2	Continuum deformation	47
	3.2	Observations		52
		3.2.1	Methods	52
		3.2.2	Characteristics of observed sea ice drift	55
		3.2.3	Strain rate and vorticity	59
		3.2.4	Deformation structures	65
	3.3	Stochastic modelling		67
		3.3.1	Two-dimensional motion using complex variables	67
		3.3.2	Mean sea ice drift field in the Arctic Ocean	69
		3.3.3	Diffusion	71
		3.3.4	Random walk	71
	3.4	Ice conservation law		72
		3.4.1	Ice states based on ice categories	73
		3.4.2	Ice thickness distribution	75

4 Sea ice rheology — 81

	4.1	General		81
		4.1.1	Rheological models	83
		4.1.2	Internal stress of drift ice	84
		4.1.3	Internal friction	87
	4.2	Viscous laws		88
		4.2.1	Linear viscous models	88
		4.2.2	Non-linear viscous models	90
	4.3	Plastic laws		91
		4.3.1	Plastic drift ice	91
		4.3.2	Mohr–Coulomb rheology	94
		4.3.3	AIDJEX elastic–plastic rheology	95
		4.3.4	Hibler's viscous–plastic rheology	97
		4.3.5	Scaling of ice strength	100
	4.4	Granular flow models		103

5 Equation of drift ice motion — 109

	5.1	Derivation of the equation of motion		109
		5.1.1	Fundamental equation	109
		5.1.2	Vertical integration	112
		5.1.3	Drift regimes	116
		5.1.4	Conservation of kinetic energy, divergence, and vorticity	116

	5.2	Atmospheric and oceanic drag forces.	118
		5.2.1 Planetary boundary layers.	118
		5.2.2 Drag force formulae	125
	5.3	Scale analysis and dimension analysis	130
		5.3.1 Magnitudes	130
		5.3.2 Dimensionless form.	133
		5.3.3 Basin scales.	137
	5.4	Dynamics of a single ice floe	138
6	**Free drift**	141
	6.1	Steady-state solution	141
		6.1.1 Classical case.	143
		6.1.2 One-dimensional channel flow	146
		6.1.3 Shallow waters.	148
		6.1.4 Linear model.	148
	6.2	Non-steady-state solution	149
		6.2.1 One-dimensional flow with quadratic surface stresses	149
		6.2.2 Linear model.	150
		6.2.3 Drift of a single floe	152
	6.3	Linear, coupled ice–ocean model	154
		6.3.1 General solution.	155
		6.3.2 Inertial oscillations	156
		6.3.3 Periodic forcing	158
		6.3.4 Free drift velocity spectrum	159
	6.4	Spatial aspects of free drift	161
		6.4.1 Advection	163
		6.4.2 Divergence and vorticity.	163
7	**Drift in the presence of internal friction**	165
	7.1	The role of internal friction.	165
		7.1.1 Examples.	166
		7.1.2 Frequency spectrum.	170
		7.1.3 Landfast ice.	171
	7.2	Channel flow	173
		7.2.1 Creep.	174
		7.2.2 Plastic flow	175
	7.3	Zonal sea ice drift.	177
		7.3.1 Steady-state velocity: wind-driven case.	178
		7.3.2 Steady state with ocean currents	182
		7.3.3 Steady-state ice thickness and compactness profiles	182
		7.3.4 Viscous models.	184
		7.3.5 Marginal ice zone.	185
		7.3.6 Circular ice drift.	187
	7.4	Modelling of ice tank experiments	191
		7.4.1 Drift ice dynamics in a tank	191
		7.4.2 Case study.	194

viii   **Contents**

8   Numerical modelling. 199
    8.1   Numerical solution 199
        8.1.1   System of equations 199
        8.1.2   Numerical technology 202
        8.1.3   Calibration/Validation 205
    8.2   Examples of sea ice dynamics models 206
        8.2.1   The Campbell and Doronin models 206
        8.2.2   Hibler model 208
        8.2.3   The AIDJEX model 211
        8.2.4   The Baltic Sea model. 212
    8.3   Short-term modelling applications. 214
        8.3.1   Research work 214
        8.3.2   Sea ice forecasting. 217
    8.4   Long-term modelling applications 222
        8.4.1   Arctic regions. 222
        8.4.2   Antarctic regions 224
        8.4.3   Baltic Sea 224

9   Use and need for knowledge on ice drift 229
    9.1   Science 229
    9.2   Practice. 232
    9.3   Final comments 236

10   Study problems 239
    Chapter 1 239
    Chapter 2 240
    Chapter 3 240
    Chapter 4 241
    Chapter 5 241
    Chapter 6 242
    Chapter 7 243
    Chapter 8 243

11   References 245

Index 261

Preface

The drift of sea ice, forced by winds and ocean currents, is an essential element in the dynamics of the polar oceans, those essential and fragile components of the world's climate system. Because its structure is almost two-dimensional, the sea ice drift problem is fascinating for theoretical research in geophysical fluid dynamics. Apart from two-dimensionality, this medium is complicated by its granularity, compressibility, and non-linearity.

The drift transports sea ice over long distances, particularly to regions where ice is not formed by thermodynamic processes. The ice transports latent heat and freshwater, and the influence of ice melting on the salinity of the mixed layer is equivalent to a considerable amount of precipitation. The ice cover forms a particular air–sea interface, which is modified drastically by differential ice motion; consequently, sea ice dynamics is a key factor in air–sea interaction processes in the polar oceans. In particular, the opening and closing of leads (open water channels) have a major impact on air–sea heat exchange. Thus the drift of sea ice plays a very important role in high-latitude weather, polar oceanography, and global climate. This is of renewed interest owing to the increasingly growing concern about man-made global warming.

The waters in the region at the ice margin are known for their high biological productivity because of favourable light and hydrographical conditions; therefore, the location of ice margins—to some degree determined by the drift of ice—is of deep concern in marine ecology.

The drift of ice is also a major environmental factor. Pollutants accumulate in the ice sheet, originating from the water body, sea bottom, and atmospheric fallout, and they are transported within the ice over long distances. A particular pollutant question is oil spills. The drift and dispersion of oil in ice-covered seas has become a very important issue; it is a difficult problem, oil being partly transported with ice and partly dragged by ice. There are three major oil exploration areas in the seasonal sea ice zone: offshore Alaska, Barents Sea, and offshore Sakhalin Island in the Sea of Okhotsk.

Sea ice has always been a barrier to winter shipping. For the purpose of sea ice monitoring and forecasting, ice information services have been in operation for about 100 years. The drift of ice shifts the ice edge, opens and closes leads, and forms pressure zones, which are all key points in navigation through an ice-covered sea. Also hummocks and sea ice ridges result in local accumulation of ice volume and strength, causing major problems to marine operations and constructions. Because of ice dynamics, the ice situation may change rapidly; therefore, ice information services work in all ice-covered seas on a daily basis, updating their ice charts and producing ice forecasts. Expansion of the Northern Sea Routes requires ongoing development of sea ice-mapping and forecasting services, where sea ice dynamics plays a key role.

This book presents the science of sea ice drift through its 100-year history to the present state of knowledge. Chapter 1 gives a brief historical overview and presents the sea ice drift problem and the subject matter of the book. The material includes geophysical theory, observations from field programs, and mathematical models. Chapter 2 describes sea ice material and how it needs to be approached from the perspective of the research and modelling of ice drift. Chapter 3 presents methodologies, data analysis, and the outcome of sea ice kinematics measurements and techniques to construct the ice conservation law, a fundamental law in the dynamics of sea ice. The equation of motion is presented and analysed in Chapters 4 and 5, including the rheology problem, derivation of two-dimensional equations, and magnitude estimation and scaling. In Chapters 6–8, solutions to the ice drift problem are presented, from simple analytical models to full numerical models. The free drift case results in an algebraic equation, while analytical solutions are easily obtained for channel and zonal flows, critically important tools to interpret the results of more complex models. Chapter 9 briefly discusses some applications of the knowledge of sea ice drift. The end matter contains a collection of study problems, the references, and the index.

The underlying idea behind the book has been to include the whole story of sea ice dynamics in a single volume, from material state through the laws of dynamics to mathematical models. There is a crying need for a synthesis of these research endeavours, as sea ice dynamics applications have been increasing and, apart from review papers, no comprehensive book exists on this topic in English.

The author has contributed to research into sea ice dynamics since 1974. In particular, his topics of interest have been (in chronological order): short-term sea ice forecasting, drift stations in the Baltic Sea, MIZEX (Marginal Ice Zone Experiment), sea ice ridging, sea ice remote sensing, seasonal modelling of Baltic Sea ice conditions, scaling problems, and finally sea ice in coastal zones. This book has grown from the author's lectures on sea ice drift, initially at the University of Helsinki, Finland and then at UNIS (Universitetsstudiene i Svalbard), Longyearbyen, Norway, and from other visits to various universities, in particular the Cold Regions Research and Engineering Laboratory, Hanover, New Hampshire, USA and Hokkaido University, Mombetsu and Sapporo, Japan.

In the progress of his research the author has learned about sea ice drift from a large number of colleagues. Especially, he wants to thank Professors Erkki Palosuo and William D. Hibler III, as well as Professors Jan Askne, Robert V. Goldstein, Sylvain Joffre, Zygmunt Kowalik, Sveinung Løset, Anders Omstedt, Sergey N. Ovsienko, Kaj Riska, Hayley H. Shen, Kunio Shirasawa, Peter Wadhams, and Wu Huiding. He is also deeply thankful to his students in sea ice dynamics, in particular Professor Zhang Zhanhai, Dr Jari Haapala, Dr Paula Kankaanpää, and Mr Keguang Wang. His home institutes, the Finnish Institute of Marine Research (1974–1992) and the Department of Geophysics (1992–; renamed in 2001 as the Division of Geophysics/Department of Physical Sciences) of the University of Helsinki, are deeply thanked for good working conditions and support. The scientists and technicians who have participated in the ice dynamics research programmes from these institutes and from several other organizations are gratefully acknowledged. Mr Keguang Wang is also thanked for a careful scientific review of the manuscript. Also, he and Ms Pirkko Numminen provided assistance with some of the graphics. Finally, the author wants to thank the Praxis team—Clive Horwood, Philippe Blondel, and Neil Shuttlewood—for professional guidance and help in getting this book completed, in particular to Dr Blondel for his very careful and excellent review of the whole book.

Matti Leppäranta
Helsinki, 18 June 2004

Figures

1.1	A section of the marine chart by Olaus Magnus Gothus (1539) showing the northern Baltic Sea	2
1.2	A drift ice landscape from the northern Greenland Sea, June 1983	5
1.3	Deployment of a corner reflector mast for deformation studies by a laser geodimeter, in the Baltic Sea in April 1978	6
2.1	The world sea's ice zones in summer and winter	12
2.2	Aerial photograph from a Finnish Air Force ice reconnaissance flight in the central Baltic, winter 1942	13
2.3	Pancake ice	13
2.4	Classification of ice concentration	14
2.5	Sea ice concentration in the Antarctic based on passive microwave SSM/I data	17
2.6	Ice chart over the Eurasian side of the Arctic Ocean, 12 March 2003	18
2.7	An optical channel *NOAA-6* image over the Barents Sea and Arctic Basin north of it, 6 May 1985	20
2.8	Open and dense locked packings of uniform, circular ice floes	21
2.9	Floe size distributions from the Arctic Ocean	23
2.10	The forms of the power law, exponential, and lognormal distributions for the floe size, scaled with a median of 500 m	24
2.11	Thin ice sheets undergo rafting in compression	29
2.12	Mean sea ice draft in the Arctic Ocean based on submarine data	31
2.13	A section of the ice draft profile in the Greenland Sea from submarine sounding	32
2.14	AEM calibration for ice thickness in the Baltic Sea, March 1993	33
2.15	Observed ice thickness distributions	35
2.16	An ice ridge shown in a field photograph (Baltic Sea, March 1988) and in a schematic cross-sectional diagram	36
2.17	The evolution of the cross-sectional profile of one ridge near Hailuoto Island, Baltic Sea, winter 1991	37
2.18	The Gulf of Bothnia in winter	38
2.19	Ridge sail height vs. ridge density in different seas	41

xiv Figures

3.1	Motion of a single ice floe	47
3.2	Strain modes and rotation, and their measures	48
3.3	Deployment of an *Argos* buoy during the Ymer-80 expedition, July 1980, north of Svalbard	53
3.4	Sea ice displacement in the Gulf of Riga, Baltic Sea, seen in consecutive optical moderate resolution imaging spectroradiometer (MODIS) images of NASA's *Terra* satellite	56–7
3.5	Mombetsu ice radar antenna on top of Mt Oyama (300 m) monitoring the sea ice of the Sea of Okhotsk	58
3.6	Sea ice distribution (ice is white) along the northern coast of Hokkaido on 3 February 2003	59
3.7	Paths of drifting stations in the Arctic Ocean	60
3.8	Sea ice buoy drifts in the Indian Ocean–eastern Pacific Ocean sector of the Southern Ocean	61
3.9	Sea ice and wind velocity time series, Baltic Sea, March 1977	61
3.10	Drift ice velocity spectra	62
3.11	Orientation of a drifter array (5 km) baseline and R/V *Polarbjørn* moored to an ice floe in the centre of the array	63
3.12	Principal strain rates during the MIZEX-83 experiment	64
3.13	Time series of 2-hourly, principal strain rates in the coastal drift ice zone, Bay of Bothnia, in March 1997	66
3.14	Principal axis strain field from consecutive *Landsat* images in the Baltic Sea	67
3.15	Drift ice in the Sea of Okhotsk shown in Yukara-ori weaving handicraft	68
3.16	Sea ice structures in the Bay of Bothnia, Baltic Sea, based on satellite remote-sensing imagery	69
3.17	Annual mean of sea ice velocity in the Arctic Ocean	70
3.18	The mean lifetime of an ice floe in the Arctic Basin based on a random walk model	72
3.19	Opening and ridging as a function of the mode of deformation	79
4.1	Stress σ on a material element	82
4.2	Basic rheology models (in one dimension) for stress σ as a function of strain ε and the strain rate	83
4.3	Schematic presentation of change in the quality of sea ice rheology as a function of ice compactness A and thickness h	86
4.4	Plastic yield curves for drift ice	93
4.5	Field experiment into the strength of sea ice ridges in the Baltic Sea, winter 1987	95
4.6	Field experiment data are used for high-quality tuning of sea ice mechanical phenomenology and for tuning rheological parameters	98
4.7	Ice pressure has captured a ship in the Baltic Sea	100
4.8	Structure of a hierarchical system	101
4.9	Sanderson's curve: the strength of sea ice vs. the loading area	102
4.10	Floe–floe interaction	104
4.11	Velocity fluctuation level according to measurements and Monte Carlo simulations based on the measured strain rate	105
4.12	Energy budget in sea ice ridging as simulated by a discrete particle model	107
5.1	Sea ice drifts past a lighthouse, northern Baltic Sea	111
5.2	The ice drift problem	111

5.3	A typical diagram of major forces in drifting sea ice (northern hemisphere)	115
5.4	A sea ice field Ω with the boundary curve Γ consisting of open water and land sections	116
5.5	A mast (height 10 m) for atmospheric surface layer measurements	121
5.6	Theoretical form of the atmospheric and oceanic Ekman layers above and beneath sea ice	122
5.7	Averaged velocities of wind, sea ice, and currents (depths 7, 10, 30, and 40 m), 8–15 April 1975 in the Baltic Sea	123
5.8	Geostrophic drag coefficient and turning angle as functions of boundary layer height scaled using the Monin–Obukhov length	129
5.9	The Soviet Union North Pole [Severnyj Poljus] drifting station programme was commenced in 1937, continuing regularly until 1991	130
5.10	Ice situation in the Bay of Bothnia, Baltic Sea on 20 and 26 February 1992	131
5.11	Observed path of a drifting station in the Baltic Sea, April 1979	136
6.1	The free drift solution	144
6.2	"In summer in the Arctic Ocean" by Gunnar Brusewitz, 1980	145
6.3	Schematic picture of open channel flow	147
6.4	Fletcher's ice island, or ice island T-3, in the Beaufort Sea, a widely used base for ice drift studies in the 1950s	152
6.5	Drift of an ice patch, 100 km across and free from any contact with solid boundaries (April 1975, Bay of Bothnia, Baltic Sea)	154
6.6	Spectra of wind, ice drift, and mixed layer in the near-free drift situation shown in Figure 6.6	155
6.7	Inertial motion of sea ice superposed on alongshore translational ice motion in the Sea of Okhotsk	157
6.8	The wind factor and lag in a wind-driven non-steady ice drift as functions of the frequency of wind forcing	159
6.9	Sea ice velocity spectra normalized by forcing	160
6.10	Ice velocity spectra from coupled, one-dimensional ice–ocean models	162
7.1	Empirical data of ice drift speed scaled with the theoretical free drift speed as a function of ice compactness	167
7.2	Sea surface slope vs. wind speed in the Bay of Bothnia, Baltic Sea in ice and ice-free cases	168
7.3	Spectra of the strain rate of ice motion and ocean currents beneath the ice in MIZEX-83 in the Greenland Sea for a 10-km area	168
7.4	Ice conditions in the Bay of Bothnia, 4 March 1985	169
7.5	Winter roads on landfast sea ice are used in the Finnish Archipelago in the Baltic Sea	172
7.6	The stability of fast ice in the Baltic Sea	173
7.7	The geometry of the channel flow problem	174
7.8	Wind-driven frazil ice in a polynya	176
7.9	Boundary-zone confituration	178
7.10	Steady-state solution of wind-driven zonal flow, northern hemisphere	181
7.11	Zonal flow speed as a function of wind direction when the coastal boundary is in the east, northern hemisphere	182
7.12	The profile of ice thickness and ice compactness from the fast ice boundary to the ice edge	184

7.13	An ice edge in the Greenland Sea		186
7.14	Configuration of zonal flow on a polar cap		188
7.15	The steady-state solution of circular ice drift		190
7.16	Sea ice cover in the Antarctic is more divergent than in the Arctic, characteristic features being the presence of polynyas and the co-existence of icebergs inside the pack ice		190
7.17	Ice tank experiment with boundary forcing		192
7.18	The force at the pusher plate in ice tank test #13		195
8.1	Spatial grid (Arakawa B type) normally employed in the numerical modelling of sea ice dynamics		204
8.2	Calibration runs for ice strength parameters against ice velocity fields obtained from *ERS-1* SAR data		206
8.3	Steady-state ice circulation in the Arctic Basin according to the linear viscous model of Campbell (1965)		207
8.4	Shipping in ice-covered seas has penetrated deeper and deeper into the ice pack		214
8.5	Simulated sea level elevation in the Gulf of Bothnia, Baltic Sea		215
8.6	The structure of the Finnish ice-forecasting system in winter 1977		218
8.7	Convoy of ships assisted by an ice-breaker in the Baltic Sea		219
8.8	Five-day ice forecast for the Liaodong Bay, Bo Hai Sea, from 13 January 1993		220
8.9	Oil spill in the Gulf of Finland		221
8.10	The average annual ice velocity and April thickness contours produced by the model of Hibler (1979)		223
8.11	Climatological sea ice velocity and sea ice thickness in the Weddell Sea		225
8.12	Ice conditions and sea surface temperature on 22 March 1984 in the Baltic		226
8.13	Average ice situation at the annual maximum in the Baltic Sea, 2050 ± 15 years		227
9.1	High productivity exists in the ocean close to the sea ice edge		233
9.2	Modern sea ice information products include navigation data (depth contours, ship routes, lighthouses, ships) overlain on ice information		234
9.3	Illustration of the geophysics–engineering scaling question		235

Tables

2.1	The main basins of the world sea's ice zones	11
2.2	Methods for sea ice thickness measurement	32
3.1	Strain rate and vorticity in AIDJEX 1975 manned array in % per day $\approx 10^{-7}\,\mathrm{s}^{-1}$	63
4.1	The viscosity of some materials	89
4.2	Parameters of the viscous–plastic ice rheology of Hibler (1979)	99
5.1	Kinetic energy budget of sea ice dynamics in the northern Baltic Sea	118
5.2	The drag coefficient for surface wind for neutral atmospheric stratification	127
5.3	The parameters of the quadratic water–ice drag laws for neutral oceanic stratification	127
5.4	Typical and extreme levels from the long-term database	131
5.5	Scaling of parameters involved in the momentum equation of drift ice	132
5.6	Comparison between the characteristics of sea ice drift in Arctic, Antarctic, and Baltic Sea waters	137
7.1	Upwelling (+) and downwelling (−) at the ice edge	188
7.2	The measured force (in Newtons) at the pusher plate at $A = 1(F_1)$ and at $h_+ = 4h_i(F_4)$	195
8.1	Parameterization of the viscous–plastic sea ice model of Hibler (1979)	209
8.2	Parameterization of the Baltic Sea viscous–plastic sea ice model	213

Symbols

A	Ice compactness
C	Drift ice strength reduction for opening
C_a	Air drag coefficient
C_{a1}	Linear air drag coefficient
C_b	Bottom drag coefficient
C_w	Water drag coefficient
C_{w1}	Linear water drag coefficient
d	Characteristic diameter of an ice floe
D	Continuum length scale
D_1, D_2	Harmonic and biharmonic diffusion coefficients
e	Aspect ratio of elliptic yield curve
E_p	Potential energy
f	Coriolis parameter
$\mathbf{F}$	Force
F	Yield function
Fr	Froude number
$\mathbf{g}$	Gravity acceleration
h	Ice thickness
h_c	Cut-off size of ridges
h_d	Thickness of deformed ice
h_h	Thickness of hummocked ice
h_i	Ice floe thickness
h_k	Ridge keel depth
h_R	Mean thickness of ridges
h_s	Ridge sail height
h_u	Thickness of undeformed ice
h_0	Demarcation thickness
h'	Freeboard of ice

Symbols

h''	Draft of ice
H	Heaviside function; ice thickness scale; water depth
i	Imaginary unit $i = \sqrt{-1}$
i	Unit vector east
I	Ice function
I	Unit tensor
j	Unit vector north
J	Ice state
k	Unit vector vertically upward
k	Thickness multiplicator in ridging
K	Diffusion coefficient
l_c	Characteristic length of ice on water foundation
L	Length; length scale
L_{MO}	Monin–Obukhov length
L_R	Length of ridges in a given horizontal area
m	Mass of ice per unit area
M	Mass transport
M_1, M_2	Elastic modulus
n	Unit vector normal to a surface
Na	Nansen number
p	Probability density; spectral density; hydrostatic pressure
p_a	Air pressure
P	Ice strength
P^*	Ice strength level constant
Q	Ridge porosity; area of floes larger than given size
r	Radius vector
r_e	Radius of the Earth
r_I	Radius of gyration
r*	Inhomogeneity vector for ice floe thickness
R	Cross-sectional area of a ridge; radius of an ice floe
Ro	Rossby number
S	Surface area of a region
S_f	Surface area of ice floe
S_R	Areal concentration of ridges
Sr	Strouhal number
t	Time
T	Timescale
T_D	Advective timescale
T_I	Inertial timescale
$\boldsymbol{u} = (u, v)$	Ice velocity
$u = u_1 + iu_2$	Ice velocity in complex form
$\boldsymbol{u}_F$	Free drift ice velocity
u^*	Friction velocity
$\boldsymbol{U} = (U, V)$	Translational velocity of an ice floe
$\boldsymbol{U}_a = (U_a, V_a)$	Wind velocity

$\boldsymbol{U}_w = (U_w, V_w)$	Water velocity
$\boldsymbol{U}_{ag}$	Geostrophic wind velocity
$\boldsymbol{U}_{wg}$	Geostrophic ocean current
v'	Velocity fluctuation
V	Displacement gradient tensor
w	Vertical velocity
x	East co-ordinate
X	Friction number
y	North co-ordinate
Y	Young's modulus of sea ice
z	Upward co-ordinate
z_0	Roughness length
Z	Zenith angle
α	Wind factor
β	Sea surface slope
γ	Ratio of keel depth to sail height of ridges; ratio of compressive strength to shear strength; free local variable
Γ	Boundary curve; ratio of frictional dissipation to potential energy production in ridging; local free variable
δ	Delta function; differential element
δ_A	Draft aspect ratio
Δ	Strain-rate invariant function connected to elliptic yield curve
Δ_0	Maximum creep rate in viscous–plastic sea ice rheology
Δx	Grid size
Δt	Time step
ε	Strain
$\dot{\varepsilon}$	Strain rate
$\dot{\varepsilon}'$	Deviatoric strain rate
$\dot{\varepsilon}_\mathrm{I}, \dot{\varepsilon}_\mathrm{II}$	Strain-rate invariants
$\dot{\varepsilon}_1, \dot{\varepsilon}_2$	Principal strain rates
ζ	Bulk viscosity
η	Shear viscosity
θ	Deviation angle between wind and wind-driven ice drift
θ_0	Deviation angle for massless ice
θ_a	Turning angle in atmosphere
θ_w	Turning angle in water
Θ_h	Ice thickness gradient correction for two-dimensional ice stress
κ	Restitution coefficient in floe collisions; von Karman constant; friction coefficient
λ	Usually local variable; longitude
λ_i	Response time of sea ice drift
λ_w	Response time of oceanic boundary layer drift
Λ	Gradient length scale
μ	Ridge density

Symbols

ν	Poisson's ratio; kinematic viscosity
ξ	Sea level elevation
π	Spatial density of ice thickness distribution
Π	Cumulative ice thickness distribution
ρ	Ice density
ρ_a	Air density
ρ_w	Water density
σ	Ice stress due to interaction between ice floes
σ_t, σ_c	Tensile and compressive strength of sea ice
σ_Y	Yield strength of sea ice
σ_1, σ_2	Principal stresses
σ_I, σ_{II}	Stress invariants
Σ	Total ice stress
τ	Shear stress
τ_a	Air stress on ice top surface
τ_w	Water stress on ice bottom surface
ϕ	Latitude; thermodynamic growth rate of ice
ϕ_A	Thermodynamic growth rate of ice compactness
Φ	Geopotential height
χ_0	Opening rate of leads
χ_d	Rate of ridging and hummocking
ψ	Mechanical deformation function, thickness redistributor derivative $d\Psi/dh$
ψ'	Redistributor in ridging (ψ'_- for loss and ψ'_+ for gain)
Ψ	Ice thickness redistributor
ω	Frequency
$\dot{\omega}$	Vorticity, rotation rate
Ω	A region in ice cover; rotation rate of the Earth
φ	Slope angle of ridge sails (φ_s) and keels (φ_k), deformation mode $\arctan(\dot{\varepsilon}_{II}/\dot{\varepsilon}_I)$
ϑ	Angle of internal friction

List of abbreviations

AARI	Arctic and Antarctic Research Institute
AEM	Airborne electromagnetic method
AIDJEX	Arctic Ice Dynamics Joint Experiment
AVHRR	Advanced very high resolution radiometer
DMSP	Defense Meteorological Satellite Program
EM	Electromagnetic
ERS	European Remote Sensing Satellite
ESA	European Space Agency
GPR	Ground-penetrating radar
GPS	Global positioning system
IABP	International Arctic Buoy Programme
MIZ	Marginal ice zone
MIZEX	Marginal Ice Zone Experiment
MODIS	Moderate resolution imaging spectroradiometer
NASA	National Aeronautics and Space Administration
NOAA	National Oceanic and Atmospheric Administration
NP	North Pole
NSIDC	National Snow and Ice Data Centre
R/V	Research vessel
SAR	Synthetic aperture radar
SIMI	Sea Ice Mechanics Initiative
SMMR	Scanning multichannel microwave radiometer
SP	*Severnyi Polyus* (North Pole)
SSIZ	Seasonal sea ice zone
SSM/I	Special sensor microwave/imager
SST	Sea surface temperature
VTT	Technical Research Centre of Finland
WMO	World Meteorological Organization

1

Introduction

Sea ice occurs in about 10% of the world ocean's surface, growing, melting, and drifting under the influence of solar, atmospheric, oceanic, and tidal forcing. Most sea ice lies in the Arctic and Antarctic Seas above 60° latitude, but seasonally freezing smaller basins exist further down toward the equator, to even below 40°N in the Bo Hai Sea, China. In sizeable basins solid sea ice lids are statically unstable and break into fields of ice floes forming *drift ice*. These fields undergo transport as well as opening and closing which altogether create the exciting sea ice landscape as it appears to the human eye. The present book is about the geophysics of the drift of sea ice. Chapter 1 gives a brief history of the research, introduces the problem with applications, and outlines the subject matter of the book.

People living by freezing seas, such as the Baltic Sea in northern Europe, have known the ice drift phenomenon for long. It has traditionally affected everyday life (Figure 1.1). Seal hunting has been based on drifting ice floes for one productive season in spring. In places the drift of ice may have stopped for a middle winter period and ice bridges have formed providing ways to cross the sea. For navigation, sea ice formation has introduced a barrier; in particular, the drift of ice has made the management of this barrier difficult. Recently the transport and dispersion of pollutants in sea ice and oil spills in ice-covered waters have become important issues.

The history of sea ice dynamics science initiates from the drift of the ship *Fram*, moored to ice between 1893 and 1896 in the Arctic Ocean. The wind-driven drift speed was on average 2% of the wind speed and the drift direction deviated 30° to the right from the wind direction (Nansen, 1902). Ekman (1902) explained this with wind forcing, ice–water drag, and the Coriolis effect. In the south, the German ship *Deutschland* drifted with the ice in the Weddell Sea between 1911 and 1912 (Brennecke, 1921). Rossby and Montgomery (1935) presented atmospheric and oceanic boundary-layer models for sea ice and discussed the role of the stability of stratification for drag forces.

Figure 1.1. A section of the marine chart by Olaus Magnus Gothus (1539) showing the northern Baltic Sea. Landfast ice bridges and seal hunting from drift ice floes are illustrated. Olaus Magnus Gothus (1490–1557), the last catholic bishop of Sweden, was deeply interested in northern nature and people. He prepared this chart and in 1555 published a book, *The History of the Nordic Peoples*.

In the Soviet Union, development of the Northern Sea Route gave rise to a large number of investigations into the movement of sea ice. First named the Northern Research and Trade Expedition, the Arctic and Antarctic Research Institute (AARI) was established in 1920. A series of *North Pole drifting stations* was commenced in 1937, where science teams drifted along with their camp sites from the North Pole to the Greenland Sea. Zubov (1945) further examined the Nansen–Ekman drift law and presented semi-empirical modifications for various particular cases. He was responsible for the isobaric drift law which states that ice drifts along the isobars of atmospheric surface pressure. In Japan, a research programme was begun in the 1940s on the drift of ice in the Sea of Okhotsk (Fukutomi, 1952), and North American scientists worked from ice islands in the Beaufort Sea, in particular from *T-3*, or *Fletcher's ice island*.

The closure of the sea ice dynamics problem was completed in the 1950s as laws for the internal friction of ice and ice conservation were established. Internal friction, already recognized by Nansen (1902), is a major factor in ice drift, and its understanding necessitates a rheological equation or a constitutive law. It smooths the drift spatially and guides the ice drift to satisfy realistic boundary conditions. The first attempt was by Laikhtman (1958), who considered drift ice as a Newtonian viscous fluid. The ice conservation law was introduced by Nikiforov (1957), allowing time integration and adjustment of the ice mass field to external forcing. This closure was then utilized in the first generation of numerical models of sea ice dynamics (Campbell, 1965; Doronin, 1970). In the 1960s remote-sensing satellites started to map the Earth's surface, providing a new look in particular over uninhabited areas. For the first time it was possible to have a large-scale daily view over the polar oceans and their sea ice cover.

In the 1970s two major steps were made in sea ice dynamics with the *AIDJEX* (*Arctic Ice Dynamics Joint Experiment*) program (Pritchard, 1980a). From a physical basis, plastic rheologies were introduced for drift ice (Coon et al., 1974), and the concept of thickness distribution was presented (Thorndike et al., 1975). Plasticity allows the presence of stationary ice under non-zero forcing and the occurrence of narrow deformation zones. Thickness distribution contains local thickness variations and adds structural information about ice into the ice conservation law. Hibler (1979) employed a viscous–plastic rheology and constructed a computationally very effective sea ice model, which has become a standard reference. In 1969 a coastal radar system was opened in Japan for continuous sea ice monitoring (Tabata, 1972), and in the 1970s sea ice dynamics investigations expanded in the Baltic Sea due to the needs of growing winter shipping.

Another approach has been derivation of a statistical theory for the drift and diffusion of ice. This was first based on simple linear models, such as the Nansen–Ekman law, which have since undergone further modifications. The *International Arctic Buoy Programme* has produced an extensive amount of data, and it is interesting that the statistics show the same numbers as Nansen's data (Thorndike and Colony, 1982). In Antarctic regions, a similar programme is also ongoing. The statistical approach was finally developed into a Markov process model by Colony and Thorndike (1984). Sea ice drift was taken as the mean field plus a random walk, and drift buoy data were used to estimate the process parameters. The statistical method complements the deterministic dynamics approach since answers can be easily provided to statistical questions, such as the probability of sea ice to drift from the Kara Sea to the Barents Sea.

In the 1980s the focus was on mesoscale ice–ocean and air–ice interaction studies. New observational data collection campaigns and process-modelling efforts were realized, in particular within the *MIZEX (Marginal Ice Zone Experiment)* programme (Wadhams et al., 1981; Muench et al., 1987). Work was begun with granular flow models to deduce the mechanical behaviour of drift ice from individual ice floe–floe interactions (Shen et al., 1986). Later, numerical discrete particle methods were developed, which resulted in some success in marginal ice zone dynamics (Løset, 1993) and the sea-ice-ridging process (Hopkins, 1994).

The launch of the *ERS-1* satellite by the European Space Agency (ESA) in 1991 initiated an intensive phase in ice kinematics mapping, with the advent of spaceborne synthetic aperture radars (SARs). These instruments became excellent research and routine tools in the 1990s (Fily and Rothrock, 1987; Kwok et al., 1990; Li et al., 1995; Kondratyev et al., 1996). Passive microwave satellite imagery was also extended to the extraction of sea ice kinematics information (Agnew et al., 1997; Kwok et al., 1998). Another landmark in the 1990s was a major increase in sea ice dynamics research in Antarctica, accompanied by wintertime expeditions and the first Antarctic drifting station, the US–Russian Ice Station Weddell-1, which was deployed in 1992 in the Weddell Sea (Gordon and Lukin, 1992; see also http://www.ldeo.columbia.edu/res/fac/physocean/proj_ISW.html).

The most recent developments in sea ice dynamics concern scaling: from local ice-engineering problems to mesoscale or large-scale geophysics (e.g., Dempsey and Shen, 2001). Anisotropic effects have been introduced into sea ice rheology due to the orientation of leads (Coon et al., 1998). A mechanics programme, called SIMI (Sea Ice Mechanics Initiative), has studied the Beaufort Sea (Richter-Menge and Elder, 1998). Russia restarted the North Pole drift station programme in 2003 after 12 years of interruption. Methods for remote sensing of sea ice thickness are slowly progressing (e.g., Wadhams, 2000). This is perhaps the most critical point for the future development of the theory and models of sea ice dynamics. Model developments also focus on deep analysis of the physics of model realizations, the improvement of coupled air–ice–ocean models, and the design of data assimilation methodology.

Drift ice is a peculiar geophysical medium (Figure 1.2). It is *granular*: ice floes are the elementary particles. The drift takes place on the scale of floes and larger. The ice moves on a horizontal sea surface plane with no vertical velocity structure; thus, ice drift can be treated as a *two-dimensional* problem. Since sea ice is located almost in a geopotential surface, packing densities of ice floes may easily change and therefore drift ice must be taken as a *compressible* medium. The rheology of drift ice shows a highly *non-linear* relation between stress and strain rate, much different from the air above and the water below. Finally, because of freezing and melting, an *ice source/sink* term is included in the ice conservation law.

The full ice drift problem includes the following unknowns: *ice state* (a set of

Figure 1.2. A drift ice landscape from the northern Greenland Sea, June 1983; the photograph was taken during MIZEX-83 (Marginal Ice Zone Experiment, 1983), onboard the R/V *Polarbjørn*. The ice field consists of ice floes with hummocks and ridges.

relevant material properties), *ice velocity*, and *ice stress*. The system is closed according to equations for the conservation of ice, the conservation of momentum, and ice rheology. Ice is driven by winds and ocean currents and responds to forcing by its inertia, internal friction, and adjustment of its state field. The Coriolis effect slightly modifies ice movement. The dynamics of sea ice is coupled with thermodynamics, since freezing strengthens and melting weakens the ice, while ice motion influences the further growth or melting of ice via transport and differential motion.

Although the ice drift problem offers quite interesting basic research possibilities, the principal science motivation has come from sea ice introducing a particular air–sea interface. The exchange of momentum, heat, and material between the atmosphere and the ocean takes place in drift ice fields in high latitudes, and this interface experiences transport as well as opening and closing due to ice drift. This is crucially important to both regional weather and global climate. Ice extent, largely influenced by ice drift, has a key role in the cryospheric albedo effect. Also, ice transports latent heat and freshwater, while ice melting gives a considerable heat sink and freshwater flux into oceanic surface layer. In the ecology of polar seas, the location of the ice edge and its ice-melting processes is a fundamental boundary condition for summer productivity. A more recent research line for sea ice dynamics is in paleoclimatology and paleoceanography (Bischof, 2000).

Marine sediments provide an archive of drift ice and iceberg data, and via its influence on ocean circulation, drift ice has been an active agent in global climate history.

In the practical world, sea ice dynamics is connected with three major questions. First, sea ice models have been applied to tactical navigation, to provide short-term forecasts of ice conditions (e.g., Leppäranta, 1981a). Second, ice forcing on ships and fixed structures is affected by the dynamical behaviour of the ice (e.g., Sanderson, 1988). Third, the question of pollutant transport by drifting sea ice has become an important issue (Pfirman et al., 1995). In particular, risk assessment of oil spills and their clean-up require proper oil transport and dispersion models for ice-covered seas (e.g., Ovsienko et al., 1999a).

Drift ice is a *geophysical fluid*, with a large horizontal scale and a very small vertical scale. Being thin, the direct Coriolis effect on the ice is rather weak, in contrast to the atmosphere and ocean. The theory and models of sea ice dynamics are applicable in general to ice dynamics in large basins (large relative to ice thickness and external forcing). The essential feature is that floating ice breaks into floes and then drifts. This is true in large lakes, such as Lake Ladoga in Russia, Lake Värnern in Sweden, and the Great Lakes of North America (Wake and Rumer, 1983). Even in large rivers (e.g., the Niagara River in North America), the drift of ice floes can be examined with the same models (Shen et al., 1993).

This book presents a treatise on the geophysical theory and empirical knowledge of the drift of sea ice. The objective is to describe ice drift by observational data (Figure 1.3), to introduce the basic ideas and laws of sea ice dynamics, to provide

Figure 1.3. Deployment of a corner reflector mast for deformation studies by a laser geodimeter, in the Baltic Sea in April 1978. A system with several reflector sites allows accurate monitoring of sea ice kinematics in a 5 km scale array.

solutions to elementary cases that help us to understand geophysics, and to present the status of sea ice dynamics models. Earlier books devoted to sea ice dynamics are only available in Russian (e.g., Timokhov and Kheysin, 1987; Gudkovic and Doronin, 2001); review articles have been written in English by Rothrock (1975b), Hibler (1986), and Leppäranta (1998). Individual chapters about ice drift can be found in sea ice geophysics books by Zubov (1945), Doronin and Kheysin (1975), and Wadhams (2000).

The general structure of the book is the following: Chapters 2–4 present the drift ice medium and its kinematics; in Chapter 5 this medium is inserted into Newton's 2nd law to obtain the equation of motion; and in Chapters 6–8 a solution to the ice drift problem is given.

Chapter 2 provides a description of drift ice material, leading to an "ice state", the set of relevant quantities for the mechanical behaviour of drift ice. In Chapter 3, ice velocity observations are presented, a theoretical framework is given for ice kinematics analysis, and the ice conservation law is derived. Chapter 4 treats drift ice rheology, a difficult but necessary subject to understand the motion of drift ice. The momentum equation of ice dynamics is derived in Chapter 5, including a discussion of principal external forcings—air and water drag force—and analysis of the equation. Free drift is examined in Chapter 6, and ice drift in the presence of internal friction is the subject of Chapter 7. In both these chapters analytical solutions are presented for particular ice flows. The modern solution to understanding sea ice drift by numerical modelling is given in Chapter 8, and the aim is to help the reader understand how this solution is arrived at, how it reflects the behaviour of real sea ice, and how one should interpret the outcome of numerical sea ice dynamics models. Finally, a brief discussion is given on some consequences of the ice drift phenomenon (Chapter 9). Chapter 10 invites the reader to tackle a collection of study problems. Chapter 11 lists the references. This is followed by the index.

2

Drift ice material

2.1 SEA ICE COVER

This chapter presents the material structure of sea ice from local scale to large scale. Ice floes form granular drift ice fields, for which continuum approximations are used when the scale of interest is much larger than the floe size. These fields are characterized by their ice type, ice compactness, floe size and shape, and ice thickness, as shown in Sections 2.1–2.3. Section 2.4 then deals with ice ridges, the thickest accumulations of mechanically deformed ice, which have a key role in the drift of sea ice. The chapter ends by introducing the concept of an "ice state"—a set of material properties of drift ice necessary to understand and model its dynamics.

In the world ocean's ice cover, ice occurs all year in the *perennial sea ice zone*, covering the inner Arctic Ocean and smaller sections in Antarctica, mainly the western Weddell Sea. Where ice occurs only in winter is an area called the *seasonal sea ice zone* (SSIZ) extending on average to 60° latitudes.

Table 2.1 and Figure 2.1 show the main sea ice basins, with sizes ranging from 200 km to 3,000 km. Southern Ocean sea ice cover is actually a ring, 20,000 km long and with a width varying from almost zero in summer to 1,000 km in winter, centred around Antarctica at 60–70°S. There are smaller sub-basins, which contain dynamically independent ice packs, such as the Gulf of Riga in the Baltic Sea (size $L = 100$ km and typical ice thickness $h = 0.2$ m). The drift ice basin closest to the equator is the Bo Hai Sea (Gulf of Chihli) off the coast of China, located between latitudes 37°N and 41°N.

The thickness of ice is 2–5 m in the Arctic Ocean and Greenland Sea, while in the SSIZ it is an order of magnitude less. For an enclosed drift ice basin there is no ice exchange with neighbouring seas, while in an open basin a large part of the ice boundary is toward open water and the mobility of the ice is therefore greater. The ratio h/L, which characterizes the stability of a solid ice sheet in a basin, ranges from 10^{-7} to 5×10^{-6} in sea ice basins where drift ice occurs (Table 2.1).

10 Drift ice material [Ch. 2

Nikolai Nikolaevich Zubov (1885–1960), a great pioneer in the science of sea ice, its physics and its geography. He was the author of the monumental book *L'dy Arktiki [Arctic Ice]*, published in 1945 in Moscow and translated into English in 1963.
© Reproduced from Collections of the Russian State Museum of Arctic and Antarctic, St Petersburg, with permission.

In a subarctic medium-size lake ($L = 10$ km, $h = 0.5$ m) the ratio is 5×10^{-5}, and the ice forms a solid stationary sheet.

2.1.1 Sea ice landscape

A "sea ice landscape" consists of leads and ice floes with ridges, hummocks, and other variable morphological characteristics. *Ice types* have been defined to provide practical standards for observers (WMO, 1970; see also http://www.aari.nw.ru/ for updates). They originate from shipping activities in ice-covered waters and are based on appearance (i.e., how the ice looks to an observer on a ship or in an aircraft: Figures 2.2 and 1.2). This ice-type classification system has worked fairly well and has not suffered from severe subjective biases. The formation mechanism, aging, and deformation influence on appearance. This gives a good idea of ice thickness, which is seldom known from direct measurements. Some ice-type names are based on their resemblance to familiar objects (Figure 2.3).

Ice concentration, or *ice compactness*, denoted by A, is normally given in percentages or tenths and further categorized into six standard classes (Figure 2.4).

Table 2.1. The main basins of the world ocean's ice zones. The types, E = enclosed, SE = semi-enclosed and O = open, refer to the ice exchange with neighbouring seas.

Basin	Size, L (km)	Type	Ice thickness, h (m)	Stability, h/L ($\times 10^{-6}$)
Central Arctic	3,000	SE	2–5	0.7–2
Greenland Sea	1,000	O	2–5	2–5
Barents Sea	1,000	O	1–2	1–2
Kara Sea	1,000	SE	1–2	1–2
White Sea	200	SE	0.1–1	0.5–5
Baltic Sea	500	E	0.1–1	0.2–2
Sea of Azov	200	E	0.1–0.2	0.5–1
Sea of Okhotsk	1,000	SE	0.1–2	0.1–2
Bohai Sea	300	E	0.1–0.3	0.3–1.0
Bering Sea	1,000	SE	0.1–1	0.1–1
Hudson Bay	500	E	0.5–1	1–2
Gulf of St Lawrence	300	E	0.1–1	0.3–3
Labrador Sea	500	O	0.5–1	1–2
Baffin Bay	500	SE	1–2	2–4
Southern Ocean	1,000	O	0.5–2	0.5–2
Weddell Sea	1,500	O	0.5–3	0.3–2
Ross Sea	500	O	0.5–1	1–2

Between ice and open water surfaces, apart from melting conditions, there are strong contrasts and satellite remote-sensing methods can detect them.

Definitions

Let us now give a brief list of the necessary sea ice nomenclature for dynamics based on internationally agreed standards (Armstrong et al., 1966; WMO, 1970):

- *Sea ice.* Any form of ice found at sea that originates from the freezing of seawater.
- *New ice.* A general term for recently formed ice.
  - *Frazil ice.* Fine spicules or plates of ice suspended in water.
  - *Nilas.* A thin, elastic crust of ice that easily bends under the action of waves and swell and rafts under pressure (matt surface and thickness up to 10 cm).
- *Young ice.* Ice in transition between new ice and first-year ice (10–30 cm thick).
- *First-year ice.* Ice with no more than 1 year's growth that develops from young ice (thickness 30 cm to 2 m). Level when undeformed, but where ridges and hummocks occur, it is rough and sharply angular.
- *Multi-year ice.* Ice of more than 1 year's growth (thickness over 2 m). Hummocks and ridges are smooth and the ice is almost salt-free.
- *Fast ice.* Sea ice that remains fast along the coast, over shoals, or between grounded icebergs (also called *landfast ice*).

12 **Drift ice material** [Ch. 2

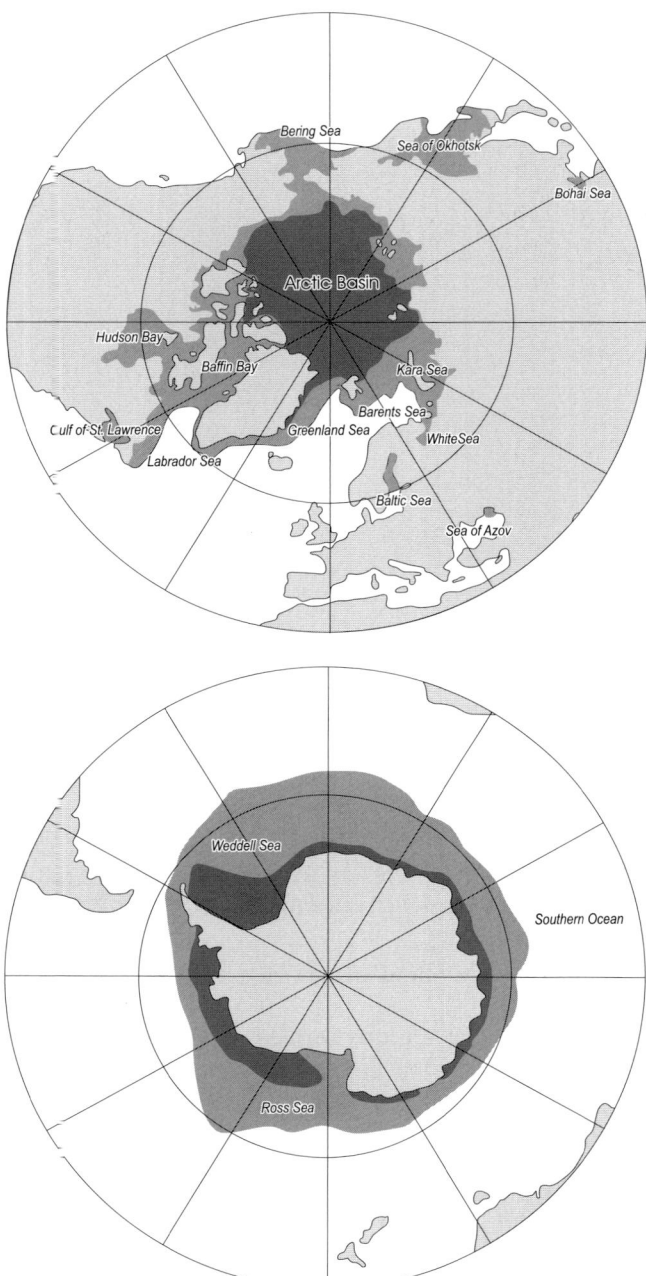

Figure 2.1. The world sea's ice zones in summer and winter. Dark area shows the perennial ice, lighter area shows the seasonal sea ice zone.
From Untersteiner (1984), with modifications for subarctic small basins.

Figure 2.2. Aerial photograph from a Finnish Air Force ice reconnaissance flight in the central Baltic, winter 1942. Airborne reconnaissance meant a huge step in understanding the morphology and drift of sea ice. The pilot was Erkki Palosuo, who later became a sea ice geophysicist and used his sea ice data from the Second World War for his doctoral thesis (Palosuo, 1953).
Reproduced with permission from Erkki Palosuo.

Figure 2.3. Pancake ice (the ice pieces shown are about 1 m across), also named lotus ice in the Far East, blini ice in Russia, and plate ice in Scandinavia and Finland.

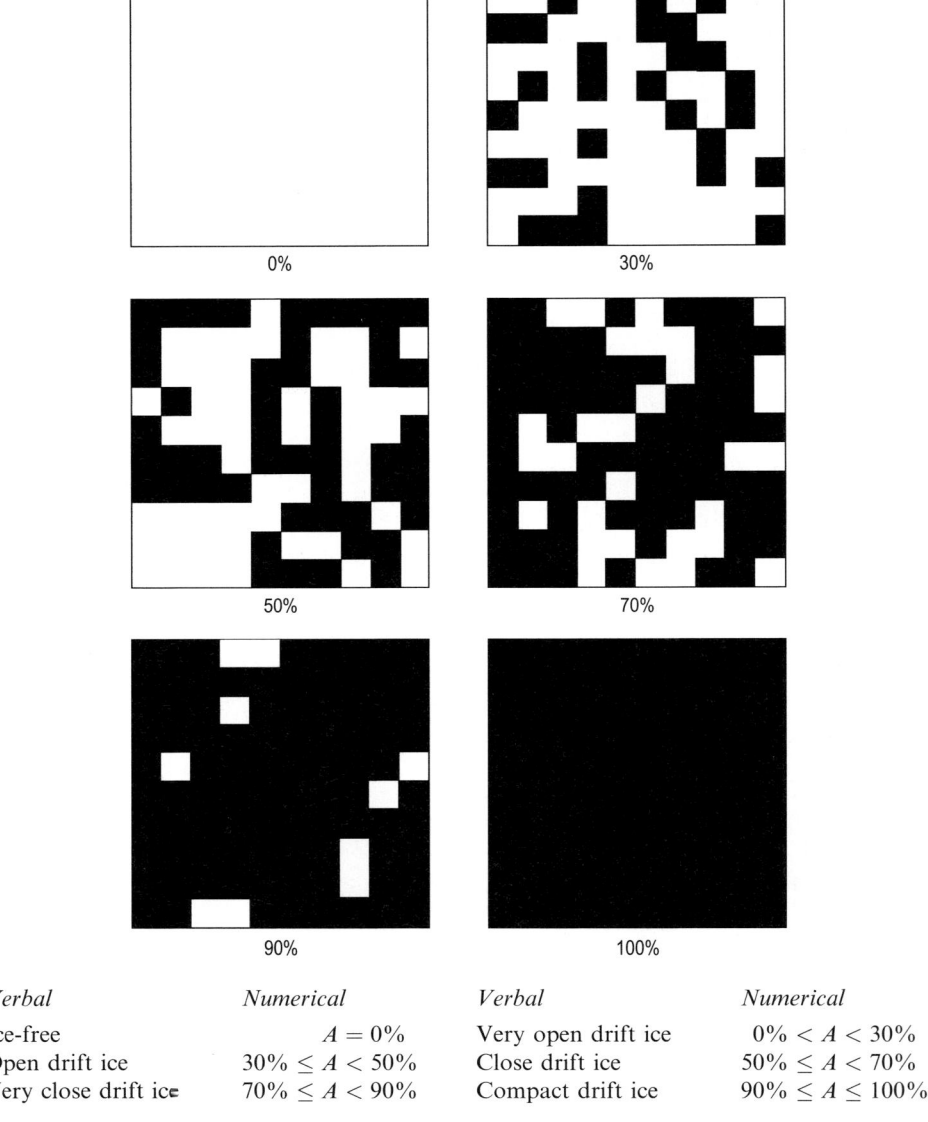

Verbal	Numerical	Verbal	Numerical
Ice-free	$A = 0\%$	Very open drift ice	$0\% < A < 30\%$
Open drift ice	$30\% \leq A < 50\%$	Close drift ice	$50\% \leq A < 70\%$
Very close drift ice	$70\% \leq A < 90\%$	Compact drift ice	$90\% \leq A \leq 100\%$

Figure 2.4. Classification of ice concentration (WMO, 1970), with the class boundaries illustrated by random binary charts.

- *Grounded ice* Floating ice that finds itself aground in shoal water.
- *Drift ice*. Term used in a wide sense to include any sea ice other than fast ice (a substitute term is *pack ice*).
- *Ice field*. Area of drift ice at least 10 km across.

- *Pancake ice.* Pieces of new ice, usually approximately circular, about 30 cm to 3 m across and with raised rims due to the pieces striking against each other.
- *Ice floe.* Any relatively flat piece of ice 20 m or more across.
- *Level ice.* Sea ice that is unaffected by deformation (a substitute term is *undeformed ice*).
- *Deformed ice.* A general term for ice that has been squeezed together and in places forced upward and downward (a substitute term is *pressure ice*).
- *Rafted ice.* A form of pressure ice in which one floe overrides another. A type of rafting common in nilas whereby interlocking thrusts are formed—each floe thrusting "fingers" alternatively over and under the other—is known as *finger rafting*.
- *Brash ice.* Accumulations of ice made up of fragments no more than 2 m across (the wreckage of other forms of ice).
- *Hummocked ice.* A form of pressure ice in which pieces of ice are piled haphazardly, one piece over another, to form an uneven surface.
- *Ridge.* A ridge or wall of broken ice forced up by pressure (the upper—above water level—part is called the *sail* and the lower part the *keel*).
- *Ice concentration.* The amount of sea surface covered by ice as a fraction of the whole area being considered (a substitute term is *ice compactness*).
- *Fracture.* Any break or rupture in ice resulting from deformation processes (length from meters to kilometres).
- *Crack.* Any fracture that has not parted more than 1 metre.
- *Lead.* Any fracture or passageway through sea ice that is navigable by surface vessels.
- *Polynya.* Any non-linear-shaped opening enclosed in ice.
- *Ice edge.* The demarcation between the open sea and sea ice.

The polar oceans also contain *ice of land origin* (i.e., ice formed on land or in an ice shelf). By the mechanism called *calving*, pieces of ice break away from land ice masses facing the ocean. These pieces are classified (WMO, 1970) according to their size: *icebergs* (top more than 5 m above sea level), *ice islands* (top about 5 m above sea level and area more than a few thousand square metres), *bergy bits* (top 1–5 m above sea level and area 100–300 m^2), and *growlers* (smaller than bergy bits). These pieces differ from sea ice floes by their thickness and three-dimensional character.

2.1.2 Sea ice zones

In a given basin, sea ice cover can be divided into zones of different dynamic character (Weeks, 1980): central pack, shear zone, fast ice, and marginal ice zone. Shear zone and fast ice form the coastal boundary zone, while the marginal ice zone is the boundary zone toward the open ocean. Very small basins only contain fast ice and the size of the boundary zones is of the order of 100 km.

The *central pack* consists of the interior ice that is free from immediate influence from the boundaries. Changes are smoother there than in the boundary zones and are caused by external forcing. The length scale is the size of the basin itself; but due to the width of boundary zones, the central pack only exists in large seas.

Fast ice, or *landfast ice*, is the immobile coastal sea ice zone, stationary for most of the ice season. The width of this zone depends on the thickness of ice, topography of the sea bottom, and the areal density of islands and grounded forms of ice. Fast ice extent develops stepwise, in an almost discontinuous manner (Jurva, 1937; Divine, 2003). Grounding of sea ice ridges creates fixed support points to stabilize the ice sheets. Because of the size of ridges and ice thickness, in Arctic seas the fast ice zone extends to depths of about 10–20 m (Zubov, 1945; Volkov et al., 2002) while in subarctic seas such as the Baltic Sea the limit is normally close to 10 m (Leppäranta, 1981b). In Antarctic waters, grounded icebergs may act as tie points for fast ice formation, and therefore the fast ice zone may extend deeper into the ocean as well (e.g., Masson et al., 2003). Off Hokkaido in the Sea of Okhotsk, there are no islands and the depth of the sea increases rapidly with distance from the shoreline. Tides "clean" the ice that forms close to the shore, and as a consequence the fast ice zone is practically non-existent.

The *shear zone* is the boundary zone of the drift ice field next to the fast ice (or coast). There the mobility of the ice is restricted by the geometry of the boundary and strong deformation takes place. The width is 10–200 km. A well-developed shear zone is found on the coast of the Beaufort Sea of the Arctic Ocean. Based on deformation data Hibler et al. (1974a) concluded its representative width was around 50 km. At the solid boundary of ice, velocity is often discontinuous, thus modifying local hydrography and ocean circulation.

The *marginal ice zone* (MIZ) lies along the boundary of open water and sea ice cover. It is loosely characterized as the area of pack ice, where the influence of the open ocean is directly observed and extends to a distance of 100 km from the ice edge (Wadhams, 1980; Squire, 1998). This distance corresponds to the penetration distance of ocean swell into a drift ice field; this distance also corresponds to the length scale below which wind fetch over ice is not long enough to build ice ridges. In the MIZ, there is a large temporal and spatial variability of ice conditions and intensive air–ice–sea interaction. Off-ice winds cause MIZ diffusion while on-ice winds drive the ice to form a sharply compact ice edge (Zubov, 1945). Well-developed MIZs are found along the oceanic ice edge of the polar oceans, located close to polar fronts. At a compact ice edge there is a discontinuity in surface velocity and roughness, and possibly a front in temperature and salinity. They affect the mesoscale circulation in the ocean, resulting in eddies and jets as well as ice edge upwelling and downwelling. A front may also form in the atmospheric boundary layer.

Well-defined marginal ice zones exist along the oceanic edge of polar sea ice caps, and their locations are largely controlled by polar fronts. In smaller, subarctic seas there is usually not enough time for a proper marginal ice zone to develop. There ice extent grows and retreats a long distance back and forth during a short ice season. The open sea influence is reflected in a narrow *ice edge zone*, up to 5 km wide.

In the past, the ice margin used to be considered the border to an unknown ocean where the ice mysteriously transported sand and drift wood from unknown

Sec. 2.1]   **Sea ice cover**   17

places. And as such drifting sea ice was strange, inspiring the Vikings to name a new land – Iceland – due to presence of drift ice in its fjords.

2.1.3 Sea ice charting

Sea ice charting began around 100 years ago for navigation in ice-covered waters. Due to the dynamics of sea ice, the charts need daily updating and therefore much is required of the mapping methods. Before the time of airborne and spaceborne remote sensing, ice charts were based on ship reports and occasional ground truth observations, which provide only limited information. Aerial reconnaissance played an important role until the 1970s, but since then satellite observation technology has been the main source of information. Sea ice information is presented according to an international standard (WMO, 2000).

Passive microwave mapping has become the principal method for regular, global sea ice charting, mainly using US Defense Meteorological Satellite Program (DMSP) special sensor microwave imager (SSM/I) data (Figure 2.5), available via the National Snow and Ice Data Center (NSIDC) in Boulder, Colorado (http://nsidc.org). This method is weather- and light-independent but limited by low

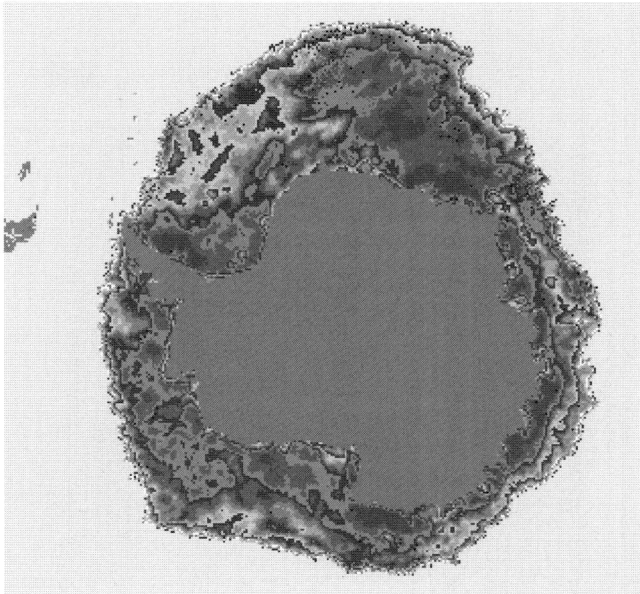

Figure 2.5. Sea ice concentration in the Antarctic based on passive microwave SSM/I data. At the ice edge, compactness increases rapidly to about 80%, while in the inner pack it is mostly 75–90% with a few areas below the 70% level only in the Weddell Sea. NRTSI Product for 2003-10-2 National Snow and Ice Data Center, Boulder, CO.
Reproduced from Cavalieri et al. (1999), with permission from National Snow and Ice Data Center, Boulder, CO.

spatial resolution (20–30 km), too small for regional ice charting. This has been addressed by polar orbiting weather satellites, such as the *NOAA* AVHRR series, with optical and infrared channels. However, both channels are limited by cloudiness. The optical channel is further limited by the lack of sunlight in polar winter.

In the last ten years satellite, synthetic aperture radar (SAR) has become an extremely useful complementary tool, first used on the satellite *ERS-1*, launched by the European Space Agency (ESA) in 1991. It is weather-independent and provides high spatial resolution but has limitations. Temporal resolution is still low, such as 5 days with the Canadian *Radarsat*, and interpretation of the radar signal for sea ice information is sometimes problematic. Ice conditions change significantly on a daily basis in the seasonal sea ice zone (Leppäranta, 1981a; Leppäranta et al., 1998), and the relation between sea ice and radar image is not one-to-one (Carsey, 1992; Wadhams, 2000) For example, ice and open water signatures are not always different, and ridges may be mixed with frost flowers on thin ice in narrow leads.

An example of a sea ice chart over the Arctic is given in Figure 2.6. Ice charts basically present ice compactness and ice type. In subarctic seas with heavy winter traffic, ice charts are invaluable in providing information about the ice thickness and

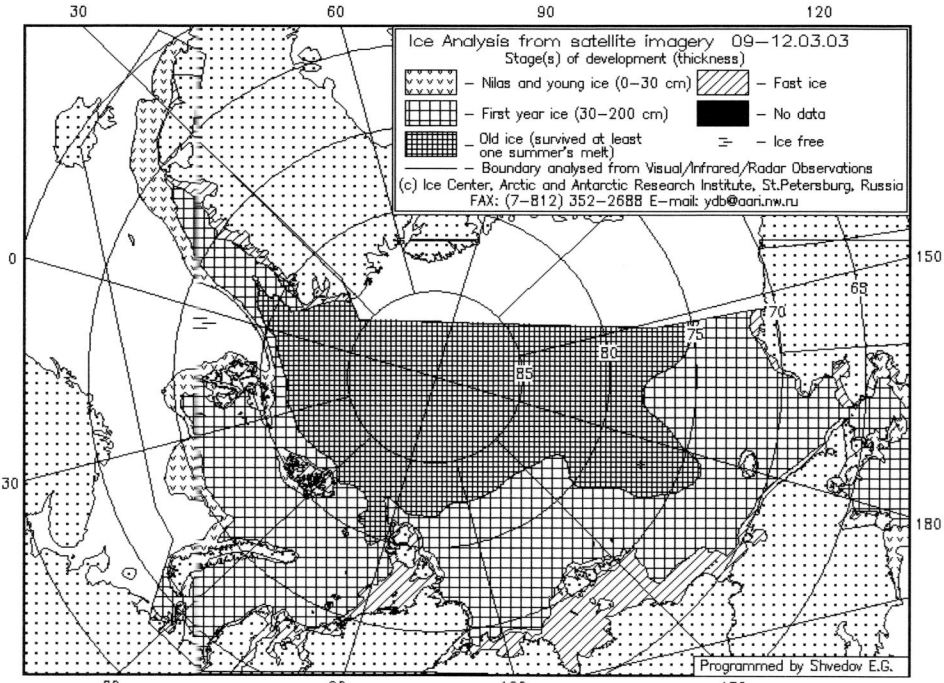

Figure 2.6. Ice chart over the Eurasian side of the Arctic Ocean, 12 March 2003. Reproduced with permission from Arctic and Antarctic Research Institute, St Petersburg, Russia: htpp://www.aari.nw.ru

floe size. In the Baltic Sea, each of the nine coastline countries has an ice service for regional ice charting (e.g., http://www.fimr.fi/ for the Finnish Ice Service). Another way of ice charting is automatic mapping of the ice concentration field, the principal ice quantity manageable by remote sensing. NSIDC provides global ice charts such as that shown in Figure 2.5 on a daily basis. In Hokkaido, Japan, a coastal radar system is used for mapping offshore ice conditions in the Sea of Okhotsk to a distance of 60 km from the coast (http://www.hokudai.ac.jp/lowtemp/sirl/sirl-e.html).

The main problem in sea ice charting is how to obtain good ice thickness information—this is also the main problem for the progress of sea ice dynamics theory and modelling.

2.2 ICE FLOES TO DRIFT ICE PARTICLES

2.2.1 Scales

Sea ice mechanics is examined over a wide range of scales. *Microscale* includes individual grains and ice impurities extending from submillimetres to 0.1 m. In the *local scale*, 0.1–10 m, sea ice is a solid sheet, a polycrystalline continuum with a substructure classified according to the formation mechanism as congelation ice, snow ice, and frazil ice (Eicken and Lange, 1989). *Ice floe scale* extends from 10 m to 10 km and includes individual floes and ice forms such as rubble, pressure ridges, and fast ice. When the scale exceeds the floe size, the sea ice medium is called *drift ice* or *pack ice*, and, as in dynamical oceanography, the scales 100 km and 1,000 km are *mesoscale* and *large scale*, respectively.

The drift of sea ice takes place on the floe scale and larger. The horizontal structure of sea ice cover is well revealed by optical satellite images (Figure 2.7). The elementary particles are *ice floes*, described by their thickness h and characteristic diameter d. The WMO nomenclature (WMO, 1970) restricts ice floes to those ice pieces with $d > 20$ m, smaller ones are termed *ice blocks*. This is convenient because for ice floes the *aspect ratio* h/d is smaller than about 0.1 and floes included in the definition are flat. The floe size ranges from the lower limit to tens of kilometres. In sea ice dynamics research we consider the drift of individual floes or the drift of a system of ice floes, called a *drift ice field*.

Ice density is taken as a constant ($\rho = 910$ kg/m^3). Therefore, armed with thickness information we can calculate the mass of ice, given as the mass per unit area by $m = \rho h$. In reality, the density varies within $\pm 1\%$ around this reference due to variations in the temperature, salinity, and gas content of the ice. Ice floes float freely, apart from shallow areas where grounding may occur. The portions of an ice floe above and beneath the sea surface are, respectively, *freeboard* (h') and *draft* (h''), $h = h' + h''$. Archimedes law states that:

$$h''/h = \rho/\rho_w \qquad (2.1)$$

where ρ_w is the density of water. Since $\rho_w \approx 1{,}028$ kg/m^3 is the density of cold seawater (salinity 35‰, temperature at freezing point), $\rho/\rho_w \approx 0.89$. The potential

20 Drift ice material [Ch. 2

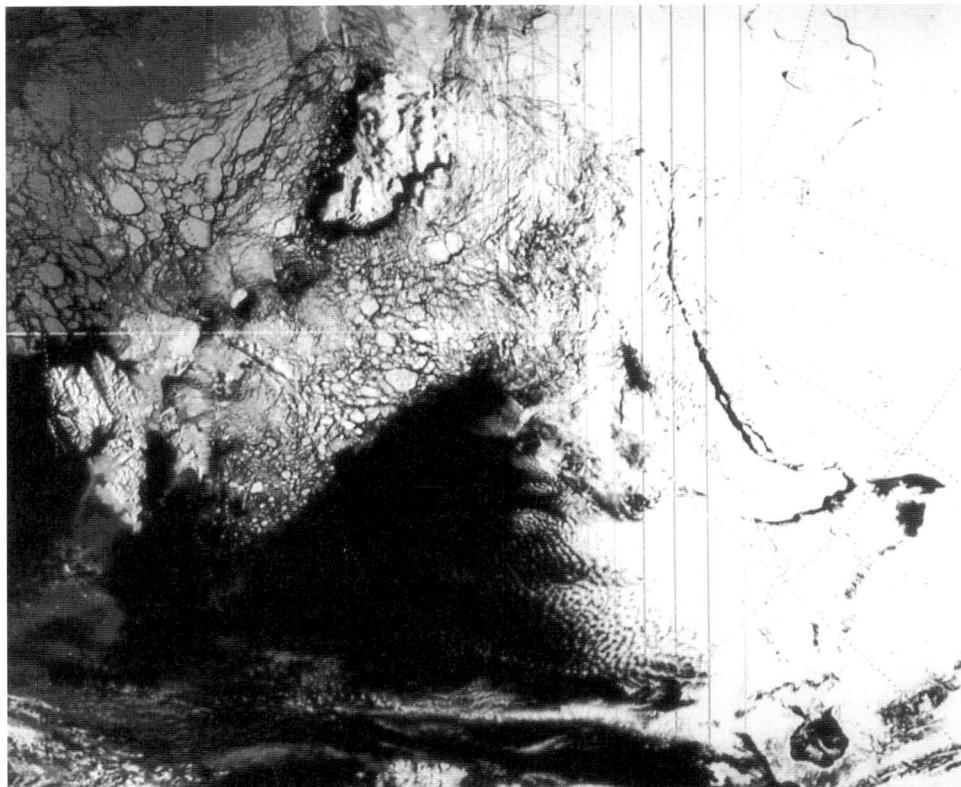

Figure 2.7. An optical channel *NOAA-6* image over the Barents Sea and Arctic Basin north of it, 6 May 1985. Svalbard is shown on the left, Franz Joseph's Land up in the middle, and Novaya Zemlya on the right. The east side is cloudy, but everywhere else drift ice floes can be seen. The image was received at the Tromsø receiving station, Norway.

energy of an ice floe per unit area consists of the freeboard and draft portions, and for floating ice it equals (Rothrock, 1975a):

$$E_F = \rho g \int_0^{h'} z\,dz - (\rho_w - \rho)g \int_{-h''}^0 z\,dz = \frac{1}{2}\frac{\rho(\rho_w - \rho)}{\rho_w} g h^2 \qquad (2.2)$$

where $g = 9.8\,\text{m/s}^2$ is acceleration due to Earth's gravity. For ice not floating but supported by the bottom the right-hand side of Eq. (2.2) is not true.

The behaviour of a drift ice field depends on its horizontal size L and the thickness and size of ice floes. This gives three length ratios: the aspect ratio h/d, granularity L/d, and stability h/L. The number of ice floes is proportional to $(L/d)^2$, and the mechanical breakage of an ice sheet mainly depends on h/L.

Drift ice particles

A drift ice material particle is a set of ice floes. A particle of size D contains $n \approx (D/d)^2$ floes. For the continuum approximation to be valid, n must be large: $n > 100$, or equivalently $D/d > 10$. Additionally, the particle size should be much less than the scale of changes or the gradient scale $\Lambda = Q/|\nabla Q|$, where Q is a property of the ice dynamics field. Summarizing, the scales should satisfy:

$$d \ll D \ll \Lambda \quad \text{for a continuum} \tag{2.3a}$$

or

$$d \sim D \sim \Lambda \quad \text{for a discrete system} \tag{2.3b}$$

In the real world, the situation is intermediate (i.e., $d < D < \Lambda$). Depending on the particular question under examination, $d \sim 10^1$–10^4 m, $D \sim 10^3$–10^5 m, and $\Lambda \sim 10^4$–10^6 m.

In the continuum theory of drift ice, field variables such as ice velocity are defined for each drift ice particle (cf. fluid parcels in fluid dynamics); but, because of the finite size of the floes their individual features may play a role in the ice motion. Thus the theory includes a basic inaccuracy. As D approaches Λ, discontinuities start to build up, and as D approaches d, a system with a single floe or a few floes appears.

Let $h = h(x, y, t)$ stand for the ice thickness at a given point and time; for open water $h = 0$. Define a function "ice", I: by:

$$I(x, y; h_0) \begin{cases} = 0, & \text{if } h(x, y) \leq h_0 \\ = 1, & \text{if } h(x, y) > h_0 \end{cases} \tag{2.4}$$

where h_0 is the *demarcation thickness*. Consider a region Ω in a sea ice pack—it may be a basin, a drift ice particle, or anything in-between—whose surface area is S. The packing density, or compactness, of ice is defined as:

$$A = S^{-1} \int_\Omega I(x, y; 0) \, d\Omega \tag{2.5}$$

For uniform circular floes, the most open and dense locked packings are $\pi/4 \approx 0.79$ and $\pi/(2\sqrt{3}) \approx 0.91$, respectively (Figure 2.8). The further below 0.79 the packing goes, so contacts between floes become fewer. Since ice floes float nearly on a geopotential surface, compactness may easily change. However, at the locked level, further compression necessitates ice breakage and pressure ice formation. The connection between compactness and mean free path l_w is for uniform circular floes $A/A_{\max} \approx (1 + l_w/d)^{-2}$, where $A_{\max}$ is the densest packing for a

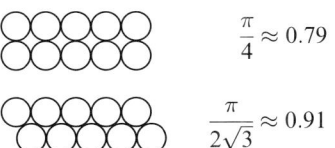

Figure 2.8. Open and dense locked packings of uniform, circular ice floes.

given set of floes. For $l_w = d$, we have $A/A_{\max} \approx \frac{1}{4}$. For distributed floe sizes, $A_{\max}$ is in general larger, $A/A_{\max} \approx \langle d^2 \rangle / \langle (d + l_w)^2 \rangle$, where $\langle \cdot \rangle$ is the averaging operator. As the distribution widens, $A_{\max} \to 1$. An important characteristic is the uniformity of floe sizes.

Mean ice thickness $\langle h \rangle$ and *mean ice floe thickness* $\langle h_i \rangle$ are defined by:

$$\langle h \rangle = S^{-1} \int_\Omega h \, d\Omega = A \langle h_i \rangle \tag{2.6}$$

Mean ice floe thickness is thus the mean thickness of actual ice floe pieces in Ω, while open water is also included in mean ice thickness. At times, these different thickness definitions have caused confusion. Clearly, we have $0 \leq A \leq 1$ as well as $0 \leq \langle h \rangle \leq \langle h_i \rangle$. Ice compactness and mean ice floe thickness are the fundamental quantities for the mechanical properties of drift ice. Other ice quantities are integrated over drift ice particles or any regions in similar ways.

The granular medium approach was introduced for drift ice in the 1970s taking ideas from soil mechanics (Coon, 1974), resulting in poly-granular continuum models. More recently, discrete particle models have been used and found to work well in the floe scale (e.g., Løset, 1993; Hopkins, 1994); but, for larger scale drift ice problems they have not overcome continuum models. A possible step between discrete and continuum models could be the "particulate medium" approach (Harr, 1977). This is composed of a complex conglomeration of discrete particles, in arrays of varying shapes, sizes, and orientations. The laws of mechanics are derived from probabilities.

2.2.2 Size and shape of ice floes

Floe size distribution

Sea ice floes break continuously into smaller pieces, and in the cold season they are at the same time frozen together into larger pieces. Floe size distributions show statistical regularity based on random floe break-up mechanisms. The characteristic floe size can be defined from its surface area S_f as $d = \sqrt{4 S_f / \pi}$. For a circle, d equals the diameter. The area of floes larger than size d is:

$$Q(d) = \int_d^\infty r^2 p(r) \, dr \tag{2.7}$$

where p is the spatial density of the floe size. Then q-fractiles d_q can be defined through $Q(d_q)/Q(0) = q$ (i.e., floes larger than d_q cover the fraction of q of the total ice area). A natural, representative floe size is $d_{0.5}$, the median. In soil mechanics (e.g., Harr, 1977), $d_{0.1}$ defines the effective size and the ratio $d_{0.6}/d_{0.1}$ is known as the uniformity coefficient. Rothrock and Thorndike (1984) introduced their mean "caliper diameter", equal to the directional average of the opening the floe goes through. It is approximately equal to π^{-1} times the perimeter. For a circle, this is therefore equal to the diameter.

Observations normally cover a range $d_0 \leq d \leq d_1$, where d_0 and d_1 are the

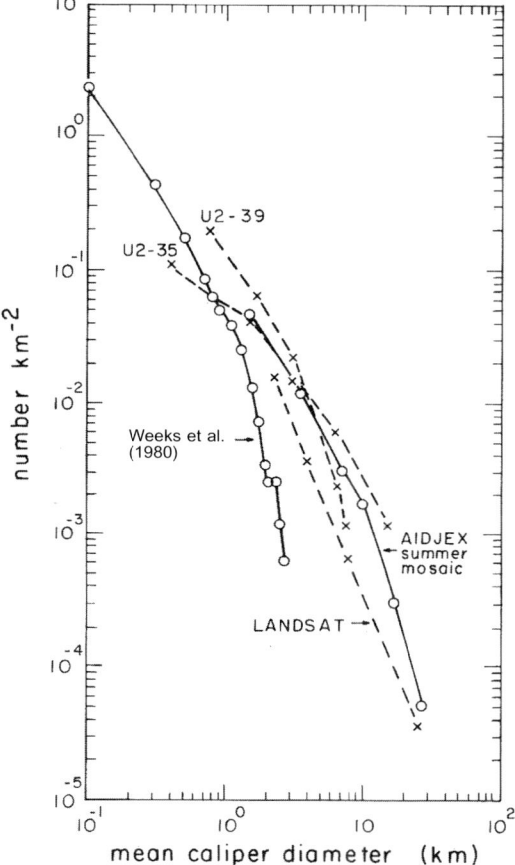

Figure 2.9. Floe size distributions from the Arctic Ocean.
Reproduced from Rothrock and Thorndike (1984), with permission from the American Geophysical Union.

minimum and maximum resolution. If d_1 is the maximum floe size, the distribution Q may be evaluated down to d_0. Anyway, the data may also be used to fit a theoretical model and consequently extended to all floes. The frequency of floe size steadily falls toward larger values with no local peaks or gaps (Figure 2.9) and their shape may be derived from rather simple fracturing principles (Figure 2.10).

Fractal geometry models lead to *power law* size distributions (e.g., Korvin, 1992). The probability density is:

$$p(d;n) \propto d^n \qquad (2.8)$$

In a d-band where $n \approx$ constant, *self-similarity* holds. Within a self-similar band the geometric structure is independent of the scale, reflected by photographs looking similar and a measure stick needing to be added for the scale information. Such scale

24 Drift ice material [Ch. 2

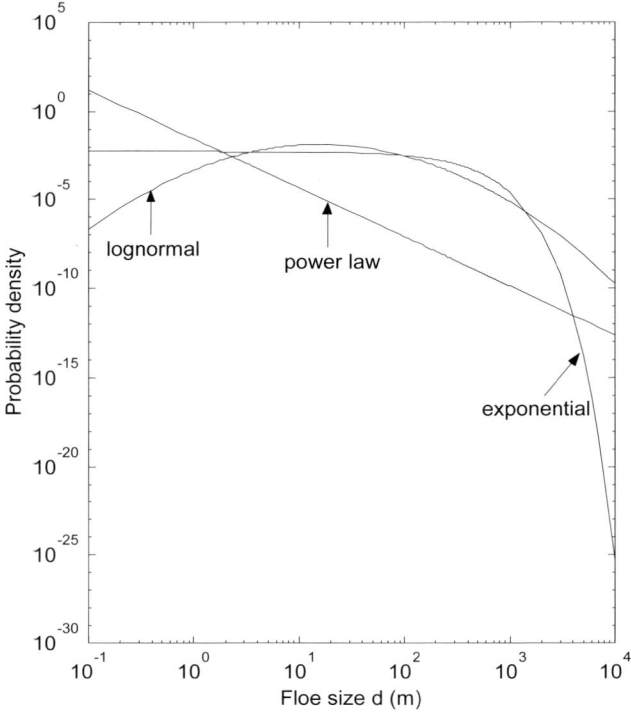

Figure 2.10. The forms of the power law, exponential, and lognormal distributions for the floe size, scaled with a median of 500 m.

invariance is quite common in geophysical data. In the data of Rothrock and Thorndike (1984) the mean caliper diameter followed the power law ($d > 100$ m) with $-2.5 < n \leq -1.7$. For the spring season in the Baltic Sea, Leppäranta (1981b) obtained $-3 < n < -2$, decreasing in the course of the spring melting season, illustrating that the area of small floes increases at the cost of large ones.

A problem with power laws is that the integral from 0 to ∞ is always infinite. Bounding from above by d_1 the integral converges for $n > -1$ and bounding from below by d_0 it converges for $n < -1$. It is clear that an upper bound exists, as the floes must be smaller than the size of the basin. Then with d_1 as the maximum floe size, $Q(d) = 1 - (d/d_1)^{n+3}$ and one needs $n > -3$ for convergence at 0. Observations usually suggest that $-3 < n < -1$ for $d > 20$ m (i.e., for ice pieces defined as floes in the WMO, 1970 nomenclature).

Example Take $n = -2$. Then $Q(d)/Q(0) = 1 - d/d_1$. The distribution of floe size in terms of areal coverage is uniform from zero to d_1, as sometimes observed (e.g., Leppäranta, 1981b). The median floe size is $d_1/2$, and any q-fractile is given by $d_q = d_1(1-q)$. The uniformity coefficient is $4/9 \approx 0.44$. For $n < -2$ small floes are areally dominant while for $n > -2$ the opposite is true. The exponent obtained from observations therefore illustrates how fractured the ice field is. ∎

A classical *random breakage* model, where the breakage probability is independent of the floe size, gives the *logarithmic normal distribution* (Kolmogorov, 1941). Then, by definition, $\log d$ is normally distributed, say, with mean M and variance s^2. Define a floe size d^* by $\log d^* = M$. The probability density then reads:

$$p(d; d^*, s^2) = \frac{1}{ds\sqrt{2\pi}} \exp\left\{\frac{-\log(d/d^*)^2}{2s^2}\right\}, \qquad d > 0 \qquad (2.9)$$

The median is then d^*, the mode and mean, respectively, are equal to $d^* \exp(-s^2/2)$ and $d^* \exp(s^2/2)$, and the variance is $d^{*2} \exp(s^2)[\exp(s^2) - 1]$ (e.g., Crow and Shimizu, 1988). The parameter s^2 is regarded as the shape parameter of the distribution. It is dimensionless since by definition $s^2 = \langle(\log d - \log d^*)^2\rangle = \langle\log^2(d/d_*)\rangle$. The median of the distribution Q for floe coverage becomes $d^* \exp(2s^2)$. The distribution has a positive mode, which is not far from zero and usually not covered in observational data; thus the existence of the mode is not clear.

If the breakage probability is proportional to the floe size, the spatial Poisson process results, leading to the *exponential distribution* of floe size:

$$p(d; \lambda) = \lambda \exp(-\lambda d) \qquad (2.10)$$

where λ is the distribution shape parameter. This is analogous to the distribution of waiting times of customers in the theory of temporal Poisson processes. Density can be directly integrated: the relative areal coverage of floes larger than d is $Q(d) = [1 + \lambda d + \frac{1}{2}(\lambda d)^2] \exp(-\lambda d)$. The median of the distribution Q for floe coverage can be numerically solved as $\lambda d_{0.5} \approx 2.67$.

Random breakage seems to explain floe size distributions well. Indeed, only in the case of small floes can physical mechanisms be found to produce a favourable floe size. An important property of floating ice is its *characteristic length*:

$$l_c = \sqrt[4]{\frac{Yh^3}{3\rho_w g(1-\nu^2)}} \qquad (2.11)$$

where Y is Young's modulus and ν is Poisson's ratio; representative values for them are $Y \approx 3\,\text{GPa}$ and $\nu = 0.3$ (e.g., Mellor, 1986). For $h = 1\,\text{m}$, we have $l_c = 18.2\,\text{m}$. The length of flexural waves of an elastic ice beam on water foundation under a point load equals $2\pi l_c$. When ice floes break under rafting, the size of pieces is $\frac{1}{4}\pi l_c$ (Coon, 1974).

At the ice edge, reflection and penetration of swell takes place (Wadhams, 1978; Squire et al., 1995; Squire, 1998). A penetrating swell becomes damped with distance from the ice edge. The wave energy as a function of distance (x) from the ice edge is:

$$E(x) = E(0) \exp(-x/l_s) \qquad (2.12)$$

where l_s is the attenuation distance. Long waves penetrate more easily. In heavy, compact ice, $l_s \approx 1\,\text{km}$ for short waves (100 m) and 5 km for long swells. Since the dominant wavelength increases exponentially from the ice edge and waves easily break floes down to one-half wavelengths, the floe size should increase exponentially from the ice edge, as long as there is enough energy in the wave field for floe breakage. This is observed in the MIZ in a qualitative sense. But no actual e-folding length scales for the floe size have been recorded, the signal being contaminated by strong mixing and the thermo-mechanical decay of ice floes.

In general, a random breakage hypothesis seems to work well. Ice floes have more or less randomly distributed defects; that is, cracks due to thermal, hydrostatic (from non-uniform ice thickness), tidal, and wind loads. The question is then how to formulate the breakage probability that would lead to the exact form of the distribution (Lensu, 2003). The size distribution of ice pieces below 20 m has not been studied to any great extent. Logarithmic normal and exponential distributions should be treated as specific cases, depending on the formulation of breakage probability.

Shape of ice floes

Sea ice floes are convex. Wintertime ice floes are typically rectangular or pentagonal, while in summer the sharp corners erode and the floes become rounded (Timokhov, 1998).

The shape of a floe is quantified using its major axis (d_{max}), minor axis (d_{min}), and characteristic diameter (d): d_{min} is the smallest opening through which the floe may penetrate and d_{max} is the length perpendicular to the direction of the minor axis. The form parameters are elongation d_{max}/d_{min} and shape factor $d^2/d_{max}d_{min}$. The latter equals unity for elliptical ice floes and $\pi/4$ for rectangular floes. In a study in the Baltic Sea (Leppäranta, 1981b), elongation was mostly 1–2 and the shape factor was 0.7–1.0. There are not enough floe shape results available to state how typical these Baltic Sea properties are.

Floe information in drift ice mechanics

Ice floe information is included in the continuum theory in that the validity of continuum hypothesis requires that the drift ice particle size should be an order of magnitude greater than the characteristic floe size (see Eq. 2.3). Otherwise, it is not yet clear how useful the floe size and shape information really are in the continuum of sea ice dynamics. In local-scale granular models both floe size and shape influence the results (and in particular, the friction coefficient between ice floes).

When ice floes reach their maximum packing density, the nature of material behaviour changes. When compactness is low, floe interactions do not transmit forces well and the level of stress is very small. When the ice pack closes up and the stress increases to a significant level, floes make contact with others, group together, and then lose many of their individual features. Nevertheless, the stress field still depends on the shape of ice floes throughout the total area of contact surfaces. This may be a reason for stress fields being lower in summer conditions. Drag forces between air and ice, and between water and ice depend on the size and shape of ice floes (Zubov, 1945; Andreas et al., 1984). When floes are smaller their number is larger and consequently there are more floe edges to provide form drag. This feature, however, disappears when the floes close up.

In granular media mechanics, horizontal grain size is an essential characteristic. In sea ice applications, floes are often assumed to be uniform and circular. Ovsienko (1976) was the first work to include floe size in a model, but without mechanical interaction. For the sea ice drift problem, Shen et al. (1986) developed a theory based on momentum transfer by floe collisions. A discrete particle model with full floe–floe interaction was studied by Løset (1993) for circular floes. Hopkins and Hibler (1991) and Hopkins (1994) examined the sea ice-ridging process with a discrete particle model.

The poor understanding of the role of horizontal floe characteristics is largely due to the fact that in sea ice dynamics the thickness of ice floes is the principal floe property, their horizontal size being given less prominence. However, floe sizes and shapes show much more regularity than the thickness field, making them more useful potentially than just a fixed background factor. The advantage of having horizontal floe data combined with their evolution laws in a model would be ease of observability. Horizontal properties are easily monitored by remote sensing, and therefore ice floe evolution laws could be calibrated and controlled. In addition, ice floes could be used as good tracers in sea ice dynamics.

2.3 THICKNESS OF DRIFT ICE

2.3.1 Basic characteristics

Sea ice can be as thick as 50 m. Its thickness is characterized by a very large variability in space and time due to thermal and mechanical processes. Thickness is defined as the distance between the upper and lower ice surfaces, any voids included. It is an irregular field and, because of fracturing and new ice growth in leads, it may have discontinuities. In the continuum approach, certain statistics of the thickness field, such as mean thickness, are considered for each continuum particle; these statistics are assumed to be smooth, as required by the theory. The thickness range normally extends down to zero, defining open water as "ice of zero thickness".

The thickness of ice determines its volume and strength. Furthermore, in ice engineering, thickness is a key parameter in estimating the force exerted by ice on structures. This is therefore the most important property of sea ice to measure and predict.

Thermal growth

The present book focuses on sea ice dynamics. Thermodynamics is not considered in any depth: instead of dealing with the heat budget directly, it is assumed that the *growth rate of ice* is a known function $\Phi = \Phi(t, h)$, with melting specified by negative growth rates. In purely dynamic events, $\Phi \equiv 0$. The reader interested in sea ice thermodynamics is directed to Maykut and Untersteiner (1971), Semtner (1976),

Makshtas (1984), Leppäranta (1993), and Kondratyev et al. (1996). For understanding scales and for simple modelling, Stefan's ice growth law says that:

$$dh/dt = a(T_f - T_s)/h \qquad (2.13)$$

where $a = k/(\rho L_F) \approx 5\,\text{cm/day}$ (k is the thermal conductivity of ice and L_F is the latent heat of freezing), T_s is the ice surface temperature, and T_f is the freezing point temperature (e.g., Leppäranta, 1993). This law assumes that $T_s \leq T_f$ and gives the solution $h^2 = h_1^2 + 2aS_F$, where h_1 is the initial thickness and $S_F = \int_0^t (T_f - T_s)\,dt'$. This is usually referred as the "sum of negative degree days" and is given in °C × day. Melting (negative growth) is determined basically by the radiation balance $Q_R = \alpha Q_s + Q_{nL}$, where α is albedo, Q_s is incoming solar radiation, and Q_{nL} is net, incoming long-wave radiation. Adding the heat flux from the water body Q_w:

$$dh/dt = -(Q_R + Q_w)/(\rho L_F) \qquad (2.14)$$

Here, the radiation balance must be positive, as is at times the case during the polar summer.

The growth of very thin ice rarely exceeds 10 cm per day, and for 1 m thick ice the growth rate is already below 1 cm per day. For the ice thickness h_1 to double in one day, Stefan's law says that $2aS_F = 3h_1^2$. If $h_1 = 10\,\text{cm}$ the daily mean surface temperature would need to be as low as $-30°\text{C}$, unrealistically low for a thin ice sheet. Melting is primarily a surface process and therefore independent of the thickness of ice. A heat flux of 30 W/m² would melt the ice vertically by 1 cm per day. Such levels are commonly observed for the radiation flux in polar summer. Oceanic heat flux is typically smaller, of the order of 5 W/m² in the Arctic Ocean, but can go up to 30 W/m² in ocean areas around Antarctica as well as in some smaller seas of the SSIZ (Shirasawa and Ingram, 1997).

In terms of scales, it is safe to conclude that, when ice thickness is 10 cm or more, it is very unlikely for the thickness to change by more than 5 cm in one day due to thermodynamic effects. In its first winter sea ice grows to 0.5–2 m thickness while in summer ice melts up to 1 m. Where summer melt is less than the first year's growth, multi-year ice fields develop. This is the case in the central Arctic Ocean and in places in the Southern Ocean, mainly in the Weddell Sea.

The sea ice nomenclature (WMO, 1970) connects thicknesses to types of undeformed ice as discussed in Section 2.1. New ice forms grow up to 10 cm, young ice is 10–30 cm, and first year ice is divided into thin (30–70 cm), medium (70–120 cm), and thick (greater than 120 cm) categories. Undeformed first-year ice can be as much as 2 m thick, while the equilibrium thickness of undeformed multi-year ice is 3–4 m (Maykut and Untersteiner, 1971). The terminology is based on ice conditions in the Arctic seas; it is not fully used in subpolar regions because of inconsistencies in the use of common language: for example, in the Baltic Sea, the thickest ice could be "thin first-year ice", and "young ice" could be several months old and as old as any ice there. It is therefore preferable to give the numerical value of thickness.

Mechanical growth

Mechanical deformation produces ice types with a range of thicknesses. The timescale of changes is short. These changes are asymmetric in that, while mechanical increase of ice thickness takes place, existing deformed ice is not undone mechanically but may disappear only due to melting. This also means that transforming kinetic energy into potential energy in deformation is irrecoverable, in contrast to ocean dynamics.

Thin ice usually undergoes rafting in compression (Figure 2.11). A theoretical formula for the maximum thickness of rafting ice is (Parmerter, 1975):

$$h_{rf} = \frac{14.2(1-\nu^2)}{\rho_w g} \frac{\sigma_t^2}{Y} \qquad (2.15)$$

where σ_t is the tensile strength of the ice sheet, a representative value being 0.65 MPa (Mellor, 1986). This gives $h_{rf} = 15$–20 cm. In single rafting the local thickness is doubled. Several layers of rafted ice have been documented (Palosuo, 1953; Bukharitsin, 1986). As the ice thickness increases, bending moments as a result of overriding become so large that small pieces break off and start to form hummocks and ridges. When thin ice is compressed against thick ice (in particular, at the fast ice boundary), jammed brash ice barriers form. Contrary to other deformed ice types, these barriers may loosen when the compressive force ceases. Working with predominantly very thin ice, the rafting process would need particular consideration.

Figure 2.11. Thin ice sheets undergo rafting in compression. The width of the interlocking fingers in the picture varies between 1 and 10 m.

Hummocking accumulates ice blocks into layers several times the thickness of the original ice sheet (Figure 1.2). The thickest forms of drift ice, ridges, may be up to 50 m thick. Fresh hummocks and ridges contain voids (20–40%) between the ice blocks. In cold conditions a consolidated layer grows down from the sea surface level. Thus the ice volume is 0.6–1.0 times the thickness, depending on the degree of consolidation.

Example (piling) Piling uniform-sized balls on top of each other results in open and closed locked packings of $\pi/6 \approx 0.52$ and $\sqrt{2} \cdot \pi/6 \approx 0.74$, respectively. Experience shows that, piling chopped firewood by randomly throwing one piece on top of another, porosity will be about 1/3. The closed packing of balls and random firewood piling are well within the range of the observed porosities of hummocks and ridges. ∎

Observations

The mean sea ice thickness is 2–8 m in the Arctic Ocean. It is lower in the Eurasian Shelf and it is at its highest off northern Greenland and the Canadian Archipelago (Figure 2.12). Distribution is strongly influenced by the drift of ice. Without dynamics the thickness would be more or less symmetric around the North Pole, but due to ice transport it is lowest (about 2 m) on the Siberian Shelf and 3 m or more in the Beaufort Sea. The highest thicknesses, averaging 8 m, result from the mechanical deformation of ice.

At a smaller scale, the mean thickness may vary a lot. A good illustration is provided by upward-looking submarine sonar data (Figure 2.13). Most of the thickness section is quite heavily deformed, which is typical for the Greenland Sea ice conditions. In this section ridge keels penetrate 10–20 m beneath the sea level (there are five ridge keels for this 2-km section). Rothrock (1986) shows a thickness section from the Beaufort Sea with a correlation length scale of the order of 1 km; the standard deviation was 2.4 m and the average was 3 m.

2.3.2 Measurement methods

To obtain thickness information is very difficult. A lot of effort has been put into this problem in sea ice remote sensing (e.g., Rossiter and Holladay, 1994; Wadhams, 2000). A good solution still does not exist—a major barrier to the progress of knowledge in drift ice dynamics. The basic techniques of sea ice thickness mapping are listed in Table 2.2.

Drilling is the traditional way of determining ice thickness, a direct measurement but not feasible for mesoscale or large-scale monitoring. One specific methodology for thickness mapping is recording from a ship (ice-breaker) the thickness of overturned ice blocks (Overgaard et al., 1983). Upward-looking sonar systems are considered the best method for ice thickness mapping by remote sensing. They measure the draft, which provides a very good estimate for total thickness. They are however restricted by their high instrumental and logistics costs, and the unavailability of

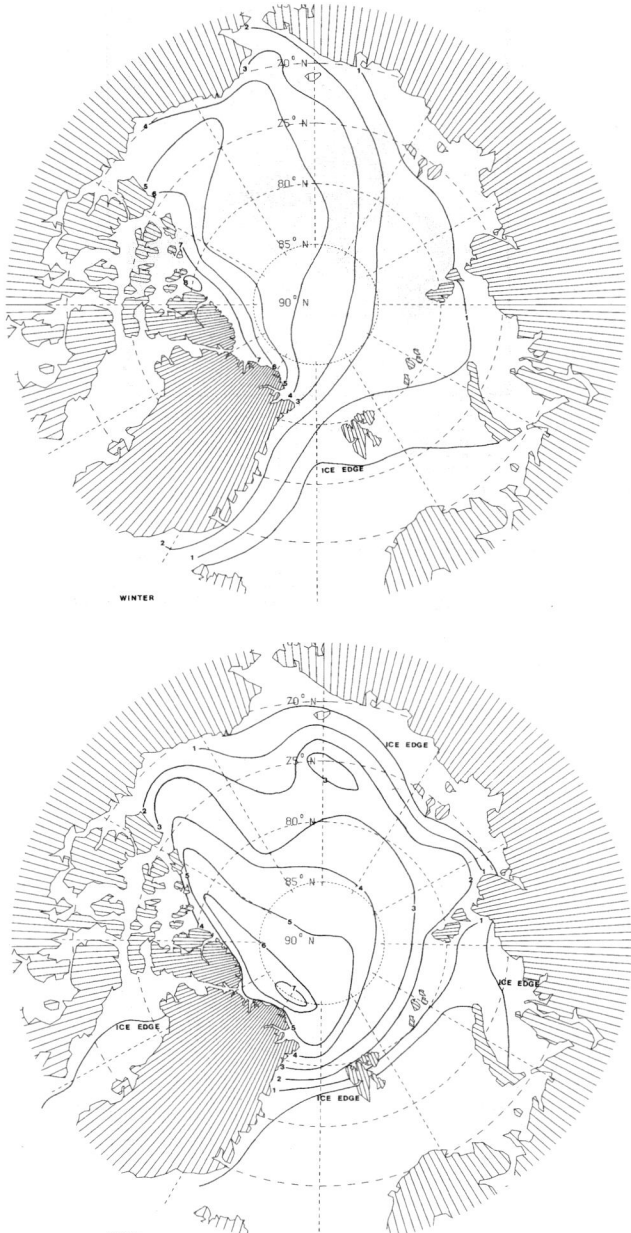

Figure 2.12. Mean sea ice draft in the Arctic Ocean based on submarine data. Since Archimedes' law states that 89% of the ice sheet is beneath the sea surface, the mean sea ice thickness is obtained by multiplying these draft numbers by 1.12.
Reproduced from Bourke and Garret (1987), with permission from Elsevier.

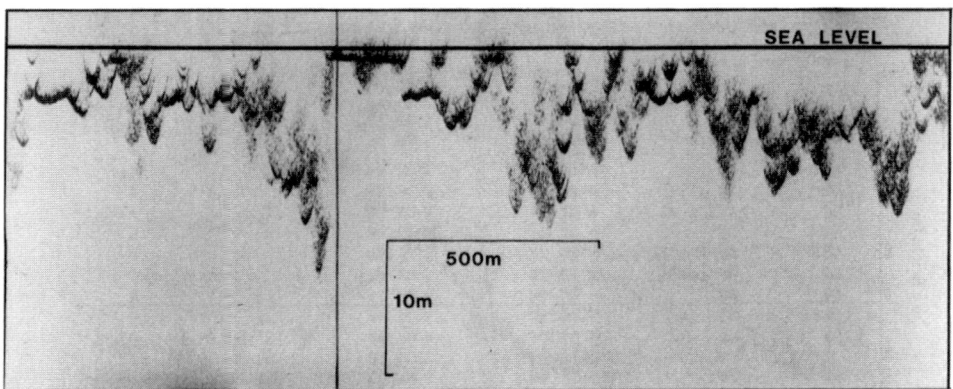

Figure 2.13. A section of the ice draft profile in the Greenland Sea from submarine sounding.
Reproduced from Wadhams (1981), with permission from the Royal Society of London.

Table 2.2. Methods for sea ice thickness measurement.

Method	Quality	Comments	Reference
Drilling	Excellent	Laborious	Traditional
Submarine sonar	Good	Access problems	Williams et al. (1975)
Bottom-moored sonar	Good	Non-real time	Vinje and Berge (1989)
AEM	Good	Large footprint	Kovacs and Holladay (1990)
Thermal mapping	Fair	Thin ice OK	Steffen and Lewis (1988)
Airborne laser	Fair	Sea ice ridges OK	Ketchum (1971)
GPR	Poor	Problems with brine	Rossiter et al. (1980)
SAR	Poor	Ice types only	Kwok et al. (1992)
Passive microwave	Poor	Ice types only	Gloersen et al. (1978)

real-time data. Submarine sonars have been routinely used since the 1950s (Lyon, 1961) but their data are normally classified (Figure 2.13 shows an example). Bottom-moored sonars were developed in the 1980s for ice thickness mapping (Vinje and Berge, 1989).

The airborne electromagnetic method (AEM), originally developed for ore prospecting and geological surveys, was introduced in the late 1980s for sea ice thickness mapping (see Rossiter and Holladay, 1994). By emitting electromagnetic fields at frequencies of 1–10 kHz, eddy currents are generated in the conductive seawater. These currents generate a secondary electromagnetic field, and, measuring this field at the receiver, the distance to seawater can be determined. Measuring the distance to the sea ice surface by an altimeter, the sea ice thickness is obtained. AEM is a logistically feasible way for reliable and quick ice thickness mapping. It also provides line data with a fairly large footprint, of the order of the flight altitude (30–100 m). The accuracy of results depends on the inversion model

Sec. 2.3]   Thickness of drift ice   33

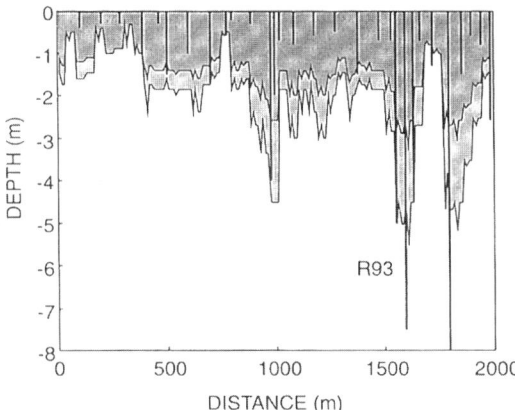

Figure 2.14. AEM calibration for ice thickness in the Baltic Sea, March 1993. The vertical bars show drilling data, dark and light-shaded areas show inversions with one- and two-layer models, respectively (the footprint is 100 m).
From Multala et al. (1996).

used (Figure 2.14), since by adding layers a more realistic conductivity distribution can be taken for the ice. Recently surface-based or shipborne EM systems have been introduced (Haas, 1998).

Thermal infrared mapping can be applied to sea ice mapping in cold weather because the surface temperature of sea ice and the eventual snow cover on top depend on the structure and thickness of the ice and snow. It works well for ice thinner than about 50 cm when the air temperature is less than $-5°C$ (Steffen and Lewis, 1988). Airborne laser profilometers give the upper surface topography (quite widely used since 1970). In principle, surface elevation can be interpreted for total ice thickness using Archimedes' law, but the measurement accuracy and snow cover cause severe problems. However, sea ice ridges are shown in surface topography profiles and constitute the principal aim of laser surveys.

Microwave sea ice mapping is carried out by ground-penetrating radar (GPR), synthetic aperture radar (SAR), and the (passive) radiometer. They are widely used for sea ice mapping but provide only poor results for ice thickness (Rossiter and Holladay, 1994; Kwok et al., 1998; Wadhams, 2000). Radar signals cannot penetrate sea ice because of brine and seawater inclusions: the connection between ice thickness and backscatter is weak. SAR can detect two to four nominal ice types: the most promising is ridged ice, because piled-up ice blocks give strong backscatter. Microwave radiometers (10–100 GHz) can differentiate between open water, first-year ice, and to some degree multi-year ice (Gloersen et al., 1978; Kondratyev et al., 1996). On the global scale they are by far the most used for ice information (see Figure 2.5).

It is clear that no single method is sufficient for ice thickness mapping; rather, a combination of instruments is needed to cover the whole range from 0-m to 50-m thicknesses. Satellite-derived thickness information results in poor quality, as it is

2.3.3 Ice thickness distribution

Sea ice thickness distribution was introduced by Thorndike et al. (1975). Let $\underline{h} = \underline{h}(x, y; t)$ stand for the actual ice thickness and h represent the ice thickness as the distribution variable. Recalling the ice function I, we have $I(\underline{h}; h) = 0$ if $\underline{h} \leq h$ or 1 if $\underline{h} > h$. In a region Ω, the area of ice thinner than or equal to h is:

$$S(h) = \int_\Omega [1 - I(\underline{h}; h)] \, d\Omega \qquad (2.15)$$

Thus $S(0)$ is the area of open water and $S(\infty)$ is the total area of Ω. The normalized form $\Pi(h) = S(h)/S(\infty)$ is the spatial ice thickness distribution, analogous to the probability distribution. The derivative (in the generalized sense) of Π, $\pi(h) = d\Pi/dh$, is the spatial density of the ice thickness. *Note*: the thickness distribution was originally (Thorndike et al., 1975) chosen to be continuous from the left (i.e., $S(h)$ equal to the area of ice thinner than h); here it is continuous from the right (i.e., $S(h)$ equal to the area of ice thinner than or equal to h). The current way is also common in probability theory.

The thickness distribution is not continuous everywhere since part of the spatial density mass is concentrated in open water and in homogenous ice patches. These discontinuities can be mathematically handled using the delta function δ, defined as $\delta(s) = 0$ for $s \neq 0$ and $\int_{-\infty}^{\infty} \delta(s) f(s) \, ds = f(0)$ for any integrable function f, and using the Heaviside function H, $H(s) = 0$ for $s < 0$ and $H(s) = 1$ for $s \geq 0$. They are connected by the generalized derivative $dH/ds = \delta(s)$. Spatial density can be written formally as a sum of a discrete part and a continuous part:

$$\pi(h) = \sum_k \pi_k \delta(h - h_k) + \pi'(h) \qquad (2.16)$$

where π_k is the probability of discrete thicknesses h_k, $k = 0, 1, \ldots$, and π' is the continuous component of the distribution. The corresponding (cumulative) distribution function $\Pi(h)$ is obtained by replacing δ by H and π' by its integral Π' in Eq. (2.16).

The thickness distribution has the following mathematical properties:

(i) $\Pi(0) = 1 - A$ and $\Pi(\infty) = 1$ by definition.
(ii) The first moment is the mean ice thickness.
(iii) The second moment is proportional to the mean potential energy of the ice (see Eq. 2.2).

There is no simple general form for the distribution function because of the very large space-time variability of its shape, and therefore discrete histogram approximations are used. However, the upper tail of the distribution drops exponentially (Wadhams, 1998). The thickness distribution includes both undeformed and

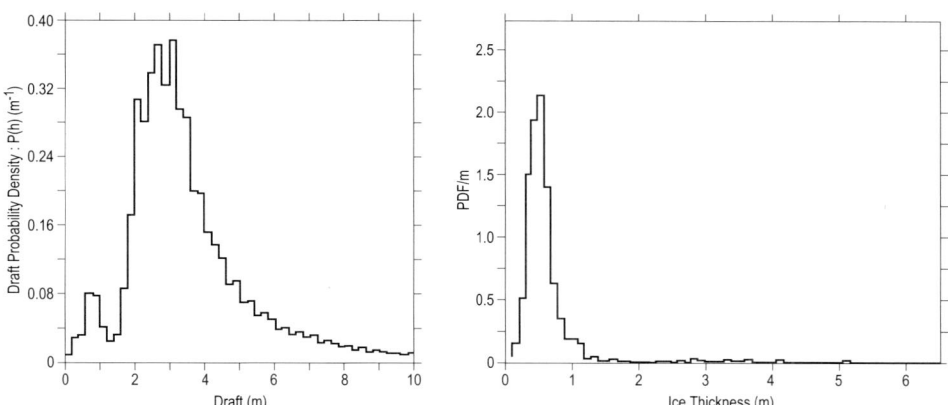

Figure 2.15. Observed ice thickness distributions. *Left*: Fram Strait (Wadhams, 1980a). *Right*: Antarctica, Atlantic sector (Wadhams et al., 1987).
Redrawn from the originals.

deformed ice but ignores spatial information about the structure of drift ice. Figure 2.15 shows ice thickness distributions from both polar regions. The multi-year peak is well developed only in the Arctic Ocean, and there is much less deformed ice in Antarctica.

2.4 SEA ICE RIDGES

Sea ice ridges are a particular form of deformed ice or pressure ice (Figure 2.16). They are the thickest ice formations, typically 5–30 m, and over large areas their volume may account for up to about one-half of the total ice volume. In sea ice dynamics, ridging is the main sink of kinetic energy in deformation due to friction and production of potential energy (Rothrock, 1975a). They are important hydrodynamic form drag elements at the air–ice and ice–water interfaces. In ice engineering, ridges are of deep concern because (i) they are connected with the highest ice loads on structures within first-year ice fields, (ii) they scour the sea bottom, and (iii) they influence on-ice traffic conditions. *In situ* field studies on the structure of ridges are made by drilling and sonar, while their spatial statistics are mapped by remote-sensing methods, primarily airborne laser profilometer, SAR, and submarine upward-looking sonar.

2.4.1 Structure of ridges

The top part of a ridge is called the *sail* and the lower part is called the *keel*. In the keel there is a *consolidated zone* that grows downward from the water surface level as the ridge ages. In the sail and lower keel the ice blocks are loose or weakly frozen

36 Drift ice material [Ch. 2

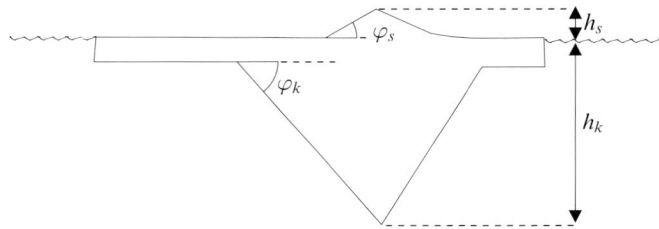

Figure 2.16. An ice ridge shown in a field photograph (Baltic Sea, March 1988) and in a schematic cross-sectional diagram.

together. A simple structural model of ridges (Figure 2.16) consists of a triangular keel and sail, described by the keel depth h_k, sail height h_s, slope angle φ, and porosity q. $\varphi \approx 30°$ (keel) or $20°$ (sail) and $q \approx 0.25$ (e.g., Timco and Burden, 1997; Kankaanpää, 1998). The cross-sectional volumes of keel and sail are $h_k^2 \cot \varphi_k$ and $h_s^2 \cot \varphi_s$. A more general model would have a trapezoidal keel. The structure of a ridge undergoes continuous evolution due to freezing, melting, and erosion, becoming smoother with time (Figure 2.17).

Ice ridges float according to Archimedes' law (see Eq. 2.1). The ratio of keel depth to sail height $\gamma = h_k/h_s$ is obtained from:

$$(\rho_w - \rho)h_k^2 \cot \varphi_k = \rho h_s^2 \cot \varphi_s \qquad (2.17)$$

We have $\gamma = 2.8$ for $\varphi_k = \varphi_s$ and $\gamma = 3.5$ for $\varphi_k = 30°$, $\varphi_s = 20°$. For multi-year ridges $\gamma \approx 3$ but for first-year ice ridges $\gamma \approx 4$–5 (Wright et al., 1978). The latter case comes from the keel being steeper than the sail. Wittman and Schule (1966) presented a triangular keel model with $\gamma \approx 3.3$ based on field data. Once the sail height is known, the total cross-section of a ridge can be estimated as $R = kh_s^2$ where

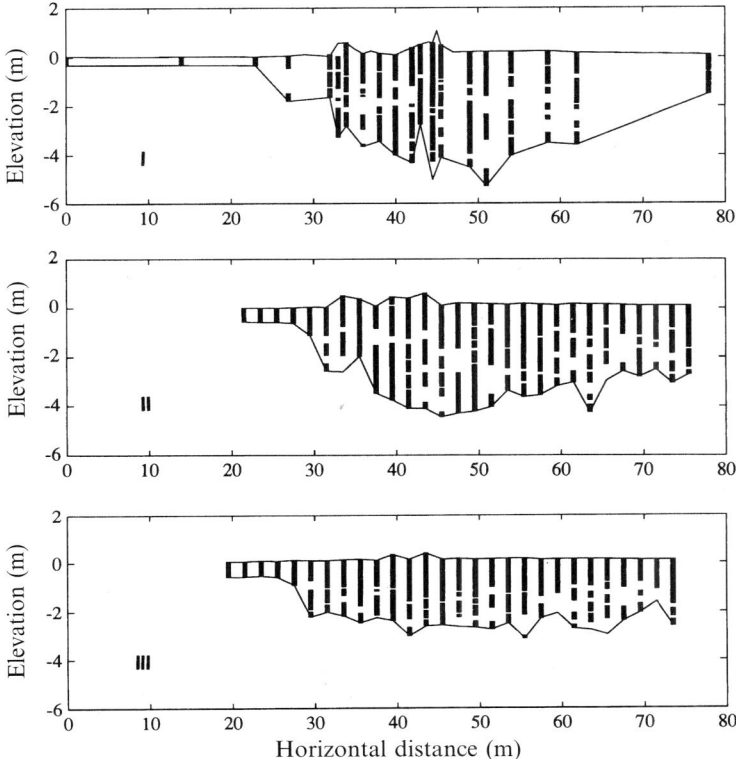

Figure 2.17. The evolution of the cross-sectional profile of one ridge near Hailuoto Island, Baltic Sea, winter 1991: I – February, II – March, III – April.
From Leppäranta et al. (1995).

the coefficient k represents all structural parameters. For the Baltic Sea, according to Leppäranta and Hakala (1992), $k \approx 17$.

A physical limitation exists for ridge growth (Parmerter and Coon, 1972; Hopkins and Hibler, 1991). When a ridge has grown to a certain vertical size, the ice sheet is too weak to penetrate into the ridge, for further growth. It breaks in front of the ridge, producing lateral growth. This limiting size mainly depends on the thickness of the parent ice sheet (very few ridges grow to this size). The record ridge sizes come from the Beaufort Sea (Wright et al., 1978): a floating ridge with a sail height of 12 m and a keel depth of 45 m, and a grounded ridge with a sail height of 18 m.

In shallow areas where the sea depth is less than the keel depth, grounding takes place and ridges anchor to fixed ice islands. This is typically observed at the landfast ice boundary. Grounded ridges serve as tie points to the ice and therefore help the fast ice boundary to extend farther away from the coast. Grounded ridges, when moving, scour the ocean bottom, and the keel may penetrate deep, depending on the bottom material (Blondel and Murton, 1997). As a consequence, pipes laid on the

38 **Drift ice material** [Ch. 2

floor of shelf waters must be buried deep enough to avoid damage from scouring keels.

2.4.2 Statistical distributions of ridge size and occurrence

Ridges give an impressive feature to the drift ice landscape (Figure 2.18). The spatial distribution of ridging is described in terms of their size and occurrence. The first occurrence data were counts made in reconnaissance flights resulting in a typical level of 5–10 ridges per km, with a maximum of 20 ridges per km (Wittman and Schule, 1966). Presently ridges are usually mapped using airborne laser profilometers for sails (Ketchum, 1971; Hibler et al., 1972; Lewis et al., 1993) and using upward-looking sonar for keels (Williams et al., 1975; Vinje and Berge, 1989).

A *cut-off size*, F_c, needs to be introduced, since the size of the chosen ridges must be well above the noise level of the measurement system. Also sails should not be mixed with snowdrifts. The cut-off sizes have been 1 m (or 3 feet ≈ 0.9 m) for sails and 5 m for the keels in the polar oceans, somewhat lower in smaller subpolar basins such as 0.4 m in the Baltic. Independent sails or keels are taken from observed

Figure 2.18. The Gulf of Bothnia in winter (Louis Belanger, according to A.F. Skjöldebrand). Travellers crossing a ridged ice field, in a romantic drawing showing the ice landscape in exaggerated linear dimensions to create a fairy-tale atmosphere.

Reproduced from Etienne Bourgelin Vialart, comte de Saint-Morys, *Voyage pittoresque de Scandinavie. Cahier de vingt-quatre vues, avec descriptions*, 1802, with permission of Collections of Museovirasto, Helsinki, Finland.

surface profiles using the *Rayleigh criterion*: ridges are taken as the local maxima, which are greater than the cut-off height and between which there is a local minimum with elevation less than half of these maxima. The cut-off is more a parameter related to observational technology than to real ice. Visual counts have given about the same number of sails as laser profilings. This confirms that the cut-off is well tuned to reflect how the ice field looks from above, a situation that is also true of the concept of the significant wave height of wind-driven surface waves.

Ridge size

Sail heights or keel depths follow the exponential distribution:

$$p(h^*; h_c, \lambda) = \lambda \exp[-\lambda(h^* - h_c)], \qquad h^* \geq h_c \qquad (2.18)$$

where h^* represents the sail height or keel depth, and λ is the distribution shape parameter. The mean size is simply $h_c + \lambda^{-1}$ and the standard deviation is λ^{-1}. This distribution was first proposed by Wadhams (1980a) for the Arctic Ocean, and it has been confirmed several times, for sails in the Baltic Sea (Leppäranta, 1981b; Lewis et al., 1993) and in the Antarctic (Weeks et al., 1989; Granberg and Leppäranta, 1999), and for keels in the Arctic by Wadhams and Davy (1986). The first size distribution was proposed by Hibler et al. (1972) who arrived at a probability density proportional to $\exp(-h^{*2})$.

The statistical background of the size distribution can be based on a certain random hypothesis concerning the probabilities of the different size arrangements (see Hibler et al., 1972). The exponential distribution comes from assuming that all height arrangements yielding the same total sum are equally probable. Such a hypothesis, however, has no clear physical background. Representative values for the mean sail height and keel depth in the central Arctic Ocean are $\langle h_s \rangle \approx 1.2$–$1.4$ m (cut-off 0.9 m) and $\langle h_k \rangle \approx 8$–$14$ m (cut-off 6.1 m) (Hibler et al., 1972). Therefore the parameter λ^{-1} is 0.3–0.5 m for sails and 2–5 m for keels. In the Baltic Sea the mean sail height was 0.5–0.6 m (cut-off 0.4 m) in Lewis et al. (1993).

Ridge spacing

Spacings between ridges relate in some way to the size of ice floes, and so we expect similar statistical laws to apply for them (see Section 2.2). In fact, the first model for ridge spacings was the exponential distribution (Hibler et al., 1972). It was later replaced by the logarithmic normal distribution (Wadhams and Davy, 1986; Lewis et al., 1993), giving the impression that new ridges would be randomly born at any point between existing ridges. In the seasonal sea ice zone, as in the Baltic Sea, this may be true. However, in the central Arctic Ocean ridges form in leads, and, in turn, the birth of ridges therefore follows the distribution of lead spacings and inherits the logarithmic normal form. The fit is in general not excellent, though, and an attempt was made in Lensu (2003) to improve the distribution model by including a clustering effect in ridging. Representative values for mean ridge spacing are 5–10 km^{-1} in the central Arctic Ocean and in the Baltic Sea, but with cut-off sail heights of 0.4 m and 0.9 m, respectively (Hibler et al., 1972; Lewis et al., 1993).

The inverse mean ridge spacing μ is called the ridge density and is equal to the mean number of ridges per length. It depends on the cut-off size. So, assuming the exponential distribution for ridge size, we have:

$$\mu(h_{c2}) = \exp[-\lambda(h_{c2} - h_{c1})]\mu(h_{c1}) \qquad (2.19)$$

With regard to the two-dimensional aspects of spacing, deviations from isotropy occur as one would expect (e.g., Mock et al., 1972; Leppäranta and Palosuo, 1983). However, isotropy has so far been the main working hypothesis. It simply leads from one to two dimensions through (Mock et al., 1972):

$$\frac{L_R}{S} = \frac{\pi}{2}\mu \qquad (2.20)$$

where L_R is the total length of ridges in the horizontal area S. For anisotropic cases this equation has been found reasonable when using the directionally averaged ridge density parameter.

2.4.3 Ridging measures

The ridge size and spacing distributions can be combined to form a measure of ridging intensity. Two natural measures arise: $\mu\langle h^*\rangle$ and $\mu\langle h^{*2}\rangle$. The first is dimensionless and describes the sum of sail heights or keel depths per unit length (with sails it is also proportional to the aerodynamic form drag of ridges: Arya, 1975). The second has the dimension of length and is proportional to the mean thickness of ridges, h_R. This quantity and the areal concentration of ridges (S_R/S) are:

$$h_R = \frac{\pi}{2}\mu\langle R\rangle, \qquad \frac{S_R}{S} = \frac{\pi}{2}\mu\langle b_R\rangle \qquad (2.21)$$

where b_R is the ridge width. If only sail data exist, the cross-section and width of ridges are approximated by $R \approx kh_s^2$ (then $h_R = \frac{1}{2}\pi\mu k\langle h_s^2\rangle$) and $b_R = 2\gamma h_s \cot\varphi_k$. Hibler et al. (1972) used the formula $h_R = 10\pi\mu\langle h_s^2\rangle (k = 20)$ to estimate the volume of ridged ice in the Arctic Ocean, while Leppäranta and Hakala (1992) obtained $k = 17$ in the Baltic Sea. The volume of ridged ice is typically 10–40% of the total ice volume (Hibler et al., 1974b; Mironov, 1987; Weeks et al., 1989; Lewis et al., 1993; Granberg and Leppäranta, 1999).

Example In approximate terms, $b_R \approx 2\sqrt{3}h_k$ and $R \approx h_k^2\sqrt{3}$. Then $S_R/S \approx 5\mu h_k$ and $h_R \approx 3\mu h_k^2$. For representative values, taking $h_k \approx 7.5$ m and $\mu \approx 5$ km^{-1} gives $S_R/S = 19\%$ and $b_R = 84$ cm. These numbers characterize a moderately ridged Arctic region. ∎

In order to describe ridging properly, one needs at least three parameters: the cut-off size, the mean size, and ridge density. The cut-off size is a free parameter that needs to be "tuned" so that ridge statistics agree with the appearance of the ice field. These parameters can be used to obtain a ridging intensity measure, but the relationship cannot be inverted to obtain the ridge size and spacing from the intensity. However,

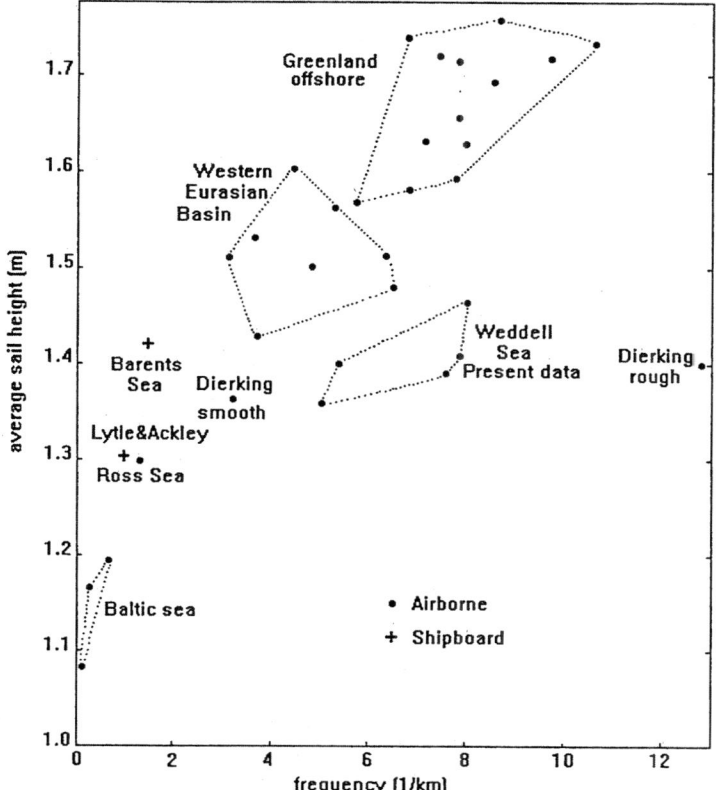

Figure 2.19. Ridge sail height vs. ridge density in different seas. Cutoff is 1 m.
From Dierking (1995), Granberg and Lepparanta (1999), Lytle and Ackley (1991), and Wadhams (1981).

there are regional regularities (Figure 2.19). In the central Arctic Ocean the size and density of ridges are correlated (Wadhams, 1981), while in the Weddell Sea the mean ridge size shows very small variations and could therefore be considered fixed (Granberg and Lepparanta, 1999). The latter case also holds in the Baltic Sea (Lepparanta, 1981b).

Example (Lepparanta, 1981b) In the Baltic Sea the average sail height is nearly constant, about 20 cm above the cut-off height of 40 cm. The ridge density alone provides a good estimator of the volume of ridged ice: in quantitative terms $h_R = 2.2 \, \text{cm}^2 \mu \cdot 10^3 \pm 27\%$ (i.e., on average one ridge per kilometre accounts for 2.2 cm in ice thickness). ∎

2.4.4 Hummocked ice

Not all mechanically deformed ice appears as ridges: in places irregular, hummocked ice fields with no regular geometry are found. Their thickness is less than that of ridges but greater than that of undeformed, level ice. Their existence is well known to ice-breaker captains, who consider them a nuisance because they may be hidden beneath the snow cover. Not many direct measurements exist about the spatial distribution of the thickness of hummocked ice. Such a field may be described by the mean and standard deviation of the thickness.

One way to estimate the volume of hummocked ice is to first consider the observation that the exponential distribution for ridge size holds for any manageable cut-off size. This suggests the following hypothesis: The exponential form can be extrapolated down to a zero cut-off height, but at some non-zero level the ridges lose their ridge-like form, corresponding to the visual appearance of the ice field. The "ridges" beneath a well-chosen cut-off height are then taken to represent hummocked ice. Without any model for the geometry of hummocks, this extrapolation allows us to determine their total volume or the mean thickness h_h:

$$\frac{h_h}{h_h + h_R} = \frac{1}{2} \exp(-\lambda h_c)[1 + (\lambda h_c)^2] \tag{2.22}$$

In most ridge observations $\lambda h_c \sim \frac{1}{2}$, and therefore $h_h/(h_h + h_R) \sim \frac{1}{4}$. Knowing the volume of hummocks is enough for ice budget calculations but it would be desirable to break it down into mean physical thickness and areal concentration. This would need a geometrical model of hummocked ice.

2.4.5 Total thickness of deformed ice

The mean thickness of deformed ice finally reads:

$$h_d = h_h + h_R \tag{2.23}$$

The spatial distribution of ridges has been traditionally described in terms of areal or volume fraction. Wittman and Schule (1966) reported that in the Canadian Basin the average areal fraction of pressure ice is 0.13–0.18 with a maximum of more than 0.5 in heavy-deformation zones, while according to Kirillov (1957) in the Kara Sea the average volume fraction is 0.28 with a maximum of 0.39.

The degree of ridging and hummocking has also been described by an index from 0 to 5. This index is roughly proportional to the total thickness of deformed ice, with the full level 5 corresponding to $h_d/h_u \approx 1.5$–2.5, where h_u is the thickness of undeformed ice and $h_i = h_u + h_d$ (Gudkovic and Romanov, 1976; Appel, 1989).

Example (Kirillov's formula) Kirillov (1957) made the following assumptions for ridges and hummocks: triangular sail and keel cross-sections, $h_k/h_s = 3$, on average $h_k = h_u$, and $q = 0.3$. Then $h_d/h_u \approx 2.0 \cdot S_R/S$. Thus observing the area of deformed ice provides an estimate for its volume. ∎

2.5 DRIFT ICE STATE

In the continuum dynamics of sea ice, an ice state J must be defined. It is the set of the material properties of drift ice necessary to solve the dynamics problem. In practice there are observational limitations: although the horizontal properties such as ice compactness and floe size can be easily mapped from satellites, ice thickness is very difficult. As the number of properties included in J is defined as N we speak of an N-level ice state. Most work in continuum sea ice dynamics has been done by assuming that the ice thickness field is sufficient to describe the state of drift ice.

The first approach is to define ice *categories* as the ice state variables. Normally these are chosen to be manageable *observables*, meaning that the ice state is largely based on the information provided by routine ice charting systems. Recall the function "ice", I, from Eq. (2.4). In a low-level approach it is necessary to define generalized ice compactness and mean ice thickness:

$$A(h_0) = S^{-1} \int_\Omega I(x,y;h_0) \, d\Omega \qquad (2.24a)$$

$$\langle h(h_0) \rangle = S^{-1} \int_\Omega h I(x,y;h_0) \, d\Omega = A(h_0) \langle h_i(h_0) \rangle \qquad (2.24b)$$

where $\langle h_i(h_0) \rangle$ is the mean ice floe thickness over the subregion $\Omega \cap \{h \geq h_0\}$. These quantities ignore all ice thinner than h_0. Very thin ice does not give significant resistance to deformation, and it is preferable to exclude this from the mean thickness for the strength estimator. On the other hand, in remote sensing very thin ice is not well detected, and consequently $A(h_0)$ corresponds better to initialization and validation data. The areal fraction $1 - A(h_0)$ is the concentration of open water and thin ice; ice only grows there thermodynamically. The argument h_0 is no longer shown below but any category may be chosen conditional on $h \geq h_0$.

$J = \{A\}$ or $J = \{\langle h \rangle\}$ are natural one-level ice states but they are not frequently used because of too limited information. The minimum feasible ice state is their union, which is in fact the very widely used *two-level ice state* (Nikiforov, 1957; Doronin, 1970):

$$J = \{A, \langle h \rangle\} \qquad (2.25)$$

or since $\langle h \rangle = A \langle h_i \rangle$, the state $J = \{A, \langle h_i \rangle\}$ has the same information.

For *three-level ice states*, the decomposition of ice into undeformed ice and deformed ice (Leppäranta, 1981a) is often used:

$$J = \{A, h_u, h_d\} \qquad (2.26)$$

The thickness of undeformed ice changes thermodynamically, while undeformed ice is dynamically transformed into deformed ice. Lu et al. (1989) used another three-level system: ice compactness, first-year ice, and multi-year ice. The timescale was less than 1 month, and therefore these ice categories were independent but very useful for modelling dynamics because of their different thicknesses. Mechanical deformation may then be correctly limited to first-year ice only.

The ice category approach can be extended to multi-level ice states $J = \{A, j_1, j_2, j_3, \ldots\}$. Additional, reasonable, new ice categories are introduced based on their dynamical significance and observability (Haapala, 2000). With a morphological model, more information can be extracted from one category.

Leppäranta (1981a) decomposed deformed ice into density and size of ridges, these being controlled visually or with a laser profilometer.

The second approach is to take the *thickness distribution* for the ice state. The question then becomes: How many levels are needed or, in other words, what is the necessary or convenient resolution of the distribution? The thickness classes are fixed, arbitrarily spaced, and their histogram contains the state variables:

$$J = \{\pi_0, \pi_1, \pi_2, \ldots\}, \qquad \sum_k \pi_k = 1 \qquad (2.27)$$

In low-level cases this is a very crude system in which the actual ice thickness information can easily disappear. Hence, the approach based on ice categories is preferable. The two-level state based on the thickness distribution is not the same as the two-level state based on the ice categories, since in the former case the thickness (or rather the thickness band) is fixed. In new, growing ice, it would need to spread evenly all over the ice band, far from what really happens.

Consequently when choosing the thickness distribution for the ice state the number of thickness classes should be large (of the order of ten). For such a number of levels, the ice category approach becomes cumbersome and the thickness distribution approach becomes preferable.

Example If there are no thermodynamic changes and mechanical changes only concern ice compactness, the two-level ice category and thickness distribution approaches overlap. The former is $J = \{A, \langle h_i \rangle = \text{constant}\}$ and the latter is $J = \{\pi_0, \pi_1 = 1 - \pi_0\}$ (i.e., only $\pi_0 = A$ changes). ∎

Also, a combination can be worked out to add more observability to the thickness distribution. One possibility is to divide the thickness distribution into two parts: undeformed and deformed ice. This would allow us to differentiate thermal ice growth from mechanical ice growth.

In principle, the horizontal properties of *ice floes* can be added to the ice state in a similar way to thickness. Floes possess size and form, and size at least shows statistical regularity (see Section 2.2). In the continuum dynamics of sea ice, floe information has been used only in floe collision models (Shen et al., 1986), there taken as uniform, circular disks.

Beginning with sea ice in the world ocean, sea ice types, and sea ice observation systems, the geophysical medium of drift ice has been introduced in this chapter. The medium consists of ice floes, and is regarded as a continuum over length scales much greater than the typical size of floes. The relevant properties of the drift ice continuum for its dynamics are mainly ice compactness and thickness, which vary largely in time and space. The last section presented the concept of an "ice state", which contains information about the drift ice field necessary to understand and model its dynamics. In Chapter 3 observations about the kinematical properties of the drift ice medium are presented.

3

Ice kinematics

3.1 DESCRIPTION OF ICE VELOCITY FIELD

This chapter deals with the methods, data, and physics that relate to sea ice kinematics. It begins with a mathematical description of the two-dimensional motion of a rigid body and a continuum. Then observations of two elements of sea ice kinematics—drift and deformation—are given and analysed using the mathematical techniques presented. The kinematics approach is unique in allowing stochastic modelling, which is treated in the third section. In the last section, the ice conservation law is derived for the ice state, linking the ice state from Chapter 2 to its mechanical changes due to kinematics.

In ice kinematics the properties of the velocity field are examined, considering also the correlation of ice velocity with wind and currents. The outcome provides answers to such questions as: How long does it take for the ice to go from point A to point B? What is the typical ice velocity range in a certain sea area? How does the separation distance between two floes evolve? What does the ice velocity spectrum look like? Kinematics data are used in stochastic drift modelling and form the basis in the construction and validation of dynamical sea ice models as well.

Sea ice floats and drifts on the Earth's sea surface, taken as the sphere of radius $r_e = 6{,}370$ km. The natural co-ordinates are the zenith angle Z (latitude is equal to $\pi/2 - Z$) and the azimuth (longitude) λ. A local Cartesian co-ordinate system with x, y, and z for the east, north, and upward co-ordinates is usually considered, with the sea surface as the zero reference level; the unit vectors in x, y, and z directions are $\boldsymbol{i}, \boldsymbol{j}$, and $\boldsymbol{k}$, respectively. The motion of sea ice takes place in the three-dimensional world, but it can be treated as a *two-dimensional* phenomenon on the sea surface. The ice velocity is then given by a two-dimensional vector function of time and the horizontal co-ordinates $\boldsymbol{u} = \boldsymbol{u}(Z, \lambda; t)$ or, as is more usual, in a local Cartesian system:

$$\boldsymbol{u} = \boldsymbol{u}(x, y; t) \tag{3.1}$$

Fridtjof Nansen (1861–1930, second from left), a Norwegian polar explorer, with *Fram* in the background. He was the leader of the *Fram* expedition (1893–1896), which drifted across the Arctic Ocean. He was the first to collect sea ice and oceanographic data from the central Arctic Ocean. He showed that the ice drift follows the wind with angular deviation in direction and that the ice flows out to the North Atlantic from the Arctic. Nansen was given the Nobel Peace Prize in 1922 for his help with refugees after the First World War. The photograph is from his book *Fram över Polarhavet* published in 1897 (Finnish translation, 1954).

with u and v, respectively, as the x- and y-components. This is true for two reasons: first, ice lays on the sea surface and does not possess a vertical velocity component; and, second, ice floes move as rigid pieces and the horizontal velocity is independent of the vertical co-ordinate. In exact terms, with respect to the Earth, the vertical velocity is not zero but given by the boundary condition $w(x, y) = \partial \xi / \partial t$, where ξ is sea level elevation. The slowly changing sea level does not cause any significant inertial effects on sea ice motion. Then there are small vertical displacements of ice blocks during pressure ice formation; they will be obtained from the ice conservation law presented in Section 3.4.

3.1.1 Motion of a single floe

The dynamics of a single ice floe is examined using classical rigid body mechanics (e.g., Landau and Lifschitz, 1976). This motion has three degrees of freedom

Sec. 3.1] Description of ice velocity field 47

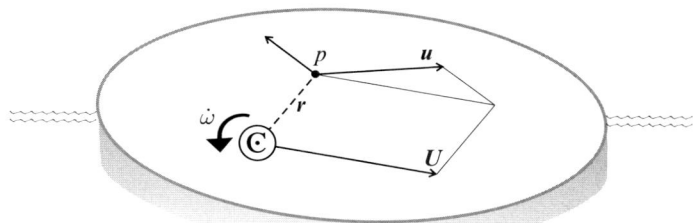

Figure 3.1. Motion of a single ice floe: translation velocity U and angular velocity $\dot\omega$ around the mass centre C.

consisting of the translational motion $U = (U, V)$ and the rotation rate $\dot\omega$ (positive anticlockwise) about the vertical axis (Figure 3.1). Let r stand for the radius vector from the mass centre of a floe. The velocity at r is:

$$\boldsymbol{u} = \boldsymbol{U} + \dot\omega \mathbf{k} \times \boldsymbol{r} \qquad (3.2)$$

For a circular floe with radius R, the translational kinetic energy is $m|U|^2/2$ and the rotational kinetic energy is $m(\dot\omega R)^2/4$, their ratio being $2|U|^2/(\dot\omega R)^2$. Therefore, if a 1-km size floe translates by 10 cm/s the rotation rate must be $\sqrt{2}\times 10^{-4}\,\mathrm{s}^{-1} = 29°\,\mathrm{h}^{-1}$ for the energies to be equal. Such rotation would be exceptionally fast. Because of the low rotational energy, it is normally sufficient to consider translational velocity only.

In a granular ice flow, each floe has its own translational and rotational velocities. As long as there are enough free paths around floes, they drift independently and each floe has its own translation and rotation. When free paths disappear, floes join into larger and larger groups. Their interaction by collisions and overriding begins, and the motion of such a set of ice floes is described using purely statistical drift models, granular flow models with individual grain interactions, or continuum flow models.

3.1.2 Continuum deformation

When the size of drift ice particles becomes one order of magnitude larger than the floe size, continuum approximation may be used for the kinematics (see Section 2.2). The continuum approach averages the movements of individual floes to the continuum length scale; the resulting velocity field needs to have continuous spatial derivatives to at least second order. The mathematics of general continuum kinematics is presented in detail by Mase (1970) or Hunter (1976), among others.

Strain and rotation

The motion of a drift ice continuum can be decomposed into rigid translation, rigid rotation, and strain. In a horizontal flow, *translation* has two horizontal components and *rotation* has one component around the vertical axis, like a rigid single floe.

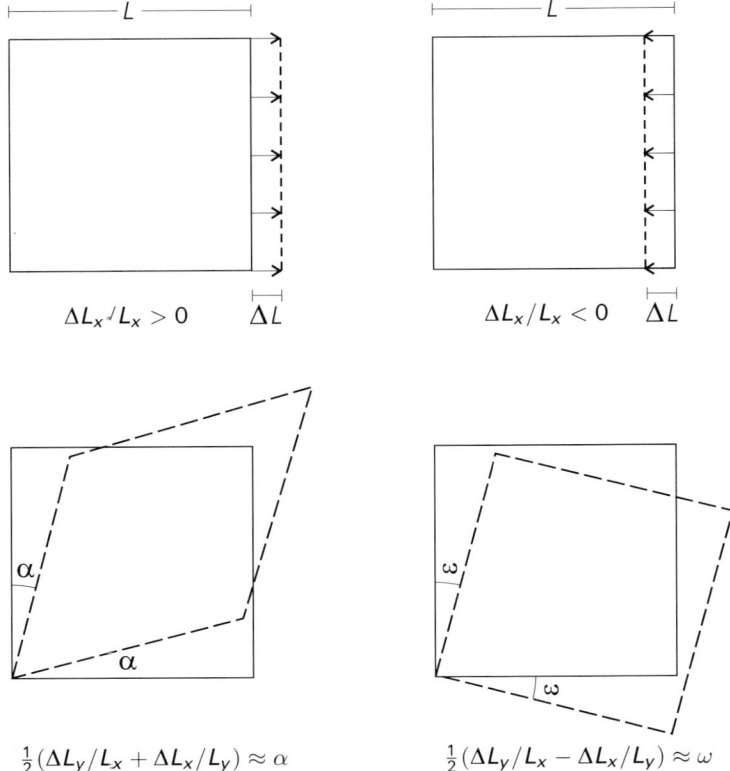

Figure 3.2. Strain modes and rotation, and their measures: *upper graphs*, extension and contraction; *lower graphs*, shear and rotation (here clockwise, $\omega < 0$).

Strain represents the physical deformation of continuum particles (Figure 3.2). There are three different modes of strain: tensile strain or extension, compressive strain or contraction, and shear strain. The first two modes are *normal strains* (with opposite signs) that change lengths, while *shear strain* changes the shape of particles. A tensile strain of 1% means that the material has lengthened by 1%, a compressive strain of 1% means that the material has shortened by 1%, and a shear strain of 1% means that a right angle in the material configuration has changed by $0.01\,\text{rad} \approx 0.57°$. Finally, a rotation of 1% corresponds to a counterclockwise rotation of the particle by $0.01\,\text{rad} \approx 0.57°$.

Strain and rotation are tensors. Let X stand for a reference configuration of a material particle and consider its change represented by a mapping $X \to x$ (i.e., $x = x(X)$). Removing the pure translation, the change is given by the *displacement gradient* $V = F - I$, where $F = \nabla x = (\partial x_i/\partial X_j)$ and $I = (\delta_{ij})$ is the unit tensor (e.g., Hunter, 1976). In small-deformation theory ($|V| \ll 1$) the displacement gradient contains the strain ε and the rotation ω, given by its symmetric and antisymmetric parts. Respectively: $\boldsymbol{\varepsilon} = \frac{1}{2}(\boldsymbol{V} + \boldsymbol{V}^T)$ and $\boldsymbol{\omega} = \frac{1}{2}(\boldsymbol{V} - \boldsymbol{V}^T)$, where the superscript T stands for the transpose.

Strain rate and vorticity

Let us consider a drift ice field moving with velocity $\boldsymbol{u}$, consisting of a rigid translation velocity and a differential velocity. The latter is given by the velocity gradient $\nabla \boldsymbol{u}$, where ∇ is the two-dimensional gradient operator $\nabla = \mathbf{i}\partial/\partial x + \mathbf{j}\partial/\partial y$. The velocity gradient is a dyadic (i.e., a second-order tensor with four components):

$$\nabla \boldsymbol{u} = \begin{bmatrix} \partial u/\partial x & \partial u/\partial y \\ \partial v/\partial x & \partial v/\partial y \end{bmatrix} \tag{3.3}$$

The displacement during a time interval δt equals $\boldsymbol{u}\,\delta t$. The normal strain along the x-axis is then:

$$\frac{\Delta L_x}{L_x} = \left[\left(u + \frac{\partial u}{\partial x}\delta x\right)\delta t - u\delta t\right]\frac{1}{\delta x} = \frac{\partial u}{\partial x}\delta t$$

Dividing by δt, the rate of normal strain along the x-axis becomes $\partial u/\partial x$. Treating the other components in a similar way, it is seen that the velocity gradient is exactly the rate of the displacement gradient. Consequently the symmetric and antisymmetric parts of the velocity gradient tensor:

$$\dot{\boldsymbol{\varepsilon}} = \mathrm{d}\boldsymbol{\varepsilon}/\mathrm{d}t = \tfrac{1}{2}[\nabla \boldsymbol{u} + (\nabla \boldsymbol{u})^T] \tag{3.4a}$$

$$\dot{\boldsymbol{\omega}} = \mathrm{d}\boldsymbol{\omega}/\mathrm{d}t = \tfrac{1}{2}[\nabla \boldsymbol{u} - (\nabla \boldsymbol{u})^T] \tag{3.4b}$$

are the *strain-rate* and rotation-rate, or *vorticity*, tensors, respectively, and $\nabla \boldsymbol{u} = \dot{\boldsymbol{\varepsilon}} + \dot{\boldsymbol{\omega}}$. Their dimension is 1/time, and their inverses thus define the timescales of strain and rotation.

The strain-rate tensor has three independent components:

$$\left.\begin{aligned} \dot{\varepsilon}_{xx} &= \frac{\partial u}{\partial x} \\ \dot{\varepsilon}_{xy} &= \dot{\varepsilon}_{yx} = \frac{1}{2}\left(\frac{\partial u}{\partial y} + \frac{\partial v}{\partial x}\right) \\ \dot{\varepsilon}_{yy} &= \frac{\partial v}{\partial y} \end{aligned}\right\} \tag{3.5}$$

The strain rates $\dot{\varepsilon}_{xx}$ and $\dot{\varepsilon}_{yy}$ are the normal strain rates in the x- and y-directions, and $\dot{\varepsilon}_{xy}$ is the shear strain rate of squares aligned with the x- and y-axes. The total strain rate, taken as the norm of the strain-rate tensor, is the square root of the sum of squares of all its components, $|\dot{\boldsymbol{\varepsilon}}| = \sqrt{\dot{\varepsilon}_{xx}^2 + \dot{\varepsilon}_{xy}^2 + \dot{\varepsilon}_{yx}^2 + \dot{\varepsilon}_{yy}^2}$. The magnitude of the drift ice strain rate is typically $10^{-6}\,\mathrm{s}^{-1}$, ranging from $10^{-5}\,\mathrm{s}^{-1}$ in active marginal ice zone (MIZ) dynamics (Leppäranta and Hibler, 1987) to $10^{-7}\,\mathrm{s}^{-1}$ in the central Arctic Ocean (Hibler et al., 1973). Note that $10^{-6}\,\mathrm{s}^{-1} = 0.36\%\,\mathrm{h}^{-1} = 8.64\%$ per day.

Strain rates of 10^{-7}–10^{-5} s^{-1} correspond to a timescale of 1–100 days. For $\dot{\varepsilon}_{xx}$ = constant, the line length becomes $L = L_0 \exp(\dot{\varepsilon}_{xx} t)$, and thus the inverse normal strain rates $|\dot{\varepsilon}_{xx}|^{-1}$ and $|\dot{\varepsilon}_{yy}|^{-1}$ are equal to the e-folding times of the corresponding line lengths. For $\dot{\varepsilon}_{xy}$ = constant, the angles of a square aligned with x- and y-axes change by $\arctan(|\dot{\varepsilon}_{xy}|t)$, and thus $|\dot{\varepsilon}_{xy}|^{-1}$ equals the time when they have changed by 45° (note that two angles decrease and two angles increase this amount).

Example For a numerical illustration, take the strain-rate magnitude as 10^{-6} s^{-1}. The deformation timescale is 10 days. Then, $\dot{\varepsilon}_{xx} = 10^{-6}$ s^{-1} means that a 10-km line opens in the x-direction by 1 cm/s $\approx$ 1 km/day and $\dot{\varepsilon}_{xy} = 10^{-6}$ s^{-1} (e.g., $\partial u/\partial y = \partial v/\partial x = 10^{-6}$ s^{-1}) means that a right angle in x- and y-axis orientations closes by 10^{-6} s$^{-1} \approx$ 5° per day. ∎

In a two-dimensional flow, vorticity has only one independent component:

$$\left.\begin{array}{c} \dot{\omega}_{xx} = \dot{\omega}_{yy} = 0 \\ \dot{\omega}_{yx} = \frac{1}{2}\left(\frac{\partial v}{\partial x} - \frac{\partial u}{\partial y}\right) = -\dot{\omega}_{xy} \end{array}\right\} \quad (3.6)$$

The component $\dot{\omega}_{yx}$, denoted usually by $\dot{\omega}$, is the vorticity about the vertical axis, and for a rigid field it is the same as the rigid body rotation rate. A typical magnitude of drift ice vorticity is 10^{-6} s^{-1} = 0.2° h^{-1} = 5° d^{-1}, which corresponds to rigid rotation progressing as $\dot{\omega} t$. At time $\dot{\omega}^{-1}$ the rotation has progressed by 1 rad and at time $2\pi\dot{\omega}^{-1} \approx 63$ days one cycle has been completed.

The principal components or the eigenvalues of the strain-rate tensor are obtained from the eigenvalue equation $\det(\dot{\boldsymbol{\varepsilon}} - \lambda \mathbf{I}) = 0$, where det is the determinant. The directions of the principal axes are obtained from the eigenvector equation $(\dot{\boldsymbol{\varepsilon}} - \lambda \mathbf{I}) \cdot \boldsymbol{\Lambda} = 0$, usually normed to $\Lambda = 1$. In two dimensions there are two eigenvalues $\dot{\varepsilon}_1$, $\dot{\varepsilon}_2$ and two corresponding eigenvectors:

$$\dot{\varepsilon}_{1,2} = \tfrac{1}{2} \operatorname{tr} \dot{\boldsymbol{\varepsilon}} \pm \sqrt{(\tfrac{1}{2} \operatorname{tr} \dot{\boldsymbol{\varepsilon}})^2 - 4 \det \dot{\boldsymbol{\varepsilon}}} \quad (3.7a)$$

$$\vartheta_1 = \arctan\left(\frac{\dot{\varepsilon}_1 - \dot{\varepsilon}_{xx}}{\dot{\varepsilon}_{xy}}\right), \quad \vartheta_2 = \frac{\pi}{2} + \vartheta_1 \quad (3.7b)$$

where tr stands for the trace, and ϑ_1 and ϑ_2 are the eigenvector directions; the eigenvalues are chosen so that $\dot{\varepsilon}_1 \geq \dot{\varepsilon}_2$. In the principal axes co-ordinate system the normal strain rates are $\dot{\varepsilon}_1$ and $\dot{\varepsilon}_2$, and the shear strain rate is zero. If $\dot{\varepsilon}_{xy} = 0$, then $\dot{\varepsilon}_{xx}$ and $\dot{\varepsilon}_{yy}$ are the principal values and the x- and y-axes are the principal axes. The particular case $\dot{\varepsilon}_{xx} = \dot{\varepsilon}_{yy}$ and $\dot{\varepsilon}_{xy} = 0$ is spherical deformation. Then, the eigenvalues are $\dot{\varepsilon}_1 = \dot{\varepsilon}_2$ and the orientations of the principal axes are arbitrary.

A two-dimensional tensor has two invariants that contain information about its frame-independent properties. The principal values are such invariants, but normally their sum and difference are introduced for the analysis of a strain-rate tensor:

$$\dot{\varepsilon}_I = \dot{\varepsilon}_1 + \dot{\varepsilon}_2 = \text{tr}\,\dot{\boldsymbol{\varepsilon}} = \dot{\varepsilon}_{xx} + \dot{\varepsilon}_{yy} \tag{3.8a}$$

$$\dot{\varepsilon}_{II} = \dot{\varepsilon}_1 - \dot{\varepsilon}_2 = \sqrt{(\text{tr}\,\dot{\boldsymbol{\varepsilon}})^2 - 4\det\dot{\boldsymbol{\varepsilon}}} = \sqrt{(\dot{\varepsilon}_{xx} - \dot{\varepsilon}_{yy})^2 + 4\dot{\varepsilon}_{xy}^2} \tag{3.8b}$$

The first invariant equals the divergence ($\text{tr}\,\dot{\boldsymbol{\varepsilon}} = \nabla \cdot \boldsymbol{u}$), while the second invariant is twice the maximum rate of shear. Note that shear deformation on a square depends on the orientation. For a square aligned with the principal axes, there is no shear, and for a square oriented 45° to the principal axes, shear is at an extremum, $\pm\dot{\varepsilon}_{II}/2$. Positive shear strain means closing the right angle, while negative shear strain means opening it.

The mode of deformation can be thought of as a vector $(\dot{\varepsilon}_I, \dot{\varepsilon}_{II})$ in the upper half-space. The proportions of normal and shear deformation are specified by the vector direction φ, $\tan\varphi = \dot{\varepsilon}_{II}/\dot{\varepsilon}_I$, $0 \leq \varphi \leq \pi$ (Thorndike et al., 1975). The angles $\varphi = 0$, $\varphi = \pi/2$, and $\varphi = \pi$ correspond to pure divergence, pure shear, and pure convergence, respectively. Note also that $|\dot{\boldsymbol{\varepsilon}}| = \sqrt{\dot{\varepsilon}_I^2 + \dot{\varepsilon}_{II}^2}$ such that the length of the vector $(\dot{\varepsilon}_I, \dot{\varepsilon}_{II})$ provides the magnitude of the total strain rate.

Another useful representation of the strain rate is its decomposition into spherical and deviatoric parts:

$$\dot{\boldsymbol{\varepsilon}} = \dot{\boldsymbol{\varepsilon}}_s + \dot{\boldsymbol{\varepsilon}}' = (\tfrac{1}{2}\,\text{tr}\,\dot{\boldsymbol{\varepsilon}})\,\mathbf{I} + \dot{\boldsymbol{\varepsilon}}' \tag{3.9}$$

The spherical part includes pure compression/dilatation, and the deviatoric part includes pure shear deformation.

There are five specific simple deformation cases:

(i) Spherical strain: $\dot{\boldsymbol{\varepsilon}} = \dot{\boldsymbol{\varepsilon}}_s$.
(ii) Pure shear or incompressible medium: $\dot{\boldsymbol{\varepsilon}} = \dot{\boldsymbol{\varepsilon}}'$, $\dot{\varepsilon}_I = 0$.
(iii) Uniaxial tension (compression): $\dot{\varepsilon}_{xx} > 0$ ($\dot{\varepsilon}_{xx} < 0$), $\dot{\varepsilon}_{yx} = \dot{\varepsilon}_{yy} = 0$.
(iv) Uniaxial shear: $\dot{\varepsilon}_{xy} = \dot{\varepsilon}_{yx} = \tfrac{1}{2}\,\partial u/\partial y$, $\dot{\varepsilon}_{xx} = \dot{\varepsilon}_{yy} = 0$.
(v) Zonal flow: $\partial/\partial y = 0$; that is, $\dot{\varepsilon}_{xx} = \partial u/\partial x$, $\dot{\varepsilon}_{xy} = \tfrac{1}{2}\,\partial v/\partial x$, $\dot{\varepsilon}_{yy} = 0$.

Sometimes it is useful to work with curvilinear co-ordinate systems, in particular with spherical co-ordinates on the Earth's surface ($r = \text{constant} = r_e$). The strain-rate tensor is then:

$$\left.\begin{aligned}\dot{\varepsilon}_{ZZ} &= \frac{1}{r_e}\frac{\partial u_Z}{\partial Z}, \qquad \dot{\varepsilon}_{\lambda\lambda} = \frac{1}{r_e \sin Z}\frac{\partial u_\lambda}{\partial \lambda} + u_Z\frac{\cot Z}{r_e} \\ \dot{\varepsilon}_{Z\lambda} &= \frac{1}{2}\left(\frac{1}{r_e \sin Z}\frac{\partial u_Z}{\partial \lambda} - u_\lambda\frac{\cot Z}{r_e} + \frac{1}{r_e}\frac{\partial u_\lambda}{\partial Z}\right)\end{aligned}\right\} \tag{3.10}$$

where Z is zenith angle and λ is longitude. In flows with zonal invariance the derivatives $\partial/\partial\lambda$ are neglected. The curvature effects are then represented by the terms with $\cot Z$.

The deformation of drift ice is particularly asymmetric: leads open and close, pressure ice forms but does not "unform". In pure shear $\dot{\varepsilon}_1 = 0$ and $\dot{\varepsilon}_1 = -\dot{\varepsilon}_2$. This results in opening along the first principal axis and closing along the second principal axis being equal in magnitude. If compactness is initially 1, the total opening will be $\dot{\varepsilon}$ and the pressure ice production will be $-\dot{\varepsilon}_2$. Therefore, it is necessary to construct a function to describe opening and ridging in a given deformation event.

Example: estimation of the strain rate and vorticity from drifter data (Thorndike and Colony, 1980; Leppäranta, 1981b) Drifters[1] are usually irregularly scattered in the study region. Assuming a continuum ice flow, the Taylor formula can be used for the velocity field:

$$u(x,y) = u(0,0) + x\frac{\partial u}{\partial x}(0,0) + y\frac{\partial u}{\partial y}(0,0) + R$$

where R is a residual of the second order. With at least three drifters, we can use linear regression and the above formula to obtain velocity derivatives. With more than three drifters, an estimate of the goodness of fit is obtained for the linear strain field. In a similar fashion, the three second-order derivatives can be added to the Taylor formula. To estimate all, three more drifters are needed. ∎

3.2 OBSERVATIONS

3.2.1 Methods

Drift ice kinematics has been mapped in three different ways. A drifter tracks the trajectory of the floe on which it has been deployed, and imaging airborne or spaceborne remote-sensing methods provide displacement fields from sequential mapping. A more recent development is acoustic remote sensing from fixed moorings beneath the ice. The drifter and imaging methods are Lagrangian, while the acoustic method is Eulerian.[2]

The *drifter method* dates back to the drift of *Fram* in the Arctic Basin (Nansen, 1902), followed by *Jermak* and *Maud* in the Arctic and *Deutschland* in the Antarctic. The Soviet North Pole station programme commenced in 1937, with annual stations drifting from the North Pole to the Greenland Sea. In the 1950s and 1960s much research was carried out on Ice Island T-3, also known as Fletcher's Ice Island, drifting in the Beaufort Sea (e.g., Hunkins, 1967). Since the 1970s, automatic tracking methods have been available. *Argos* buoys (Figure 3.3) have produced an

---

[1] Sea ice kinematics is traditionally mapped using drifters, buoys or research stations anchored into the ice.
[2] In the Lagrangian approach the paths of physical ice floes are monitored, while in the Eulerian approach instantaneous velocities are observed at fixed points in space. This terminology comes from the Lagrangian and Eulerian formulations in fluid dynamics (e.g., Li and Lam, 1964).

Figure 3.3. Deployment of an *Argos* buoy during the Ymer-80 expedition, July 1980, north of Svalbard. *Argos* buoys have provided fundamental knowledge about sea ice drift climatology in the polar oceans.

extensive data set of the ice kinematics in the Arctic Basin, with a positioning accuracy of 500 m (Thorndike and Colony, 1982). The temporal resolution of the traditional drifter method was at best one to five cycles per day, but spatial information was quite limited.

In the 1960s, the mapping of differential ice velocity began at distances of 10–100 km around the ice camps. The first field experiment on sea ice deformation was made in 1961 in the central Arctic Ocean by monitoring a rectangle with sides 75–100 km in length (Bushuyev et al., 1967). By closing up the scales to less than 10 km, a high degree of accuracy was needed. The methods used included theodolites (Legen'kov, 1977), laser geodimeters (Hibler et al., 1974b; Leppäranta, 1981b), and microwave transponder systems (Leppäranta and Hibler, 1987). Currently, the GPS (global positioning system) is the best way to monitor ice kinematics. The accuracy is excellent, down to metres, and the only limiting factor is spatial coverage, or how many drifters are deployed. Leppäranta et al. (2001) reported using GPS in a coastal sea ice dynamics experiment, where buoy monitoring and data transmission were performed using the local mobile phone network.

The principal properties of a drifter system are positioning accuracy (δx) and the measurement time interval (δt). From a practical point of view the cost of a single drifter and the realization logistics must also be considered. Taking the positioning error as white noise with variance σ^2, the velocity error in first-order differentiation becomes blue noise[3] with the variance spectrum $\sigma^2(1-\cos\lambda)/\pi$, $0 \leq \lambda \leq \pi$ (where λ is given in radians and δt is the time unit). The temporal resolution of a drifter system may be taken as the frequency at which real variance is twice the noise level. At the 10-km scale the variance magnitude of sea ice velocity is 1 (cm/s)^2. Therefore, to resolve the deformation up to 1 c.p.h. (cycle per hour) the positioning accuracy must be 10 m or better.

Example The *Argos* system is simple to use and reliable, but the disadvantages are the cost of drifters, low accuracy (500 m), low data transmission rate (a few hours time interval), and as a consequence low temporal resolution. Let us take $\delta t = 3$ h and let the variance of the position error be $s^2 = (250\,\text{m})^2$ (accuracy represented as 2s). Estimating the ice velocity by first-order differentiation of the drifter positions, the total variance of velocity error is then $2s^2/\delta t^2 \approx 10.7$ (cm/s)^2 and the Nyquist frequency is $(6\,\text{h})^{-1}$. Integration of the spectral density shows that, for frequencies lower than 1 c.p.d. (cycle per day), the total variance is 0.27 (cm/s)^2 (i.e., the standard error in these low-frequency displacements is 450 m per day). ∎

Sea ice velocity can be extracted from *sequential remote-sensing imagery* by identifying ice floes and then determing the distance travelled (Figure 3.4). This was first done manually using optical imagery (e.g., Gorbunov and Timokhov, 1968; Hibler et al., 1975; Muench and Ahlnäs, 1976). Imaging radars are very good at ice velocity mapping because of their high resolution and weather independence. In particular, along with data obtained by synthetic aperture radar (SAR) satellites, automatic algorithms have been developed based on cross-correlation or feature tracking (Fily and Rothrock, 1987; Vesecky et al., 1988; Kwok et al., 1990; Sun, 1996). Ice kinematics products have been among the most useful outcomes from SAR satellites.

More recently algorithms have been developed to extract sea ice motion from satellite passive microwave data using the cross-correlation method (Agnew et al., 1997; Kwok et al., 1998; Liu and Cavalieri, 1998; Kwok, 2001). This approach has the great advantage of the availability of daily data from the *Nimbus*-7 scanning multichannel microwave radiometer (SMMR) and DMSP special sensor microwave imager (SSM/I) ever since 1978 for both polar regions (htpp://nsidc.org/). Therefore, the results are excellent for climatological investigations.

The drifter method provides temporally dense Lagrangian velocity time series for a limited number of floes, and the satellite or airborne imaging method produces spatially dense Lagrangian displacement fields between consecutive overflights, which are infrequent. Neither is preferable; rather, they are complementary. A high space-time resolution can be provided by fixed remote-imaging stations. A system of three coastal radar stations (Figure 3.5) has been used in Hokkaido, Japan since 1969, allowing continuous mapping of the ice cover along a 250-km section of coastline out to 60 km from the shore (Tabata, 1972). Figure 3.6 gives an

---

[3] In white noise the spectral density is constant, while in blue (red) noise there is more variance at higher (lower) frequencies.

example of sea ice distribution. As the ice moves along the coast, sometimes eddies can be traced in the coastal current.

Acoustic Doppler current profilers are normally used for water velocity profiling. They are based on the acoustic backscatter of suspended particles in natural waters. The ice bottom serves as a good backscatterer, and therefore the method can also be used beneath the ice cover. It has been shown to work in shallow seas (less than 200 m deep) for the ice velocity in the Gulf of St Lawrence (Belliveau et al., 1990). The advantages are the safety of the velocity measurement arrays, a good temporal resolution, and truly Eulerian data series.

3.2.2 Characteristics of observed sea ice drift

Drift ice speed is of the order of 1–100 cm/s. Figure 3.7 shows the long-term paths of several drifting stations in the Arctic Ocean. They follow the general sea ice circulation. The *Transpolar Drift Stream* takes ice from the Siberian Shelf across the Eurasian side and through the Fram Strait into the Greenland Sea, while on the American side ice rotates in the *Beaufort Sea Gyre*. These two flows meet in the central parts of the Arctic Basin where mixing of ice floes may take place. Average velocities are 1–5 cm/s, with the higher levels reached when approaching the Fram Strait. The drift pattern follows the ice–ocean linear response to the average atmospheric pressure field over the Arctic Ocean.

In the Antarctic, the governing features are the westward flow zone close to the continent and eastward drift zone farther out, driven by easterly and westerly winds (Figure 3.8). In places there are meridional movements, which interchange ice floes between the two zones. In particular the Antarctic Peninsula forces the westward flow toward the north in the Weddell Sea.

Figure 3.9 shows a 1-week time series of hourly sea ice velocity in the Baltic Sea together with wind data. As the ocean current is weak there, the ice followed the wind with essentially no time lag. On average the drift velocity was 2.5% of the wind speed and the direction of ice drift was 20° to the right of the wind direction. However, the ice velocity experienced some remarkable changes, which cannot be explained by a linear wind drift rule: occasionally, the ice nearly stopped during a moderate wind. This behaviour is typical of drifting sea ice: in general there is a good connection with the wind field (or ocean currents), but sometimes the ice takes "unexpected" steps, which are due to its internal friction.

The *frequency spectra* of ice velocity reach the highest levels on synoptic timescales, and a secondary peak appears at the inertial period (Figure 3.10). In the Arctic Ocean the variance level is lower and the inertial signal weaker in winter than in summer. Apart from the inertial peak, the slope of the spectral density for frequencies higher than 1 c.p.d. is about -3, meaning that the spectral density is proportional to the frequency to the power of -3. The reported spectra of sea ice motion go up to around 5 c.p.d. To obtain very high frequencies the positioning accuracy would need to be of the order of metres.

In special cases very high ice velocities (more than 1 m/s) have been observed. Zubov (1945) reported ice velocities up to 2.5 knots as a 1-day average in the

56 Ice kinematics

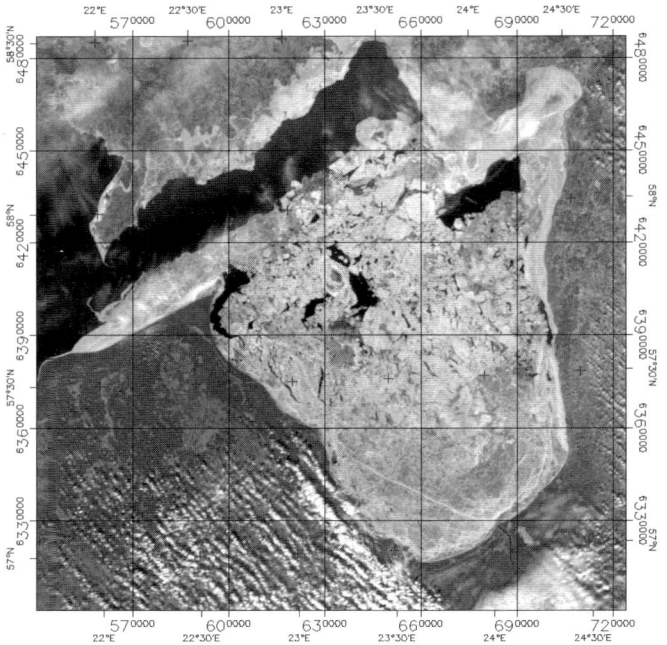

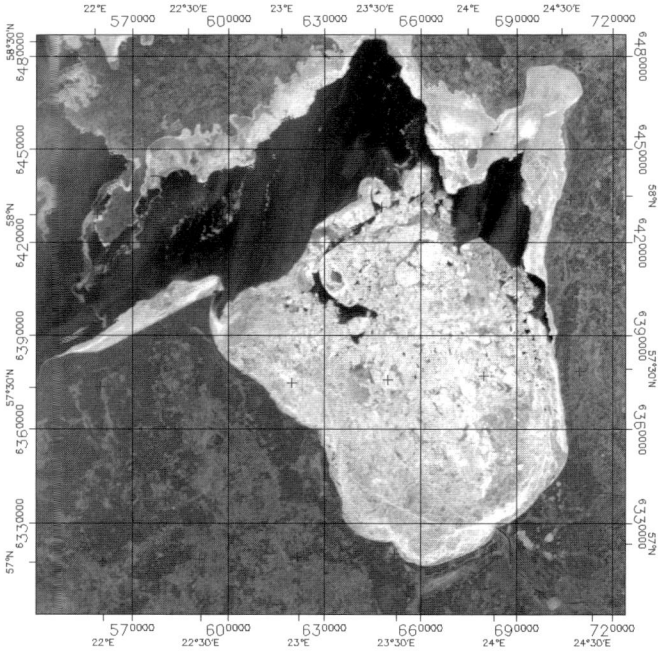

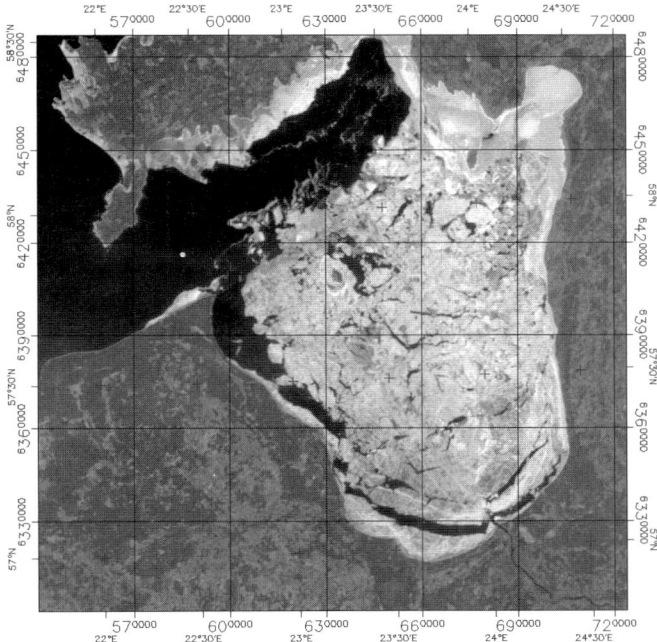

Figure 3.4. Sea ice displacement in the Gulf of Riga, Baltic Sea, seen in consecutive optical moderate resolution imaging spectroradiometer (MODIS) images of NASA's *Terra* satellite. The size of the basin is 120 km across; note an island at 57°50′N 23°15′E. Dates: March 17, 21 and 28, 2003.

Chukchi Sea; at the time there was wind and a 1-knot current driving the ice. Kuznetsov and Mironov (1986) reported rapid ice movements in transient currents, forced by surface pressure gradients in straits and along the coastline. An extreme and rare case is the *ice river* phenomenon: in coastal regions a narrow ($\approx \frac{1}{2}$ km) band of close pack ice flows at a very high speed, up to 3 m/s, much higher than the surrounding ice field. Kuznetsov and Mironov (1986) described three ice river cases from 1967 to 1977 from the Arctic coast of Russia, based on ship reports. They were sudden and brief; two of them took place during a strong southerly wind.

An excellent illustration of the ice motion in the Arctic Ocean, tracked by drift buoys, is provided on the website of the International Arctic Buoy Programme (IABP) (http://IABP.apl.washington.edu/Animations/). Animations of high-frequency coastal ice motion in the Sea of Okhotsk are available at the Hokkaido University web page (http://www.hokudai.ac.jp/lowtemp/sirl/sirl-e.html).

The magnitude of the *rotation rate of ice floes* is usually $10^{-6}\,\text{s}^{-1} = 0.2°\,\text{h}^{-1} = 5°\,\text{d}^{-1}$. According to observations, typically in the Arctic Ocean central pack, $\dot{\omega} \sim 10^{-7}\,\text{s}^{-1}$, $R \sim 10$ km, and therefore $\dot{\omega} R \sim 10^{-3}$ m/s. In the MIZ, $\dot{\omega} \sim 10^{-5}\,\text{s}^{-1}$, $R \sim 100$ m, and again $\dot{\omega} R \sim 10^{-3}$ m/s. Sverdrup (1928) reported a floe rotation rate of $2 \times 10^{-6}\,\text{s}^{-1}$ in the Eurasian Basin. The results of Legen′kov et al. (1974) from the

Figure 3.5. Mombetsu ice radar antenna on top of Mt Oyama (300 m) monitoring the sea ice of the Sea of Okhotsk.

Soviet North Pole station NP-19 showed that the rotation rates of several floes were almost always less than $10^{-6}\,\mathrm{s}^{-1}$, ice compactness was 0.85–1.00, the floes rotated as groups, and the rate increased with decreasing compactness. Leppäranta (1981b) reported a 1.5-km floe in the Baltic Sea rotating at rates below $3 \times 10^{-6}\,\mathrm{s}^{-1}$, ice compactness was about 0.9, and floe rotation correlated well with the local vorticity. Figure 3.11 shows the rotation of a floe together with a 5-km array in the drift ice around the Greenland Sea MIZ. The rates fall within the range 1–10°/day and correlate well (ice compactness was 0.8–0.9 and floe size was 50–100 m).

Occasionally, floe rotation may be fast. Gorbunov and Timokhov (1968) studied

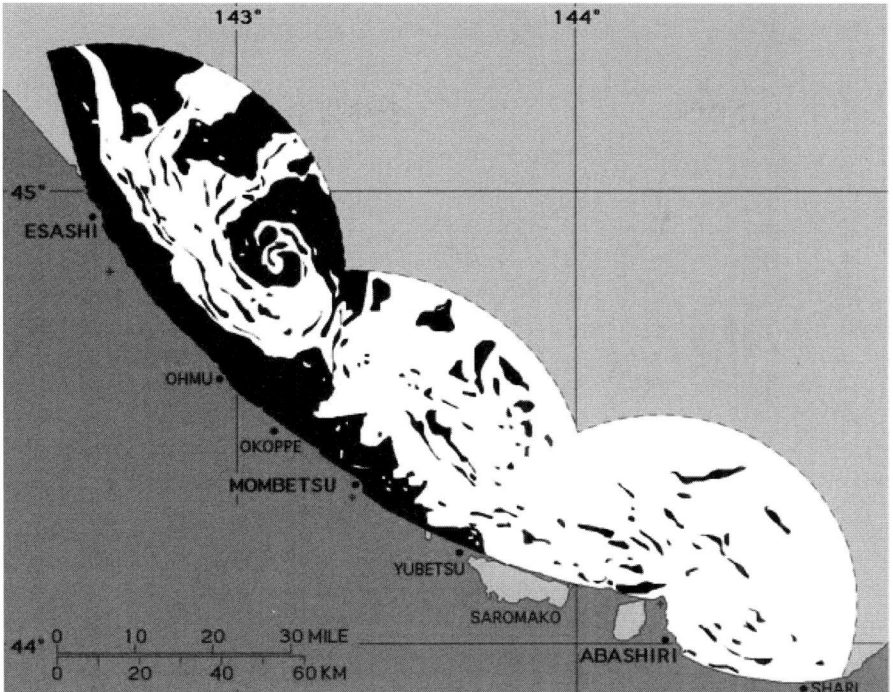

Figure 3.6. Sea ice distribution (ice is white) along the northern coast of Hokkaido on 3 February 2003. The chart is based on three coastal radars mapping the ice up to 60 km from the coastline.
Reproduced with permission from the Sea Ice Research Laboratory, Institute of Low-temperature Science, Hokkaido University.

the rotation of floes in the Chukchi Sea. Compactness was rather small, less than 0.8, and floe size was mostly less than 200 m. Their rotation showed very high values: the typical range was ±30° per hour, but on occasion went as high as 100° per hour. Smaller floes had higher rotation rates. The interaction between ice floes had a positive effect on rotation since the rate tended to be higher in close drift ice than in open drift ice. Also floes usually rotated in groups rather than as individuals. During the Marginal Ice Zone Experiment (MIZEX-84) in the Greenland Sea, rotation rates higher than 100° per day were noted by the author when the pack was opening. The very high rotation rates may be caused by ocean eddies or the morphology of the ice floes.

3.2.3 Strain rate and vorticity

The 100-km scale

Continuum deformation observation campaigns commenced in the 1960s, along with serious work on sea ice rheology that necessitated knowledge of strain and

60 Ice kinematics [Ch. 3

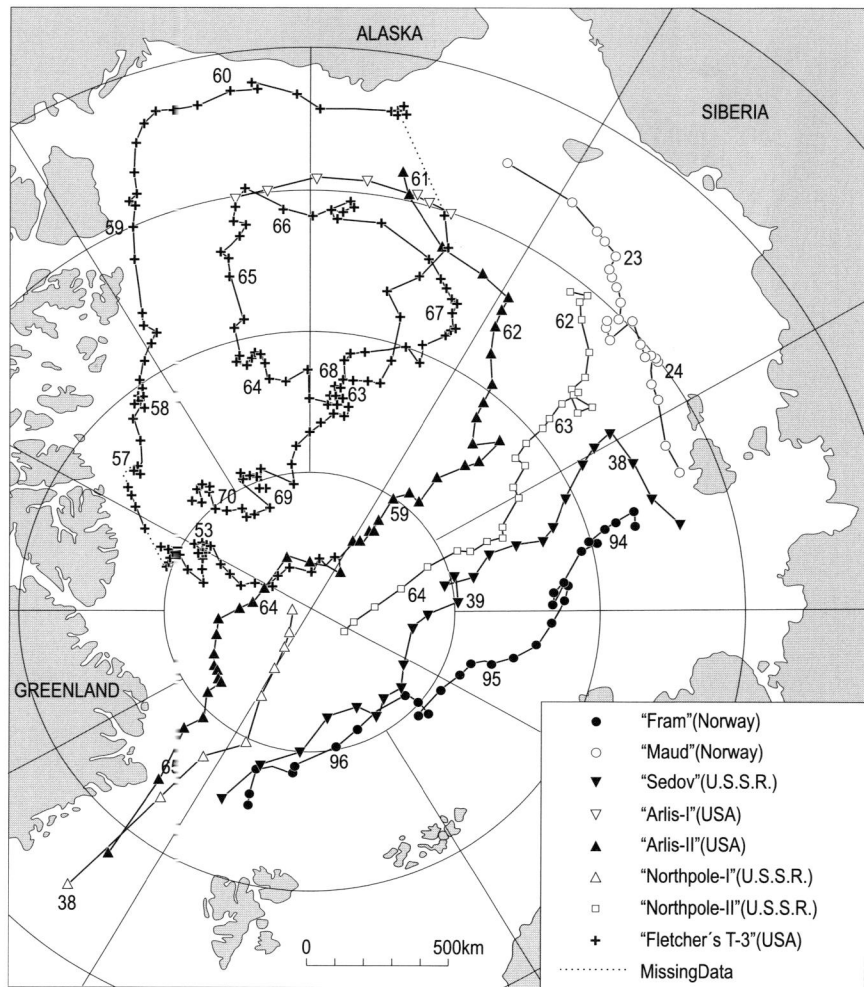

Figure 3.7. Paths of drifting stations in the Arctic Ocean. The numbers show the year (1897–1970) and marks between are at monthly intervals.
Redrawn from Hibler (1980b).

strain rate. Bushuyev et al. (1967) observed the motion of seven points, four of which formed a rectangle with sides 75–100 km in the central Arctic Ocean. In 22 days, the total translation of the square was about 90 km, while the normal strains along the sides of the square were within 3% in 22 days (i.e., the mean strain rate was $10^{-8}\,\mathrm{s}^{-1}$, or 0.1% per day). They also measured rotation on an approximately daily basis, and the magnitude was below $10^{-7}\,\mathrm{s}^{-1}$ with two general rotation schemes:[4] the Beaufort

---

[4] See Figure 3.17 for a general idea of these rotation fields.

Sec. 3.2] Observations 61

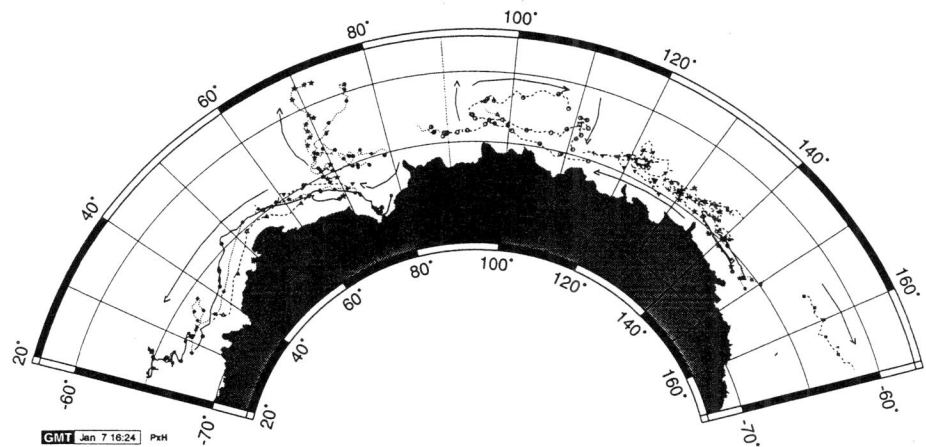

Figure 3.8. Sea ice buoy drifts in the Indian Ocean–eastern Pacific Ocean sector of the Southern Ocean: 1985 (short dashed lines with solid symbols), 1987 (solid line with solid symbols), 1992/93 (solid line with open symbols), 1995/96 (dotted lines) and 1996 (dashed lines). The symbols are plotted at 10-day intervals; arrows show the net drift.
Reproduced from Worby et al. (1998), with permission from the American Geophysical Union.

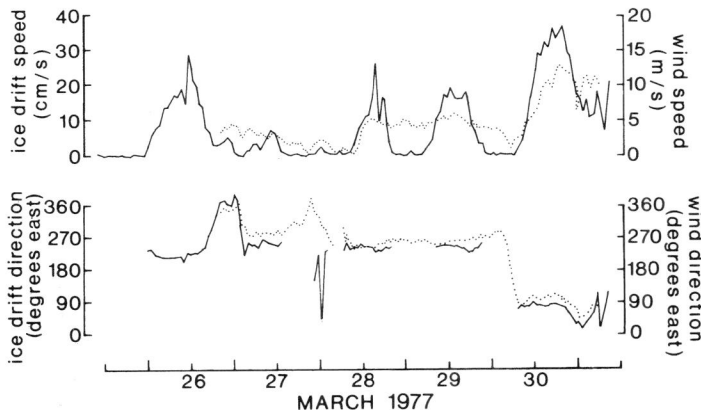

Figure 3.9. Sea ice (solid lines) and wind velocity (dotted lines) time series, Baltic Sea, March 1977.
From Leppäranta (1981a).

Sea Gyre ice rotated more or less as a rigid disc; and in the Transpolar Drift Stream there was a maximum speed in the centre and rotation was connected with shear on both sides of the centre.

Long-term sea ice deformation was measured during the main Arctic Ice Dynamics Joint Experiment (AIDJEX) in 1975 (Thorndike and Colony, 1980; Thorndike, 1986) (Table 3.1). The magnitude of strain rate and rotation was $10^{-7}\,\text{s}^{-1}$; this level increased by 50–100% in summer. In winter the divergence was

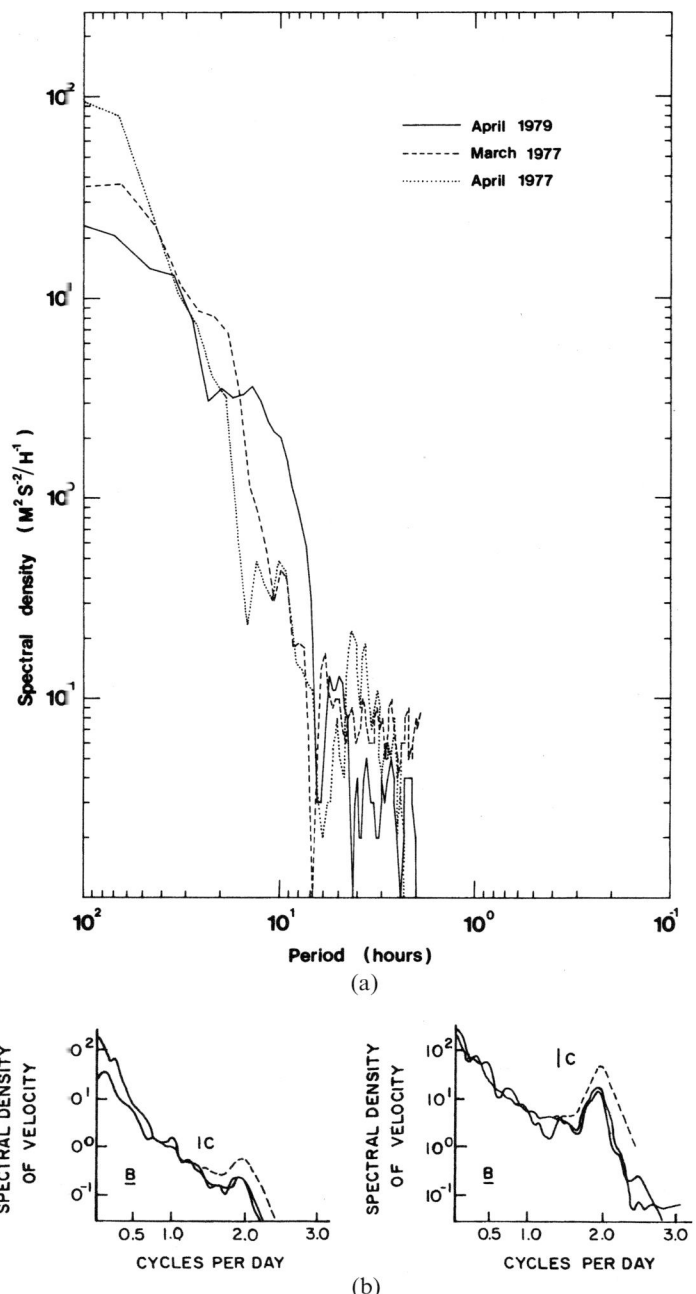

Figure 3.10. Drift ice velocity spectra: (a) Baltic Sea, 1979; (b) Arctic Ocean.
(a) From Leppäranta (1981b). (b) Reproduced from Thorndike and Colony (1980), with permission of University of Washington Press.

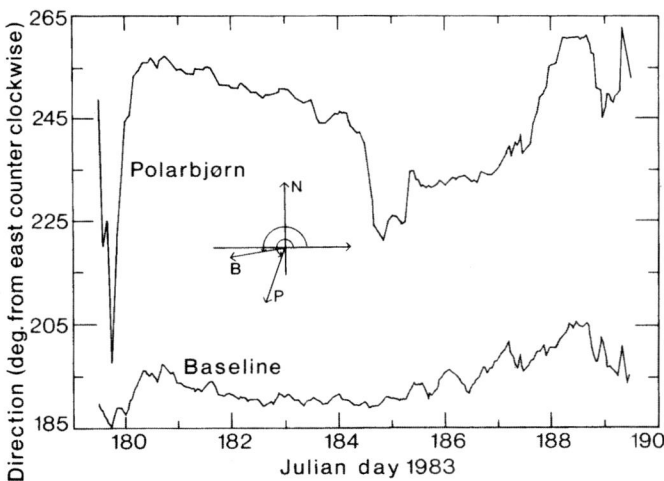

Figure 3.11. Orientation of a drifter array (5 km) baseline and R/V *Polarbjørn* moored to an ice floe in the centre of the array (MIZEX-83 in the Greenland Sea). The large bump for the data from R/V *Polarbjørn* in days 184–185 is non-natural, caused by a necessary correction of the ship's mooring.
From Leppäranta and Hibler (1987).

Table 3.1. Strain rate and vorticity in AIDJEX 1975 manned array in % per day $\approx 10^{-7}\,\mathrm{s}^{-1}$: length scale is 200 km (Thorndike, 1986); —, no data provided in the reference.

	Divergence		Shear		Vorticity	
	Mean	Standard deviation	Mean	Standard deviation	Mean	Standard deviation
Winter	0.07	1.0	1.6	1.6	−0.52	2.0
Summer	−0.03	1.6	3.5	2.2	—	—

on average 0.07% d^{-1}, which means a total opening of 7% over 100 days, while its standard deviation was 1% d^{-1}. Shear measure is positive and therefore its average is much more than that of divergence, but note that the standard deviations are close. The mean rotation was anticlockwise in winter (as in the Beaufort Sea Gyre) and summed to 30° over 100 days.

The 10-km scale

Hibler et al. (1973) monitored mesoscale sea ice deformation in the Beaufort Sea manually, using tellurometers[5] in a 10-km triangle. Short-term ($\approx 6\,\mathrm{h}$) deformation

---

[5] A tellurometer is a geodetic distance meter based on the phase modulation of transmitted radio waves.

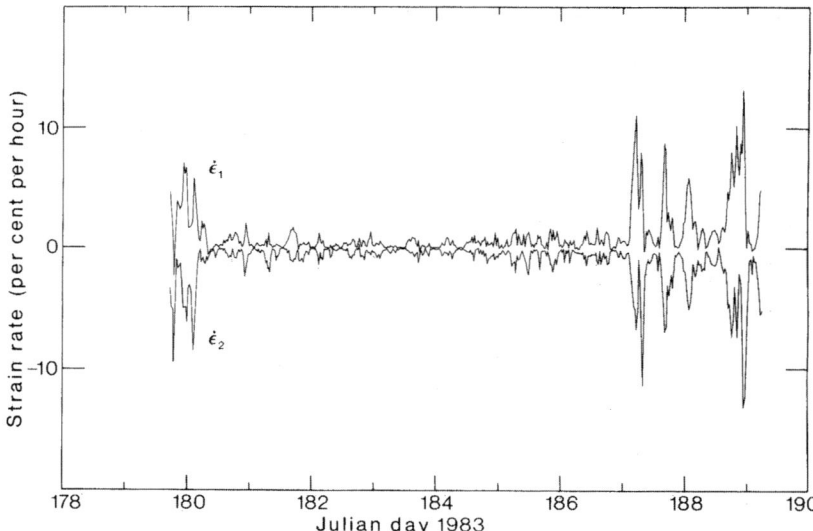

Figure 3.12. Principal strain rates during the MIZEX-83 experiment. The size of the measurement region was 5 km.
From Leppäranta and Hibler (1987).

events were observed with strain-rate invariants of $10^{-7}\,\text{s}^{-1}$, but over intervals of 1 day or more the level was one order of magnitude less. They also reported good correlation between the strain ellipse and fracturing of the ice. In the next phase a laser geodimeter was used for more accurate time series data (e.g., Hibler et al., 1974b). In the shear zone the level of strain rates was three to five times as much as in the central pack (Hibler et al., 1974a). Therefore, as found by Bushuyev et al. (1967), the Beaufort Sea Ice Gyre is mostly a cohesive wheel with intensive deformation at the boundary of landfast ice. The width of this shear zone was shown to be 50 km.

Legen'kov (1977) analysed measurements of the Soviet NP-19 drifting station in the Arctic Basin, collected from an ice island using a theodolite and a fixed baseline. The size of the study area was about 5 km. His strain-rates were also $\sim 10^{-7}\,\text{s}^{-1}$. The temporal stability of the sign of strain rates was almost always below 24 hours, averaging to about 6 hours. Therefore, in wind-driven ice dynamics, deformation events are much shorter than the velocity time scale, which is at the synoptic level. The spatial correlation length was also short; although not quantified this was estimated as less than the whole study region for 3-hour strain events. It was also shown that deformation activity increased with decreasing compactness (range of compactness was 0.85–1.0).

In the MIZ the strain-rate level is much larger than in the central pack. Data from the MIZEX-83 experiment in the Greenland Sea, in June 1983, showed that the level is as much as $10^{-5}\,\text{s}^{-1}$ (5% h^{-1}) in short, intensive periods and normally one order of magnitude less (Figure 3.12). For a 1% h^{-1} strain rate, the e-folding time is

100 hours for ice compactness and the shear strain on a right angle is 45° in 100 hours. However, the direction of deformation alternated, and the total change of the drifter array remained small during the 10-day period of the experiment. In a study by Leppäranta (1981b) in the Baltic Sea, the magnitude of deformation rates was $10^{-6}\,\text{s}^{-1}$ and major deformation events lasted 5–10 hours. The statistics over a 1-week period for a 5-km array were the following (unit $10^{-6}\,\text{s}^{-1}$): $\dot{\varepsilon}_1 = 0.44 \pm 0.66$, $\dot{\varepsilon}_2 = -0.50 \pm 0.55$, with a maximum magnitude of 3; the vorticity was 0.79 ± 0.77, with a maximum of 3.4.

In all reported time series the principal components have mostly opposite signs and are of the same magnitude, which means that their sum (opening/closing of the ice pack) is small and their difference (twice the maximum rate of shear) is large.

In winter 1997 a special coastal boundary zone experiment was performed in the Baltic Sea (Leppäranta et al., 2001; Goldstein et al., 2001). Sea ice dynamics was examined using GPS drifters in a 10 km size array. Compared with central basins (Leppäranta, 1981b), the strain had more uniaxiality and was more stepwise (Figure 3.13). The deformation signal was seen to travel at $\sim 5\,\text{m/s}$ across the array, and the time response indicated separate local- and basin-wide (100 km size) responses.

With sequential remote imaging of the ice pack, it is now possible to determine spatial deformation fields. A coastal radar network has been utilized for research in sea ice kinematics in the Sea of Okhotsk since 1969 (Tabata, 1972). The system tracks deformed sea ice floes such as ridges, and consecutive radar images can be utilized to obtain sea ice kinematics information. Tabata (1971) and Tabata et al. (1980) followed several quadrangles in the coastal ice drift. The spatial scale of the deformation was about 20 km, and the magnitude of hourly strain and rotation was 1–5% or $10^{-5}\,\text{s}^{-1}$.

Another case from the northern Baltic Sea is shown in Figure 3.14, based on *Landsat* images. Ice velocities were obtained manually by identifying shifts in a number of ice floes, and these were then averaged into an 18.5-km grid. The resulting deformation field showed an overall level of 5% per day for both divergence and maximum shear (the length scale of the deformation field was some 50 km).

3.2.4 Deformation structures

As a result of mechanical deformation, ordered structures appear in sea ice cover. Figure 3.15 shows banding of drift ice as seen from the hills of the northern coast of Hokkaido. Deformation structures are identifiable in airborne and spaceborne remote-sensing imagery and can be used to observe kinematic events and to interpret their effect on the mechanical behaviour of the ice. For example, Erlingsson (1988) examined fracture geometry and the mesoscale plastic properties of drift ice.

Goldstein et al. (2000) examined Baltic Sea imagery and identified (a) a parallel system of faults under unidirectional tension, (b) a concentric system of radial faults, (c) coupled compression–tension structures, (d) broom-like shear structures, and (e) vortices (Figure 3.16). Point (c) refers to a ridge–lead structure, a typical and

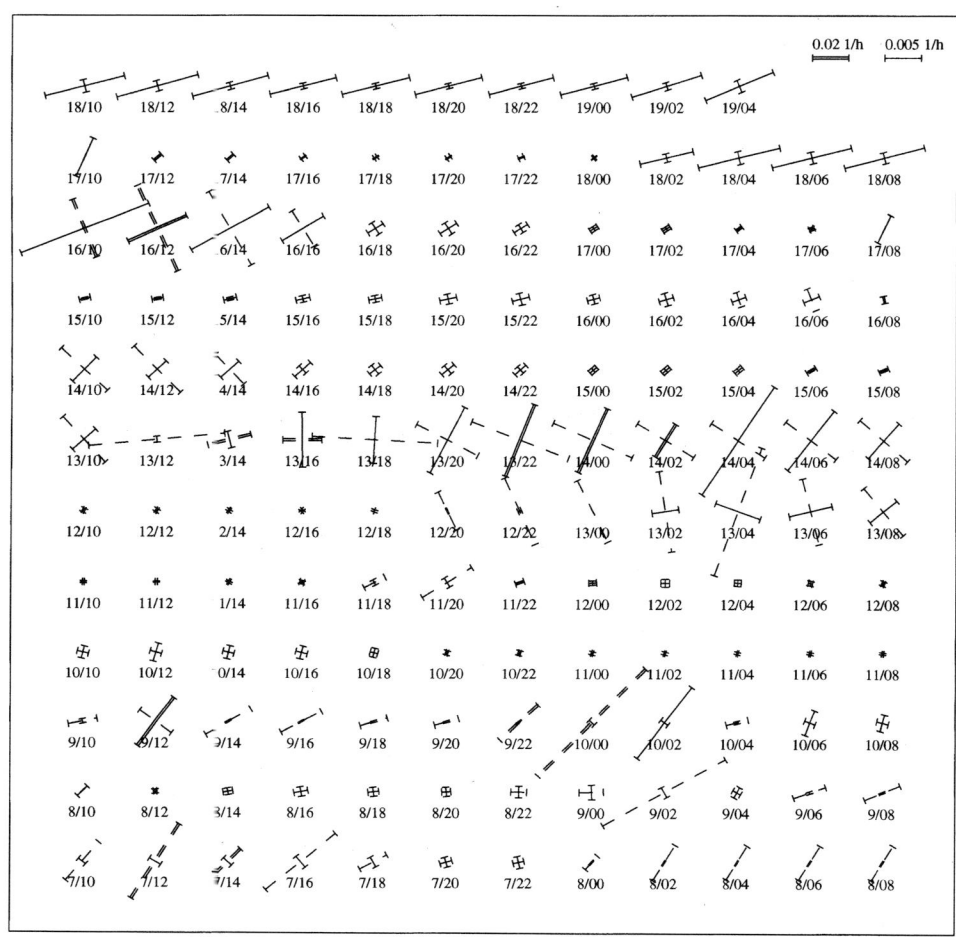

Notation 10/15 means 10 March 1997, 15:00 UTC

Figure 3.13. Time series of 2-hourly, principal strain rates in the coastal drift ice zone, Bay of Bothnia, in March 1997. The coastline is aligned vertically, and the notation 10/14 means 10 March at 14:00 hours.
From Leppäranta et al. (2001).

important deformation process, illustrating how in pure shear simultaneous lead opening and pressure ice formation occurs. This is further indicated by principal strain rates having normally opposite signs as shown earlier (Section 3.2.3).

Sea ice cracks and leads are easily recognizable in remote-sensing imagery. Classification of cracks, and their formation and structure was deeply examined by Volkov et al. (2002). Lead formation structures in the Arctic were analysed by Hibler (2001) for anisotropy of lead pattern and for development of an anisotropic sea ice model.

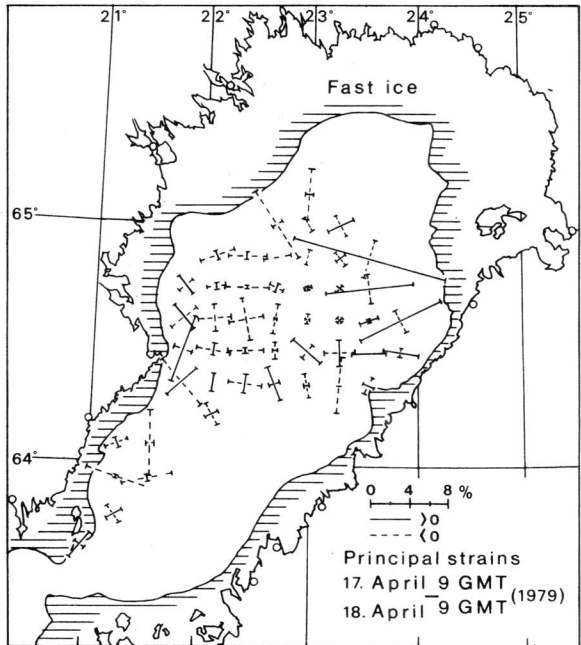

Figure 3.14. Principal axis strain field from consecutive *Landsat* images in the Baltic Sea. Satellite passes were 1 day apart.
From Leppäranta (1982).

3.3 STOCHASTIC MODELLING

3.3.1 Two-dimensional motion using complex variables

Let us now carry out time series analyses of two-dimensional vectors using complex variables. We denote them like scalars, as done in mathematics; they possess real and imaginary components oriented along the x- and y-axes, respectively (e.g., Ahlfors, 1966). Ice velocity is then expressed as $u = u_1 + iu_2$, where $i = \sqrt{-1}$ is the imaginary unit, and $u_1 = \operatorname{Re} u$ and $u_2 = \operatorname{Im} u$ are the real and imaginary components. For a complex variable u, the magnitude, or *modulus*, is $\sqrt{uu^*} = \sqrt{u_1^2 + u_2^2}$, where $u^* = u_1 - iu_2$ is the complex conjugate of u. Its direction, or *argument* (counter-clockwise angle from the x-axis), is denoted as $\arg u$. The following elementary properties are very useful. For u and q complex variables and a real variable φ, we have:

(i) $|uq| = |u||q|$.
(ii) $\arg(uq) = \arg u + \arg q$.
(iii) $\exp(i\varphi) = \cos\varphi + i\sin\varphi$ (Euler's formula).

Figure 3.15. Drift ice bands in the Sea of Okhotsk shown in Yukara-ori weaving handicraft. This style was started by Ms Aya Kiuchi in the early 1960s; thereafter, she developed it with her son Kazuhiro.
Reproduced with permission from Yukara-ori Kougeikan [Yukara Weaving Museum], Asahikawa, Hokkaido.

Example The covariance between two complex random variables u and q is given by $\mathrm{Cov}(u,q) = \langle u - \langle u \rangle \rangle \langle (q - \langle q \rangle)^* \rangle$. The correlation coefficient is $r = \mathrm{Cov}(u,q)/[\mathrm{Cov}(u,u)\,\mathrm{Cov}(q,q)]^{1/2}$. $|r|$ gives the total correlation between u and q, and $\arg r$ gives the angle from u to q at which the quantities are correlated. Since the directions of ice drift, wind, and ocean current are different, it is convenient to study their correlations in complex form. In this way, the correlation coefficient provides both total and directional correlation. ■

A simple, stochastic sea ice drift model is written:

$$\mathrm{d}u/\mathrm{d}t = -\lambda u + F \tag{3.11}$$

where λ is the "memory" of the system and F is the forcing function. The spectrum of velocity is then

$$p(\omega) = \frac{p_F(\omega)}{|\lambda + i\omega|^2} \tag{3.12}$$

For $|\omega| \ll |\lambda|$, ice velocity follows the forcing as $u = F/\lambda$. Conversely, for $|\omega| \gg |\lambda|$ the ice velocity spectrum is ω^{-2} times the forcing spectrum. The spectrum has a local

Action type	Boundary condition	Type of structure		Structure scale (km²)	Region of probable occurrence
1. Uniform uniaxial tension caused by wind	Free boundaries	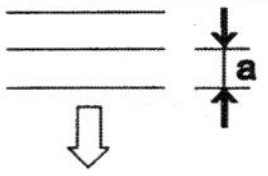	System of parallel faults a ~ const.	~10^4	Central Bay regions
	Displacement restrictions at the region boundary	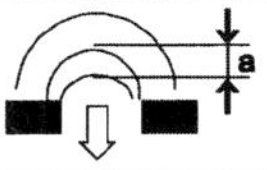	Arc-shaped faults a ~ const.	~10^3	Shore regions, islands of north part of the Bay
2. Bi-axial tension caused by wind	Free boundaries	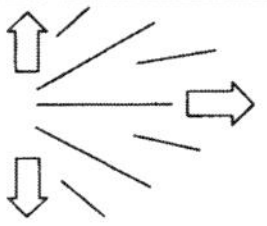	Concentric system of radial faults	~10^4	Region adjusting the central part of the Bay
3. Uniform compression caused by wind	Displacement restrictions in the compression direction	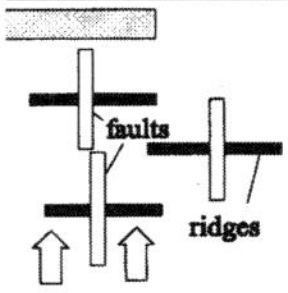	System of conjugated structures of compression-tension ridges (rafts) faults	~10^4	Region near the banks of the Bay
4. Local compression under motion of large mass or ice push on an obstacle	Displacement restrictions in the compression	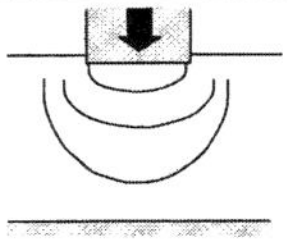	Arc-shaped structures of ridges	~10^2	The north region of the Bay
5. Local shear caused by wind	Displacement restrictions at the fast ice boundary	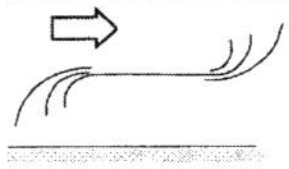	Shear with feathering "broom"-type structure	~10^3	East and north-east regions of the Bay
6. Tracing of flows			Rotational structures of faults and ridging regions related to the structure of underice flows	~10^4	East and north regions of the Bay

Figure 3.16. Sea ice structures in the Bay of Bothnia, Baltic Sea, based on satellite remote-sensing imagery.
From Goldstein et al. (2000).

maximum at $\omega = -\mathrm{Im}\,\lambda$, which is a singularity if $\mathrm{Re}\,\lambda = 0$. This can be expanded by coupling Eq. (3.11) with the similar model used to ascertain ocean mixed layer velocity and then force the system by geostrophic wind and ocean currents (a model we will return to in Chapter 6).

3.3.2 Mean sea ice drift field in the Arctic Ocean

If we consider sea ice velocity as a random field, observations such as drifter data can be used to determine the field structure; this leads to interpolation of data into the mean ice velocity field. The first mean field was presented by Gordienko (1958) based on the drift of ships and manned stations. It was revised by Colony and Thorndike

70 Ice kinematics [Ch. 3

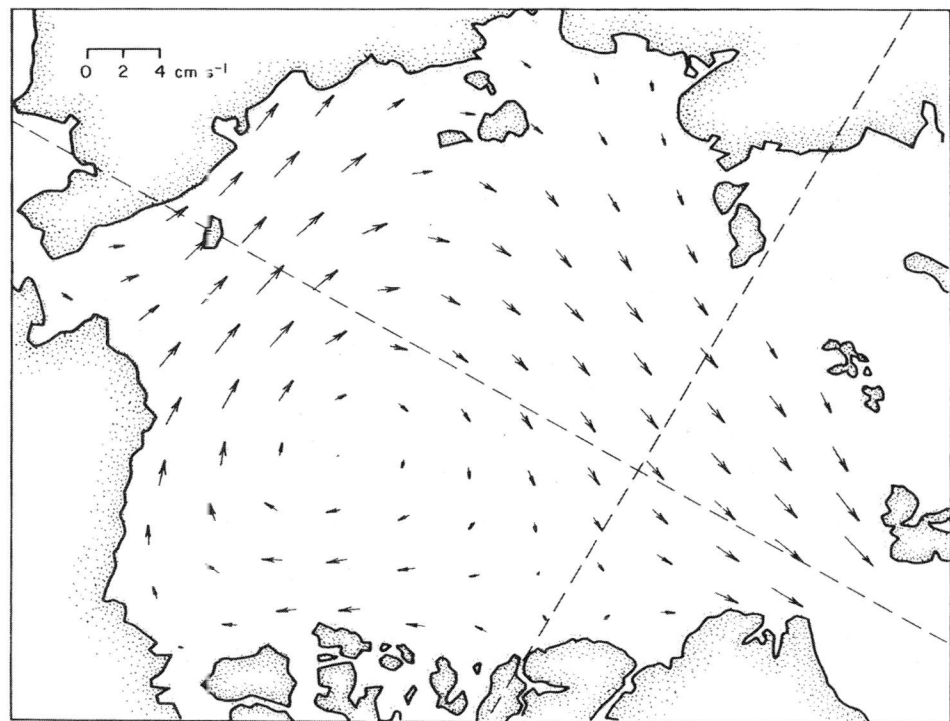

Figure 3.17. Annual mean of sea ice velocity in the Arctic Ocean.
Reproduced from Colony and Thorndike (1984), with permission of the American Geophysical Union.

(1984) with large amounts of drift buoy data added and using optimum linear interpolation (Figure 3.17). The detailed structure of the Transpolar Drift Stream, the Beaufort Sea Gyre, and outflow through the Fram Strait were then revealed. According to Colony and Thorndike (1984), daily velocities average about 2 km/d, with a standard deviation of 7 km/d, and the integral timescale (time integral of the correlation coefficient) is 5 days. They also demonstrated that displacement variance increases with time, following a 1.4 power law. Inter-annual variations appear in annual mean fields as reflections of the variations in atmospheric pressure over the Arctic Ocean.

Thorndike (1986) examined spatial correlation structures in the large-scale longitudinal and transverse components of ice velocity, a view often taken in turbulence studies. Correlation lengths (the distance after which correlation has vanished) became 2,000 km for the longitudinal component and 800 km for the transverse component, close to the situation in the geostrophic wind field. This reflects that the ice drift follows atmospheric forcing in the large scale. The variance level over distances above the correlation length is ~ 100 $(cm/s)^2$.

3.3.3 Diffusion

One way of examining the spatial variability of velocity is by looking at it from the diffusion viewpoint. For tracer spots in the ocean, the diffusion coefficient K depends on the length scale L as $K \propto L^{4/3}$ (Okubo and Ozmidov, 1970). Okubo and Ozmidov also showed that there are two separate regimes with a transition zone at 1–10 km; and $K \sim 1\,\text{m}^2/\text{s}$ and $K \sim 10\,\text{m}^2/\text{s}$ at 1 km and 10 km, respectively. In pure diffusion the standard deviation of displacement (s) increases as $s = 2(Kt)^{1/2}$ (e.g., with $K = 10\,\text{m}^2/\text{s} \approx 1\,\text{km}^2/\text{d}$ and $t = 1\,\text{d}$, we have $s = 2\,\text{km}$).

Gorbunov and Timokhov (1968) made a case study in the Chukchi Sea based on aerial photography. They estimated the diffusion coefficient $K = \langle \Delta l^2 \rangle / \Delta t$ for $A < 0.8$, resulting in $K \approx 8.41 \times 10^{-4} L^{5/8}\,\text{m}^{11/8}\,\text{s}^{-1}$. Compared with oceanic tracer spots, the magnitude was close at $L = 1\,\text{km}$, illustrating that ice floes become well mixed when their compactness is lower than 0.8; however, length-scale dependence was weaker. Gorbunov and Timokhov (1968) further showed that the diffusion coefficient increased with decreasing compactness and decreasing floe size. Legen'kov et al. (1974) studied a 5-km area in which variation in ice velocity was within about 1 cm/s from the mean, with high sensitivity for ice compactness, which is in good agreement with Gorbunov and Timokhov (1968).

Diffusion models can be used to describe the spread and mixing of ice floes, but they have severe limitations. In the ice field itself there are no self-diffusive mechanisms, but the cause lies in atmospheric and oceanic forcing. Ice floes have a more or less random component in their motion due to their individual properties, but these components are not always diffusive. Forcing may also collect floes together (i.e., cause "negative diffusion").

3.3.4 Random walk

A Markov chain model for a two-dimensional random walk states that a particle moving in a grid has normalized arbitrary probabilities of stepping on one of the neighbouring cells (e.g., Feller, 1968). Dividing a sea surface into N cells, transition probabilities can be obtained either from direct observations or using a Monte Carlo method[6] to define a Markov chain. This idea was used by Colony and Thorndike (1985). They divided the Arctic Basin into 111 cells. Cell sizes ranged within some hundreds of kilometres; although the authors did not explain how the cells were fixed, in practice a good choice must be based on data availability and resolution of interest. *Argos* buoy data were used for mean sea ice motion as a deterministic background and for displacement variance. The latter was estimated as $2.5 \times 10^4\,\text{km}^2$ over 90-day time steps. The purpose of this study was to examine long-term statistical problems. Colony and Thorndike (1985) concluded that the model works well for questions such as the probability distributions for the path

---

[6] Monte Carlo methods estimate ensemble properties by means of large-number simulations generated by random numbers (e.g., Korvin, 1992). Here they can be used for transition probabilities in a given space-time grid.

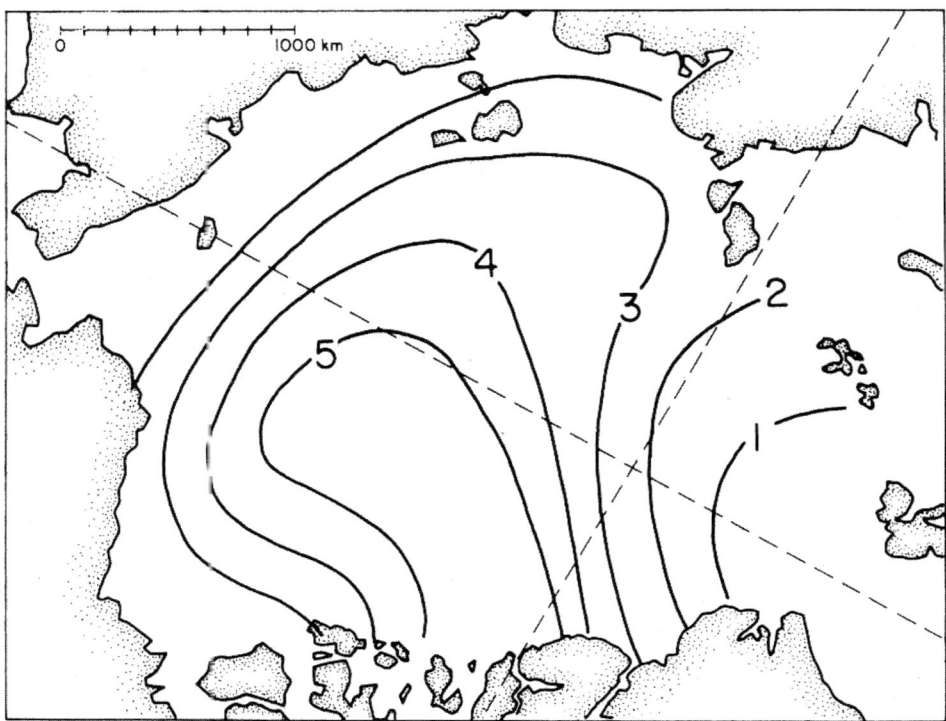

Figure 3.18. The mean lifetime (years) of an ice floe in the Arctic Basin based on a random walk model.
Reproduced from Colony and Thorndike (1985), with permission from the American Geophysical Union.

of pollutants from a given source, ice renewal times in different basins, and statistical variability of ice conditions in a given climate. Figure 3.18 shows the resulting mean lifetime of an ice floe as a function of its location of birth.

3.4 ICE CONSERVATION LAW

Ice conditions in a given region, Ω, are modified by thermal and mechanical processes. The ice volume as well as ice-state variables in this region must obey conservation laws. For any property of the ice field (say, Q) the conservation law is in general form:

$$\partial Q/\partial t + \boldsymbol{u} \cdot \nabla Q = \psi_Q + \phi_Q \tag{3.13}$$

where left-hand side terms are the local rate of change and advection, and on the right-hand side ψ_Q represents mechanical deformation and ϕ_Q stands for thermodynamic changes. The left-hand side can be expressed briefly as DQ/Dt, where $D/Dt = \partial/\partial t + \boldsymbol{u} \cdot \nabla$ is the material or total time derivative operator, the rate of change in a particle following motion. Mechanical deformation occurs as opening and closing of leads, rafting, hummocking, ridging, etc. Equation (3.13) includes two dynamical timescales: one for advection, $T_Q \sim L_Q/U$, and one for deformation, $|\dot{\varepsilon}|^{-1} \sim L_U/U$, where L_Q and L_U are the length scales of the property Q and velocity, respectively. Since $L \sim 100$ km and $U \sim 10$ cm/s, the deformation timescale is typically 10 days, ranging from 1 day to 100 days.

Thermodynamics is not the topic of the present book, and therefore the heat budget is left out. The scales of ice growth and melting were briefly discussed in Section 3.3. Therefore, we assume for property Q the rate of thermodynamic change is provided as an external forcing. The property of concern is the ice state J (i.e., basically the thicknesses of different ice categories or the thickness distribution of ice).

The conservation of ice mass $m = \rho h$ is a necessary condition for any ice state. Since $\rho =$ constant, this is equivalent to the conservation of mean ice thickness, which may change due to the divergence of ice motion, and freezing or melting. Consequently:

$$\partial h/\partial t + \boldsymbol{u} \cdot \nabla h = -h\nabla \cdot \boldsymbol{u} + \phi(h) \qquad (3.14)$$

where $\phi(h)$ is the thermal growth rate of ice of thickness h. Apart from the thermodynamic term, the conservation law of mean ice thickness is derived as the mass conservation law in shallow water models. These models give an equation for sea level that exactly matches Eq. (3.14), with $\phi = 0$ (e.g., Pond and Pickard, 1983). If $h = 0$, the mixed layer must first be cooled to freezing point for ice growth to start; if $h > 0$, it is assumed that the mixed layer is at freezing point. The mechanical increase of ice thickness due to ridging is more than the thermal increase. If $h \sim 1$ m and $\nabla \cdot \boldsymbol{u} \sim -10^{-6} \text{s}^{-1}$, the mechanical growth rate would be ~ 10 cm/day, while thermal growth rates are usually less than 1 cm/day (thin ice excluded). Mechanical growth events are very short, so that in the long run thermal production of ice volume usually exceeds mechanical production. In regions of intensive ridging, such as off the northern coast of Greenland, the mean thickness of ice is more than twice the thermal equilibrium thickness of multi-year sea ice in the Arctic.

The ice conservation laws presented below deal with ice categories and ice thicknesses. A conservation law can be derived in a similar manner for floe size (Ovsienko, 1976). The advection and thermal growth of floes is straightforward, thermal growth being mainly due to lateral melting in summer. The break-up of floes as well as their merging together may be parameterized in terms of floe-size distribution parameters.

3.4.1 Ice states based on ice categories

The two-level ice state obeys the conservation law of ice thickness and compactness. The mean thickness law is as above (Eq. 3.14). For ice compactness, the conservation law can be derived similarly to that of thickness. The divergence of ice motion changes compactness mechanically, the one limitation being that compactness cannot be more than 1. The result can be expressed as:

$$\partial A/\partial t + \boldsymbol{u} \cdot \nabla A = -|\dot{\varepsilon}|\chi_0(\varphi)A + \phi_A \qquad (0 \leq A \leq 1) \qquad (3.15)$$

where $\chi_0(\varphi)$ gives the opening of leads under different deformation modes φ, and ϕ_A is thermodynamic change. For example, $\chi_0(0) = -1$ (pure divergence) and $\chi_0(\pi) = 1$ (pure convergence). When the mixed layer temperature is at freezing point, any heat loss would potentially freeze the surface over. Very thin ice is, however, insignificant in sea ice dynamics, and therefore a demarcation thickness h_0 is introduced to account only for ice thicker than that (see Eq. 2.24). When the mixed layer temperature is at freezing point (as always when $A > 0$), it is assumed that, in the fractional area $(1 - A)$, ice thickness is uniformly distributed between zero and the demarcation thickness h_0. Then:

$$\phi_A = \frac{\phi(h_0)}{h_0}(1 - A) \qquad (3.16)$$

When $A < 1$, lateral melting may take place via absorption of solar radiation in leads (see Rothrock, 1986). This can be immediately added to Eq. (3.16) by distributing the heat evenly over vertical floe surfaces. When the mean ice thickness and compactness are known, the thickness of ice floes is obtained from $h_i = h/A$.

Example For a numerical illustration, take $h_0 = 10$ cm and assume that $\phi(h_0) = 2$ cm/day. For ice compactness $A(h_0) = 0.8$, we have $\phi_A = 0.04$/day. It would take $h_0/\phi(h_0) = 5$ days to reach full compaction $A(h_0) = 1$. ∎

In multi-level cases different ice types and their volumes are included. The main question is then how the mechanical redistribution of ice is arranged in the case of compression of compact ice (i.e., how deformed ice is produced). Thin ice sheets (less than 10–20 cm) undergo rafting, and the consequence is local doubling of the ice thickness. By overriding thicker ice, the ice breaks into blocks of size from a few centimetres up to a few metres to form pressure ice: rubble fields, hummocks, and ridges.

The three-level case (A, h_u, h_d) treats ice compactness in the same way as the two-level case (Leppäranta, 1981a). Undeformed ice thicknesses only change by thermodynamics and advection, while also mechanical deformation may increase the thickness of deformed ice:

$$\partial h_u/\partial t + \boldsymbol{u} \cdot \nabla h_u = \phi(h_u) \qquad (3.17a)$$

$$\partial h_d/\partial t + \boldsymbol{u} \cdot \nabla h_d = -|\dot{\varepsilon}|\chi_d(\varphi, A)h + \phi_d(h_d) \qquad (3.17b)$$

where $\chi_d(\varphi)$ is the production of deformed ice as a function of the deformation mode, and ϕ_d is the growth rate of deformed ice. Deformed ice production depends on compactness since it occurs only in compact ice: for example, $\chi_d(\pi, 1) = -1$ (pure convergence of compact ice). The term $-|\dot{\varepsilon}|\chi_d(\varphi, A)h$, the gross term for the production of deformed ice, could be named the "packing rate". Thermal change in the thickness of undeformed ice is displayed as growth or melting at the upper and lower boundaries (ignoring brine pockets). For deformed ice, however, not only may the thickness change but the voids between ice blocks in the keels may freeze as well. Therefore, the thermodynamic growth of deformed ice depends on porosity.

The mechanical deformation terms and Eqs (3.15) and (3.17b) must be consistent with the conservation of total ice mass. In earlier approaches (Doronin, 1970; Hibler, 1979; Leppäranta, 1981a) lead opening and pressure ice formation in shear were ignored, meaning that these terms were given by $-A\nabla \cdot \boldsymbol{u}$ and $-h\nabla \cdot \boldsymbol{u}$ (if $A = 1$ and $\nabla \cdot \boldsymbol{u} < 0$). This likely leads to a small underestimation of the volume of deformed ice, but there is no real data available to show exactly how much.

Example With $h \sim 1$ m and a pure convergence of $-\nabla \cdot \boldsymbol{u} \sim U/L \sim 10^{-6}\,\mathrm{s}^{-1}$ in compact ice, the packing rate is ~ 0.36 cm/h. Such rapid packing events do not last long. In the Baltic Sea the thickness of ice that participates in ridging is 20–40 cm. Then for $\nabla \cdot \boldsymbol{u} \sim -10^{-6}\,\mathrm{s}^{-1}$, the packing rate would be 1.7–3.5 cm per day. If there were one such deformation event every 10 days, the thickness of deformed ice would grow to ~ 25 cm in one winter season, closely comparable with what has been actually observed (Lewis et al., 1993). The magnitude 1.7–3.5 cm can also be compared with the mean equivalent thickness of 2.2 cm for one ridge per one kilometre in the Baltic Sea. ∎

It is known that typically $h_d \sim h_u$ in mesoscale area averages (see Section 2.4). Thus, deformed ice is built up in short, intensive deformation events, while undeformed ice grows slowly but steadily through the whole cold season. An alternative formulation would be to replace h by h_u in the mechanical deformation term of Eq. (3.17b), to allow only undeformed ice to build ridges. But then the mass conservation law (Eq. 3.14) would need to use $-h_u \nabla \cdot \boldsymbol{u}$ as well for the first term on the right-hand side.

The mass conservation law (Eq. 3.14) also allows us to examine vertical velocity. As the top and lower ice surfaces move farther from the sea surface with velocities w_t and w_b, respectively, we can ascertain the movement of a drift ice particle by:

$$w_t - w_b = \mathrm{D}h/\mathrm{D}t = -h\nabla \cdot \boldsymbol{u} + \phi(h) \tag{3.18}$$

And if isostasy is assumed, $w_t/w_b = -(1 - \rho/\rho_w)$. The thermodynamic contribution to vertical motion is less than $5\,\mathrm{cm/day} \approx 5 \times 10^{-7}\,\mathrm{m/s}$. For $h \sim 1$ m and $\nabla \cdot \boldsymbol{u} \sim 10^{-6}\,\mathrm{s}^{-1}$, we have $w_t - w_b \sim 10^{-6}\,\mathrm{m/s}$. All in all, the vertical motion of ice is weak and its integral is equivalent to changes in mean ice thickness (its details are not relevant here). Consequently, as stated at the beginning of this chapter, for the problem of sea ice drift it is sufficient to solve the horizontal velocity on the sea surface plane and the vertical dimension is taken care of by the mass conservation law.

3.4.2 Ice thickness distribution

The ice conservation law is obtained for thickness distribution in a similar fashion (recall Figure 2.15 for examples of thickness distributions). Freezing and melting advect the distribution in the thickness space, but additional assumptions are needed to specify how mechanics change the form of the thickness distribution. In general, thin ice is crushed and piles up into thicker, deformed ice. In discrete form the distribution is an N-category histogram $\{\pi_k, \Delta h_k\}$, and for any category a conservation law like that given in Eq. (3.13) is established. These categories interact as ice is transferred from one category to another by mechanics and thermodynamics. Formally one may take the limit $\Delta h_k \to 0$ to obtain an almost continuous distribution, with delta peaks taking care of the singularities. This form of the conservation law is presented in Thorndike et al. (1975) and Rothrock (1986).

The way of working with the thickness distribution is not immediately clear. Therefore, the logic behind the derivation of the conservation law is shown below in detail for a simple two- to three-class distribution; the generalization should then become more understandable.

Derivation

The derivation of the conservation law is straightforward when using the thickness distribution function Π. In general, consider a material (Lagrangian) element with area thickness distribution $S(h; t)$ in normalized form $\Pi(h; t) = S(h; t)/S_\infty$, $S_\infty = S(\infty; t)$. The change in Π due to mechanics is for a drift ice particle:

$$\frac{\partial \Pi}{\partial t} = \frac{\partial}{\partial t}\left(\frac{S(h;t)}{S(\infty;t)}\right) = \frac{1}{S_\infty}\frac{\partial S}{\partial t} - \frac{S}{S_\infty^2}\frac{dS_\infty}{dt} \qquad (3.19)$$

This is the basic form in which the mechanical evolution of sea ice thickness distribution is normally presented. The reason is that this form splits the evolution into rearrangement and net change parts. Indeed, the first term on the rightmost side of Eq. (3.19) states how the area coverages of different thickness categories change, defining the *thickness redistributor* as $\Psi = S_\infty^{-1} \partial S/\partial t$. Including the second term, noting that $S_\infty^{-1} dS_\infty/dt = \nabla \cdot \boldsymbol{u}$ and $S/S_\infty = \Pi$, we have $\partial \Pi/\partial t = \Psi - \Pi \nabla \cdot \boldsymbol{u}$ from Eq. (3.19). In general the redistributor depends on thickness distribution and strain rate, $\Psi = \Psi(\pi, \dot{\varepsilon})$.

The thermodynamic change in thickness distribution also becomes clear from the distribution function Π. Since, clearly, by thermal growth:

$$\Pi(h; t) = \Pi[h + \phi(h)\delta t; t + \delta t]$$

we have:

$$\Pi(h, t) = \Pi(h, t) + \frac{\partial \Pi}{\partial h}\phi(h)\delta t + \frac{\partial \Pi}{\partial t}\delta t + O(\delta t^2) \qquad (3.20)$$

Dividing by δt and letting $\delta t \to 0$, the change in the distribution function becomes $\partial \Pi/\partial t = -\phi(h)\,\partial \Pi/\partial h$. This can be considered as advection of the spatial density by the growth rate $\phi(h)$ in the thickness space.

Finally, by combining dynamics and thermodynamics together and taking the Eulerian frame, the conservation law of thickness distribution is (Thorndike et al., 1975):

$$\frac{\partial \Pi}{\partial t} + \boldsymbol{u} \cdot \nabla \Pi = \Psi - \Pi \nabla \cdot \boldsymbol{u} - \phi(h)\frac{\partial \Pi}{\partial h} \qquad (3.21)$$

or, taking the derivative with respect to h, we have in terms of the spatial density of thickness:

$$\frac{\partial \pi}{\partial t} + \boldsymbol{u} \cdot \nabla \pi = \psi - \pi \nabla \cdot \boldsymbol{u} + \frac{\partial \phi(h)\pi}{\partial h} \qquad (3.22)$$

where $\psi = \mathrm{d}\Psi/\mathrm{d}h$. The functions Π and π must satisfy the general conditions of spatial distribution and density (see Section 2.3), and the conservation law of ice volume or the mean ice thickness (Eq. 3.14). This means that for spatial density:

$$\int_0^\infty \psi \, \mathrm{d}h = \nabla \cdot \boldsymbol{u}, \qquad \int_0^\infty h\psi \, \mathrm{d}h = 0 \qquad (3.23)$$

The first equation follows from the conservation of area, and the second states that rearrangement of thicknesses gives no net change.

Under the convergence of compact ice, the thinnest ice is broken and transformed into deformed ice. The thickness of the deformed ice may be spread into a certain range or simply transformed to a multiple of the original (say, from h to kh, $k > 1$). In purely mechanical processes there is a limitation that ice ridges have a maximum thickness of around 10–50 m depending on the thickness of the ice sheet used in building the ridge, but these thicknesses are very seldom reached.

In the following the formulation of mechanical deformation is examined from particular cases to a general case and a simple three-class redistributor:

$$\Psi = \frac{1}{S_\infty}\frac{\partial S_0}{\partial t}H(h) + \frac{1}{S_\infty}\frac{\partial S_1}{\partial t}H(h-h_1) + \frac{1}{S_\infty}\frac{\partial S_2}{\partial t}H(h-h_2) \qquad (3.24)$$

is used as an example. Here S_0, S_1, and S_2 are the areas of open water and ice of thicknesses h_1 and h_2, respectively; $S_0 + S_1 + S_2 = S_\infty$; and $h_2 = kh_1$ represents the deformed ice produced from the undeformed ice of thickness h_1. Equation (3.24) transforms to the function ψ when the Heaviside functions are replaced by delta functions.

Pure divergence and convergence

In pure divergence, the open-water fraction increases while the relative coverage of non-zero thicknesses decreases (i.e., $\psi(0) = \nabla \cdot \boldsymbol{u}$ and $\psi(h) = 0$ for $h > 0$). The converse is true in pure convergence as long as the compactness of ice is less than unity (or less than a specified maximum). Consequently:

$$\frac{\partial \pi}{\partial t} + \boldsymbol{u} \cdot \nabla \pi = [\delta(h) - \pi]\nabla \cdot \boldsymbol{u} \qquad (3.25)$$

Since $\pi_0 = 1 - A$, this equation is for π_0 exactly the same as Eq. (3.15) is for mechanical changes in ice compactness under pure divergence or convergence.

Example Assume that $\nabla \cdot \boldsymbol{u} = \text{constant} = \lambda$, $|\lambda| = 0.1/\text{day}$, $A(t=0) = 0.8$, and ignore thermodynamics and advection. Then $\mathrm{d}\pi_0/\mathrm{d}t = (1-\pi_0)\lambda$, which gives $\pi_0 = 1 - 0.8\mathrm{e}^{-\lambda t}$. If $\lambda > 0$, the open-water fraction approaches zero with an e-folding timescale $\lambda^{-1} = 10$ days; if $\lambda < 0$, the ice closes up and compactness is one at the time $-\lambda^{-1}\log(1/0.8) \approx -0.22 \times \lambda^{-1} = 2.2$ days. ∎

Convergence in compact ice leads to redistribution of thicknesses by transforming thin ice into thick ice. The area of thin ice decreases at the rate $\nabla \cdot \boldsymbol{u}$ as a result of convergence itself. As the ice piles up, the thin-ice area decreases at the rate $(k-1)^{-1}\nabla \cdot \boldsymbol{u}$, while the deformed ice area increases by the same rate.

Consider the distributor of Eq. (3.24). We have $S_\infty^{-1}\partial S_1/\partial t = k/(k-1)\nabla \cdot \boldsymbol{u}$ and $S_\infty^{-1}\partial S_2/\partial t = -1/(k-1)\nabla \cdot \boldsymbol{u}$, and then:

$$\frac{\partial \pi}{\partial t} + \boldsymbol{u} \cdot \nabla \pi = -\left[\left(\pi_1 - \frac{k}{k-1}\right)\delta(h - h_1) - \left(\pi_2 + \frac{1}{k-1}\right)\delta(h - h_2)\right]\nabla \cdot \boldsymbol{u} \quad (3.26)$$

Example Assume that $\nabla \cdot \boldsymbol{u} = $ constant $= \lambda = -0.1$/day, ignore thermodynamics and advection (as in the previous example), and take initially $\pi = \delta(h - h_1)$. This ice is transformed into deformed ice with thickness kh_1. Then $d\pi_1/dt = -[\pi_1 - k/(k-1)]\lambda$, which gives $\pi_1 = (k - e^{-\lambda t})/(k-1)$. Since $\lambda < 0$, the fractional area of thin ice decreases, and at time $-\lambda^{-1}\log k$ the thin ice has totally disappeared; if $k = 5$, the time for this total deformation is ≈ 16 days. ∎

Pure shear

Although divergence in shear deformation is zero, opening takes place in the first principal axis and closing in the second principal axis. In non-compact ice there is room enough for the ice floes to perform shear deformation without interaction, and the area density of ice is unchanged. This case corresponds to shear deformation of incompressible fluids. But in the case of compact ice, pressure ice forms and leads open along the principal axes, and being of different nature they cannot cancel each other. Consequently, thickness distribution may also change in pure shear.

Consider the redistributor in Eq. (3.24). By opening leads at the rate of $\dot{\varepsilon}_0$, the same area of undeformed ice is taken and piled up as deformed ice. Clearly, the deformed ice area increases by $(k-1)^{-1}\dot{\varepsilon}_0$. The undeformed ice area must decrease by $k/(k-1)\dot{\varepsilon}_0$ to balance this. In pure shear $\dot{\varepsilon}_{II} = 2|\dot{\varepsilon}_{xy}|$, and one could then expect that $\dot{\varepsilon}_0 = \chi\dot{\varepsilon}_{II}$, $0 \leq \chi \leq \frac{1}{2}$, and χ decreases with decreasing compactness. Consequently:

$$\frac{\partial \pi}{\partial t} + \boldsymbol{u} \cdot \nabla \pi = \left[\delta(h) - \frac{k}{k-1}\delta(h - h_1) + \frac{1}{k-1}\delta(h - h_2)\right]\chi\dot{\varepsilon}_{II} \quad (3.27)$$

Thus, open water is formed here at the rate $\chi\dot{\varepsilon}_{II}$ even though divergence is zero.

General strain rate

In general the mode of deformation is described by the angle φ, defined by $\tan\varphi = \dot{\varepsilon}_{II}/\dot{\varepsilon}_I$ (see Section 3.1). The pure divergence, convergence, and shear correspond to $\varphi = 0$, π, and $\pi/2$, respectively. For any mode φ, the amount of opening is given by a function $\chi_0(\varphi)$ and the amount of pressure ice formation is

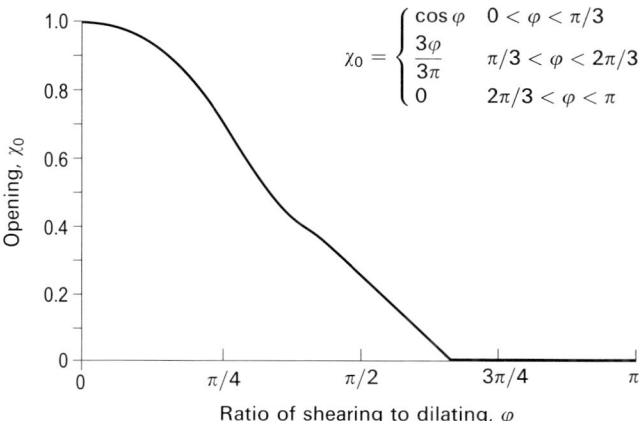

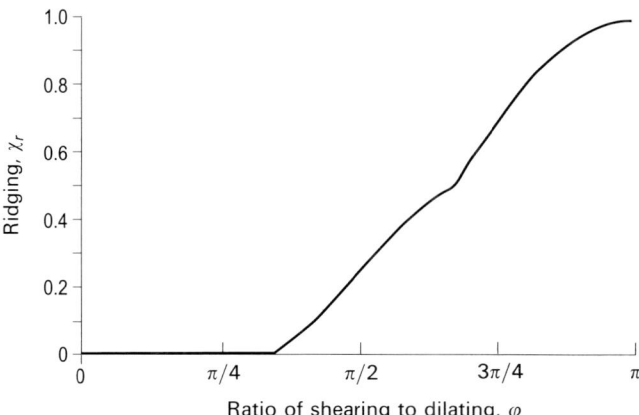

Figure 3.19. Opening and ridging as a function of the mode of deformation φ, $\tan\varphi = \dot{\varepsilon}_{II}/\dot{\varepsilon}_{I}$. Redrawn from Rothrock and Hall (1975), according to Pritchard (1981).

given by a function $\chi_r(\varphi)$ (Thorndike et al., 1975). Figure 3.19 shows the AIDJEX form of lead opening and ridging functions. Then:

$$\psi = |\dot{\varepsilon}|[\chi_0(\varphi)\delta(h) + \chi_r(\varphi)\psi'(h)] \qquad (3.28)$$

where the function $\psi'(h)$ specifies the loss and gain of different thicknesses in ridging. It is commonly assumed that ψ' picks up the 0.15 lower tail from the thickness distribution (Thorndike et al., 1975) and transfers it to k-multiple thicknesses. The original value was suggested as $k = 5$ (Thorndike et al., 1975), while later $k = 15$ was considered better at producing realistic thickness distributions (Pritchard, 1981; Rothrock, 1986). Flato and Hibler (1995) showed that $k = 5$ produces realistic thickness distributions up to 10-m thicknesses, but concluded that ψ' should rather pick up 0.05 thin-ice fractions and with a fixed k the resulting thickness distribution has a sawtooth shape.

It has also been suggested that the rafting process is significant in thin ice. This would be described by $k = 2$ (Leppäranta, 1981b; Bukharitsin, 1986; Babko et al., 2002), or more generally rubble fields or hummocked ice fields where the parameter k is more than 2 but much less than for ridged ice.

Sea ice kinematics has been examined in this chapter for methods, theory, and data. Ice motion was taken as a continuum flow and analysed using two-dimensional theory. The movement of a drift ice field was shown to consist of rigid displacement, rotation, and strain. Stochastic models for ice drift were also introduced. In Section 3.4 the ice conservation law was derived for the ice state defined in Chapter 2. Chapter 4 is about the rheology of drift ice, examining how the internal ice stress is determined by the ice state and strain. The geophysical interpretation of stochastic models is given in Chapter 6, and the ice conservation law is combined with the equation of motion for fully closed ice drift models in Chapters 7 and 8.

4

Sea ice rheology

4.1 GENERAL

This chapter first explains the general concepts underlying the science of rheology. Then the usual drift ice rheology models are presented: viscous, plastic, and granular medium. There are two principal reasons for the inclusion of this chapter. First, drift ice is a very complicated medium and its rheology changes drastically as a function of ice state and deformation. The picture of this rheology is still far from complete. Second, for students and scientists in oceanography or meteorology, rheological problems are not so familiar, because the ocean and atmosphere are linear Newtonian fluids that obey the well-established Navier–Stokes equation.

To derive the two-dimensional drift ice rheology, it is necessary to begin in the three-dimensional world. In this way it is possible to analyse the three-dimensional effects of the rheology problem and determine the conditions when the two-dimensional approach is allowed for formulation of drift ice rheology. For clarity, three-dimensional vectors and tensors are underlined. The internal stress $\underline{\sigma}$ in a medium specifies the internal force field over its arbitrary internal surfaces. In three dimensions the stress has nine components, arising from three independent surface orientations and three independent force directions (Figure 4.1):

$$\underline{\boldsymbol{\sigma}} = \begin{bmatrix} \underline{\sigma}_{xx} & \underline{\sigma}_{xy} & \underline{\sigma}_{xz} \\ \underline{\sigma}_{yx} & \underline{\sigma}_{yy} & \underline{\sigma}_{yz} \\ \underline{\sigma}_{zx} & \underline{\sigma}_{zy} & \underline{\sigma}_{zz} \end{bmatrix} \qquad (4.1)$$

Stress is thus a second-order tensor. Analogous to the strain and strain-rate tensors, the diagonal components are *normal (compressive or tensile) stresses* and off-diagonal components are *shear stresses*. The component $\underline{\sigma}_{xx}$ gives the normal stress in the x-direction, the component $\underline{\sigma}_{yx}$ gives the shear stress across the y-axis in the x-direction, etc. The stress tensor is necessarily *symmetric*, since otherwise

82 Sea ice rheology [Ch. 4

Tadashi Tabata (1923–1981), professor at the Institute of Low Temperature Science of Hokkaido University, founded the Sea Ice Research Laboratory in Mombetsu in 1967. He designed a coastal radar system, which has been operational since 1969, and largely advanced the understanding of the material properties and motion of sea ice. He also made major contributions to the study of the small-scale structure and mechanics of sea ice.
Reproduced with permission from the Institute of Low Temperature Science, Hokkaido University.

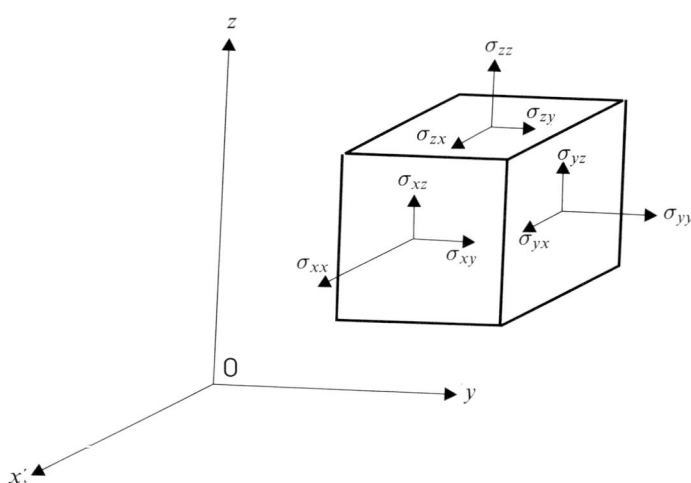

Figure 4.1. Stress $\underline{\sigma}$ on a material element.

there would be net internal torques that would not disappear when particle sizes approach zero (e.g., Hunter, 1976). Thus $\underline{\sigma}_{xy} = \underline{\sigma}_{yx}$, $\underline{\sigma}_{xz} = \underline{\sigma}_{zx}$ and $\underline{\sigma}_{yz} = \underline{\sigma}_{zy}$ which reduces the number of independent stress components to six. At a given point, the force exerted by the stress on a unit area with normal $\mathbf{n}$ is $\mathbf{n} \cdot \boldsymbol{\sigma}$. This vector has a normal component in the direction $\mathbf{n}$ and a shear component parallel to the surface.

Example Hydrostatic pressure is a stress $\boldsymbol{\sigma} = -p\mathbf{I}$ where $\mathbf{I}$ is the unit tensor; it is spherical (i.e., equal in all directions) and compressive ($p > 0$). In seawater $p = p_0 - \rho_w g z$, where p_0 is the surface atmospheric pressure and z is the vertical co-ordinate (negative downward). In any direction, $\mathbf{n} \cdot \boldsymbol{\sigma} = -p\mathbf{n}$ (i.e., hydrostatic pressure compresses the particle normal to the surface). ∎

4.1.1 Rheological models

The science of rheology examines how the stress in a medium depends on its material properties and strain (with strain rate and possibly higher order strain derivatives). The basic models are *linear elastic* or Hooke's medium, *linear viscous* or Newton's medium, and *ideal plastic* or St Venant's medium. Linking scientists' names to the rheology he or she has introduced is a common practice (e.g., Mase, 1970; Hunter, 1976). One-dimensional cases are illustrated in Figure 4.2. The linear elastic model assumes that stress is proportional to strain, while in the linear viscous model stress is proportional to the strain rate (proportionality coefficients are, respectively, the elastic modulus or Young's modulus and viscosity). Rubber and water (in laminar flow) are good material examples of linear elastic and viscous media. An ideal plastic medium collapses once stress achieves yield strength (children's modelling wax serves as an example of a plastic medium). Mechanical analogues for these models include spring balances for linear elasticity, dashpots[1] for linear viscosity, and static friction for plasticity. Further analogues can be found in the refrigerator: sausage is elastic, jam is viscous, and gelatine is plastic.

Basic rheological models can be expanded to more complex ones (e.g., Mellor, 1986). First, linearities can be changed to general non-linear laws in elastic and viscous models. Second, models can be combined (e.g., linear elastic and linear

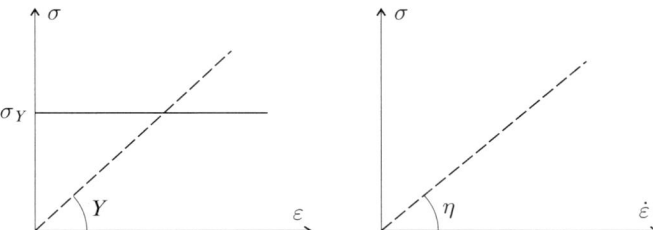

Figure 4.2. Basic rheology models (in one dimension) for stress σ as a function of strain ε and the strain rate $\dot{\varepsilon}$ (Y is elastic modulus, η is viscosity, and σ_Y is yield strength).

---

[1] Such as a door stroke compressor, which dampens or softens the movements of a door.

viscous models in series give a Maxwell medium while combining them parallel gives a Kelvin–Voigt medium). For a constant load, a Maxwell medium has an immediate elastic deformation and then flows in a linear viscous manner, while a Kelvin–Voigt medium flows in a viscous manner toward an asymptotic determined by the elastic part.

An ideal plastic medium collapses at infinite strain to yield stress. The rate of this strain then depends on other forces acting in the system. In compressive deformation usually strain hardening takes place. The medium becomes denser and more stress is needed for further deformation. To continue the modelling wax example, more and more force would be needed to make a ball smaller (for a constant load a stationary steady state would result).

Example (bearing capacity). For short-term loading floating ice is considered as an elastic plate on an elastic foundation (e.g., Parmerter and Coon, 1972). The bearing capacity can be approximated by load (in kg) $\sim 5\times$ ice thickness (cm) squared. ∎

4.1.2 Internal stress of drift ice

On a small scale, sea ice behaves in a linear elastic manner for short-term loading and in a general viscous manner for long-term low loading (Mellor, 1986). A typical Young's modulus is 2 GPa and the compressive elastic strength is 2 MPa. The viscous regime is more complicated and consists of three consecutive regimes: I— sublinear, II—linear and III—superlinear (in that order).[2] Viscosity is dependent on loading rate and ice temperature, with the strain inflection point for constant load at about 1% of strain. As a comparison, Young's modulus and compressive strength are one order of magnitude larger for wood and two orders of magnitude larger for steel.

Floating at the air–sea interface, drift ice actually experiences hydrostatic pressure from seawater and air in addition to the stress $\boldsymbol{\sigma}$ due to the interaction between ice floes. The total internal stress is thus:

$$\Sigma = \boldsymbol{\sigma} - p_a \mathbf{I} + \rho_w g \min(0, z) \mathbf{I} \qquad (4.2)$$

where p_a is atmospheric pressure at the sea surface. The hydrostatic pressures of air and water have a role in the creation of the state of drift ice. But, they can also have horizontal gradients that act as external forces on the ice, in much the same way as the horizontal pressure gradient works in dynamical oceanography. The influence of these surrounding media is straightforward and will be introduced in the derivation of the equation of motion in Chapter 5.

In this chapter the focus is on stress $\boldsymbol{\sigma}$, the most difficult and the least known

---

[2] Law $y = cx^a$ is sublinear for $a < 1$, linear for $a = 1$, and superlinear for $a > 1$.

factor in sea ice dynamics. The stress given in Eq. (4.1) is first split into three subfields:

$$\underline{\sigma} = \begin{bmatrix} HH & HH & HV \\ HH & HH & HV \\ VH & VH & VV \end{bmatrix}$$

The left upper 2×2 stress (HH), denoted by $\underline{\sigma}_H$, is the horizontal stress due to horizontal interactions between ice floes. The left lower 1×2 stress (VH) is the vertical shear stress, which transfers air and water stress into the ice. The right side 3×1 stress (HV and VV) takes care of ice floating on the sea surface. These 1×2 and 3×1 subfields will be attended to in the vertical integration of the momentum equation in the next chapter.

Integration of $\underline{\sigma}_H$ through the ice thickness gives two-dimensional stress:

$$\boldsymbol{\sigma} = \int_h \underline{\sigma}_H \, dz \qquad (4.3)$$

this is normally referred as *internal ice stress* in sea ice dynamics. The notation $\int_h$ stands for integration through the thickness of the ice. Note that the dimension of $\boldsymbol{\sigma}$ is force/length. The components σ_{xx} and σ_{yy} are normal stresses, and $\sigma_{xy} = \sigma_{yx}$ is the xy shear stress. The principal components, invariants, etc. for the stress tensor are found similarly to the method used for the two-dimensional strain-rate tensor in the previous chapter.

The rheology of drift ice is particularly complicated because qualitatively and quantitatively different laws apply to different ice states (particularly to the different packing densities and thicknesses of ice floes). The stress-generating mechanisms are:

 (i) Floe collisions.
 (ii) Floe break-up.
 (iii) Shear friction between floes.
 (iv) Friction between ice blocks in pressure ice formation.
 (v) Potential energy production in pressure ice formation.

From observational evidence, we know the qualitative features of drift ice rheology are as follows (the numbering has no significance):

(1) Stress level ≈ 0 for $A < 0.8$ (weak floe contacts).
(2) Tensile strength ≈ 0 (non-zero but small for compact consolidated ice).
(3) Shear strength is significant.
(4) Shear strength $<$ Compressive strength.
(5) Yield strength > 0 for $A \approx 1$.
(6) No memory.

Items (1) and (2) mean that significant stresses are present only under compression of very close or compact drift ice. Item (6) means that stress may depend on both strain and strain rate (i.e., the zeroth- and first-order time derivatives of the strain) but not

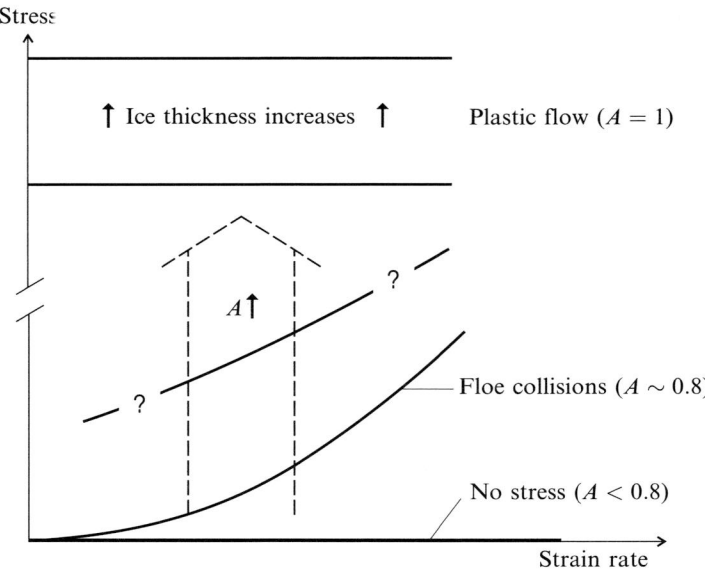

Figure 4.3. Schematic presentation of change in the quality of sea ice rheology as a function of ice compactness A and thickness h. The cut in the ordinate axis corresponds to a jump of several orders of magnitude.

on higher order time derivatives. Consequently, the rheological law of drift ice in general form is:

$$\boldsymbol{\sigma} = \boldsymbol{\sigma}(J, \varepsilon, \dot{\varepsilon}) \qquad (4.4)$$

The material properties are included in the ice state J, and the deformation is specified by the strain ε and strain rate $\dot{\varepsilon}$. The form of the rheological equation is normally assumed, and its parameters are found from tuning a full large-scale or mesoscale model against ice state and kinematics data.

A one-dimensional illustration of drift ice rheology is given in Figure 4.3. The simplest rheology is the no-stress case ($\boldsymbol{\sigma} \equiv 0$) known as *free drift*. The main drawback of the free drift system is that coupling with the ice conservation law is lost and ice floes may accumulate without limitations. But it is applicable for $A < 0.8$ when stress levels are very small. In a way this is analogous to using the pure Ekman drift in ocean hydrodynamics. In such a case, water could "pile up" or go onshore uncontrolled by hydrostatic pressure from sea level elevation or geostrophic adjustment (see Gill, 1982).

With increasing ice compactness, the significance of *floe collisions* increases and stress becomes a quadratic function of strain rate (Shen et al., 1986). But still the level of stress is low. With compactness increasing above 0.7–0.8, shear friction between ice floes becomes important. With compactness approaching unity, friction and potential energy increase due to pressure ice formation, resulting in plastic sea ice behaviour. How rheology changes from superlinear collision rheology to a plastic law is not known. In the plastic regime, yield strength increases with increasing ice thickness. Discrete particle models with detailed floe–floe interaction processes have been used to examine the closing up of drift ice (Løset, 1993; Hopkins, 1994; Savage, 1995), and the results are guiding the development of drift ice continuum rheologies.

In two dimensions shear strain and shear stress are also important. Their dependence is qualitatively similar to the case of normal stresses in the one-dimensional case, but the shear strength of drift ice is lower than its compressive strength. When shear strength is ignored, this becomes a *cavitating fluid* model (Flato and Hibler, 1990).

4.1.3 Internal friction

Spatial stress differences give rise to forces that step into the equation of motion. They smooth the velocity field, force the flow to satisfy boundary conditions, and may transmit momentum over large distances. The irrecoverable part of these forces is called *internal friction*. Apart from small elastic deformations, all changes due to stress differences are irrecoverable in drift ice mechanics.

Consider a square $\delta x \times \delta y$ in a two-dimensional internal stress field $\boldsymbol{\sigma}$. In the same way as in general continuum mechanics (e.g., Hunter, 1976), the net force is in the x-direction $[\sigma_{xx} + (\partial \sigma_{xx}/\partial x)\delta x - \sigma_{xx}]\delta y = (\partial \sigma_{xx}/\partial x)\delta x \delta y$ from normal stress and $[\sigma_{yx} + (\partial \sigma_{yx}/\partial y)\delta y - \sigma_{yx}]\delta x = (\partial \sigma_{yx}/\partial y)\delta x \delta y$ from shear stress; and similarly for the y-direction. Thus the total net force per unit area due to the interaction between ice floes gives us the divergence of the stress tensor $\nabla \cdot \boldsymbol{\sigma}$; that is:

$$[\nabla \cdot \boldsymbol{\sigma}]_x = \frac{\partial \sigma_{xx}}{\partial x} + \frac{\partial \sigma_{yx}}{\partial y}, \qquad [\nabla \cdot \boldsymbol{\sigma}]_y = \frac{\partial \sigma_{xy}}{\partial x} + \frac{\partial \sigma_{yy}}{\partial y} \qquad (4.5)$$

The three-dimensional case is analogous (i.e., stress divergence gives the force per unit volume).

Example Let p stand for hydrostatic pressure, $\boldsymbol{\sigma} = -p\mathbf{I}$. Evaluating the pressure differences for an elementary volume gives $\nabla \cdot \boldsymbol{\sigma} = -\nabla p$, which is known as the pressure gradient term in fluid dynamics. In general, internal stress differences lead to similar forcing of the medium. ∎

Here, internal ice stress $\boldsymbol{\sigma}$ is actually the vertically integrated three-dimensional stress $\boldsymbol{\sigma} = \int_h \boldsymbol{\sigma}_H \, dz$, and how it forces ice dynamics needs special attention (Nye, 1973). Namely, the partial derivative of a stress component σ_{ij} for a co-ordinate x_k is ($i, j, k = 1, 2$):

$$\frac{\partial \sigma_{ij}}{\partial x_k} = \frac{\partial}{\partial x_k} \int_{-h''}^{h'} \boldsymbol{\sigma}_{H_{ij}} \, dz = \int_{-h''}^{h'} \frac{\partial \boldsymbol{\sigma}_{H_{ij}}}{\partial x_k} \, dz + \boldsymbol{\sigma}_{H_{ij}}(h') \frac{\partial h'}{\partial x_k} + \boldsymbol{\sigma}_{H_{ij}}(-h'') \frac{\partial h''}{\partial x_k} \qquad (4.6a)$$

and assuming isostatic balance, the thickness gradient effect can be expressed as:

$$\boldsymbol{\sigma}_{H_{ij}}(h')\frac{\partial h'}{\partial x_k} + \boldsymbol{\sigma}_{H_{ij}}(-h'')\frac{\partial h''}{\partial x_k} = \left(\boldsymbol{\sigma}_{H_{ij}}(h')\left(1 - \frac{\rho}{\rho_w}\right) + \boldsymbol{\sigma}_{H_{ij}}(-h'')\frac{\rho}{\rho_w}\right)\frac{\partial h}{\partial x_k} \quad (4.6b)$$

In other words, the vertical integral of forcing by the stress $\int_h \partial \underline{\sigma}_{H_{ij}}/\partial x_k \, dz$ equals forcing by the two-dimensional stress $\partial \sigma_{ij}/\partial x_k$ minus the ice thickness gradient effect. If the thickness gradient is very small, its influence can be neglected, as usually implicitly done in sea ice dynamics. In general we could replace $\boldsymbol{\sigma}$ by $\boldsymbol{\sigma} - [\boldsymbol{\sigma}_H(h')h' + \boldsymbol{\sigma}_H(-h'')h'']$, and then the divergence of this stress would give the forcing correctly. However, to know the correction due to the thickness gradient, one would need a full three-dimensional rheology of the ice sheet. Nye (1973) analysed the linear viscous case and showed that for ice thickness gradients of 1 m per 20 km the thickness gradient correction becomes important. Such gradients would be exceptional in mesoscale averages, but the level is actually one order of magnitude less. Anyway, in this book the condition for the pure two-dimensional approach is assumed to be valid.

4.2 VISCOUS LAWS

In most drift ice dynamics research the elastic regime has been neglected, hence the fluid rheology $\sigma = \sigma(J, \dot{\varepsilon})$ results. A general viscous model is provided by the *Reiner–Rivlin fluid model* (e.g., Hunter, 1976):

$$\boldsymbol{\sigma} = \alpha \mathbf{I} + \beta \dot{\boldsymbol{\varepsilon}} + \gamma \dot{\boldsymbol{\varepsilon}}^2 \quad (4.7)$$

where the coefficients α, β, and γ may depend on the state variables of the medium and on strain-rate invariants (e.g., Hunter, 1976). In drift ice dynamics the last term has not been accounted for, but since α and β may depend on strain-rate invariants non-linear rheologies are possible.

4.2.1 Linear viscous models

Linear viscous models are not very representative of drift ice. However, in the history of sea ice dynamics they were the first class of applied rheologies and were used throughout the 1960s. The first was the Newtonian fluid model by Laikhtman (1958): $\boldsymbol{\sigma} = 2\eta\dot{\boldsymbol{\varepsilon}}' = 2\eta(\dot{\boldsymbol{\varepsilon}} - \frac{1}{2}\operatorname{tr}\dot{\boldsymbol{\varepsilon}}\mathbf{I})$, where η is the shear viscosity. This law is for linear, viscous, incompressible fluids such as water in laminar flow (viscosity gives the resistance of the fluid to shear deformation). For a compressible fluid, bulk viscosity or the second viscosity ζ is included to give the resistance to spherical deformation through compressive stress $\zeta(\nabla \cdot \boldsymbol{u})\mathbf{I}$. In principle, shear and bulk viscosities are independent. In the linear case they may depend on the material properties of the fluid but not on the strain rate. Internal friction is then:

$$\nabla \cdot \boldsymbol{\sigma} = \zeta\nabla(\nabla \cdot \boldsymbol{u}) + \eta\nabla^2\boldsymbol{u} + \nabla\zeta(\nabla \cdot \boldsymbol{u}) + \nabla\eta \cdot [\nabla\boldsymbol{u} + (\nabla\boldsymbol{u})^T - (\nabla \cdot \boldsymbol{u})\mathbf{I}] \quad (4.8)$$

Table 4.1. The viscosity of some materials.

Water (laminar flow, 0°C)	1.8×10^{-3} kg m^{-1} s^{-1}
Wax, shoemakers (8°C)	4.7×10^{5} kg m^{-1} s^{-1}
Pitch (0°C)	5.1×10^{9} kg m^{-1} s^{-1}

Source: Hunter (1976).

The third and fourth terms on the right-hand side are due to changing viscosity. Laikhtman (1958) assumed a constant shear viscosity and consequently $\nabla \cdot \boldsymbol{\sigma} = \eta \nabla^2 \boldsymbol{u}$.

The magnitude of the linear viscosities of drift ice is 10^8–10^{12} kg/s (Campbell, 1965; Doronin, 1970; Doronin and Kheysin, 1975; Hibler and Tucker, 1977; Leppäranta, 1981b). If the thickness of ice were 1 m, the corresponding three-dimensional viscosities would be 10^8–10^{12} kg m^{-1} s^{-1}. Hibler and Tucker (1977) further showed that there is a clear seasonal cycle in viscosities with winter and summer extrema at 10^{11} kg/s and 10^9 kg/s, respectively, in the Arctic Basin. Since deformation rates are of magnitude 10^{-6} s^{-1}, the stress magnitude is 10 kPa for viscosity 10^{10} kg/s; thus at the geophysical scale the stress is about two orders of magnitude less than at the local scale. An early improvement (Campbell and Rasmussen, 1972) was to introduce stepwise linear viscosities to separate opening and closing flows based on then-current qualitative knowledge (see Section 4.1): ζ, $\eta = 0$ for $\dot{\varepsilon}_I > 0$ and ζ, $\eta > 0$ for $\dot{\varepsilon}_I < 0$.

Kinematic shear or bulk viscosity K is obtained from the corresponding dynamic viscosity ζ or η by dividing by ρh: this includes time- and space-scale information, $K \sim L^2/T$. Thus in the viscous spreading of velocity we have $L^2 \sim KT$; for ζ, $\eta \sim 10^{10}$ kg/s this gives $L \sim 200$ km at $T \sim 1$ h.

For comparison, viscosities of some materials are shown in Table 4.1. These are, however, small-scale viscosities. Similar spatial scales to those in the drift ice problem exist in large-scale turbulence in ocean dynamics. There, horizontal turbulent stresses are often treated analogously to Newtonian viscous stresses (e.g., Gill, 1982), and then the internal friction is proportional to the Laplacian of velocity as in Laikhtman's (1958) drift ice model. The turbulent shear viscosity in ocean dynamics models is of the order of 10^7–10^9 kg m^{-1} s^{-1}, at the lower end of the range for the linear viscosity of drift ice. For glacier flows non-linear viscous laws are used (see Section 4.2.2); a linearized form for typical strain rates of 10^{-9} s^{-1} gives a linear viscosity of the order of 10^{14} kg m^{-1} s^{-1}.

Linear models are too crude for quantitative ice velocity analysis and modelling. For example, Hibler et al. (1974a) found that in the Beaufort Sea the ice velocity could fit a linear model only with a stepwise viscosity or with a slip boundary. The stepwise model would need viscosities of 10^{12} kg/s in the central pack but only 10^8 kg/s in the shear zone under convergence.

Russian authors have also used a hydrostatic pressure term in linear viscous models (e.g. Doronin and Kheysin, 1975). This has been formulated as $p = k_p \delta A$, with $k_p = 0$ for $\delta A > 0$ and $k_p > 0$ for $\delta A < 0$, δA being compactness change. Thus, pressure becomes active under compression and, according to Kheysin and Ivchenko (1973), at $A \approx 1$ and $\delta A \sim 10^{-3}$–10^{-2}, $k_p \sim 10^5$ N/m. The existence of this hydrostatic pressure term shows the general possibility of compressional waves in the system. Assuming $k_p = $ constant for small disturbances in compactness, with symmetry for opening and closing, the speed of compression waves is $c_p = \sqrt{k_p/m} \approx 10$ m/s. Doronin and Kheysin (1975) also provide evidence for the existence of transverse waves that propagate at speed $c_p/2$ perpendicular to the propagation of shear ridging. The existence of signal propagation in drift ice at speeds ~ 10 m/s has been reported by other scientists as well (Legen'kov, 1978; Goldstein et al., 2001). The physical mechanism for these waves is, however, not known. An interesting parallel can be made with underwater acoustics with waves propagating in the seabed (see chap. 2 of Lurton, 2002).

4.2.2 Non-linear viscous models

In non-linear viscous laws, ζ and η are functions of strain-rate invariants. The general form of Glen (1970), with the hydrostatic pressure p explicitly included, is:

$$\boldsymbol{\sigma} = \boldsymbol{\sigma}(\dot{\boldsymbol{\varepsilon}}, p; \zeta, \eta) = (-p + \zeta \operatorname{tr} \dot{\boldsymbol{\varepsilon}})\mathbf{I} + 2\eta \dot{\boldsymbol{\varepsilon}}' \qquad (4.9)$$

This rheology assumes isotropy and homogeneity.

Since our two-dimensional drift ice lies on, or nearly on, a surface with constant, or nearly constant, geopotential,[3] it is difficult to imagine any hydrostatic pressure build-up in the system. Therefore we can assume $p = 0$. However, in some rheological models the internal ice stress uses a formal hydrostatic pressure, and so this term is included in the general rheological equation.

It is normally assumed that the ratio of viscous compressive strength to viscous shear strength, ζ/η, is constant and greater than 1. This ratio is defined as the ratio of viscous stress in pure compression to viscous stress in pure shear (i.e., $-\zeta\dot{\varepsilon}_\mathrm{I}/(\eta\dot{\varepsilon}_\mathrm{II})$ with $-\dot{\varepsilon}_\mathrm{I} = \dot{\varepsilon}_\mathrm{II}$). Compressive strength $\zeta = \zeta(\dot{\varepsilon}_\mathrm{I}, \dot{\varepsilon}_\mathrm{II})$ can be taken as a homogeneous function of power n. Since $\sigma \propto \dot{\varepsilon}^{n+1}$, negative n results in a sublinear law while positive n gives a superlinear law. In close or compact ice the power is negative ($-1 < n < 0$), but in more open ice where floe collisions are important $n = 1$ (as will be seen below).

Non-linear shear viscous laws have been much used for glacier flow (e.g., Paterson, 1995). The basic model can be written in one-dimensional form:

$$\sigma = \eta \left| \frac{du}{dz} \right|^n \frac{du}{dz} \qquad (4.10)$$

where the z-coordinate is the local normal of the glacier surface. Normally $n = -\frac{2}{3}$, known as Glen's law (Glen, 1958), and then the viscosity is $\eta \approx 10^5$ kPa s$^{1/3}$. In three

---

[3] Geopotential surfaces contain the same potential energy and are perpendicular to local gravity.

dimensions, ice is incompressible, and the three-dimensional glacier rheology is expressed in tensor form as:

$$\underline{\sigma}' = \eta \dot{\varepsilon}_0^{-n} \underline{\dot{\varepsilon}} \qquad (4.11)$$

where $\underline{\sigma}' = \underline{\sigma} - \frac{1}{3} \mathrm{tr}\, \underline{\sigma} \underline{\mathbf{I}}$ is deviatoric stress, and $\dot{\varepsilon}_0^2 = \dot{\varepsilon}_{xx}^2 + \dot{\varepsilon}_{yy}^2 + \dot{\varepsilon}_{zz}^2 + 2(\dot{\varepsilon}_{xy}^2 + \dot{\varepsilon}_{xz}^2 + \dot{\varepsilon}_{yz}^2)$ is a strain-rate invariant equal to the sum of all squared strain-rate components. Thus in glacier flow, stress is proportional to the strain rate to the power $\frac{1}{3}$.

4.3 PLASTIC LAWS

4.3.1 Plastic drift ice

The most common approach to the rheology of drift ice has been the plastic model. The medium is stable when stress is below *yield stress*, while at *yield stress* it fails. The plastic law comes as the limit of sublinear viscous laws as $n \to -1$ in $\sigma \propto \dot{\varepsilon}^{n+1}$. The yield function is a property of the medium. In a one-dimensional system the situation is simple: the yield function consists of two points: there is one yield stress for compression and another for tension (see Section 4.1.1). For a two-dimensional medium, a *yield curve* needs to be specified for the yield function. In the principal stress space, we have:

$$F(\sigma_1, \sigma_2) = 0 \qquad (4.12)$$

A yield curve is a closed curve in the principal stress space. If $F < 0$, stress is small and the medium acts like a rigid body; this corresponds to stress being below yield stress in the one-dimensional case. At $F = 0$ the principal stress combination reaches the yield level and the medium fails. Values $F > 0$ are not allowed. Yield curves are always convex.

Inside the yield curve, no deformation takes place and stress varies between zero and yield stress (i.e., there is no one-to-one relation between stress and deformation). For numerical modelling such a situation is not feasible, and therefore the rigid (i.e., no deformation) mode is replaced by a "nearly rigid" mode. Two approaches have been taken: the Arctic Ice Dynamics Joint Experiment (AIDJEX) model (Coon et al., 1974) used a linear elastic model, while Hibler (1979) chose a linear viscous model. Combined with yield behaviour, elastic–plastic and viscous–plastic rheologies result. There is no good observational evidence to qualify and quantify the correct model for drift ice stresses below the yield level.

How plastic deformation takes place is specified by a *flow rule*. The one-dimensional case is again simple: compressive strain for compressive stress and tensile strain for tensile stress. For higher dimensional cases, *Drucker's postulate* for stable materials states that the yield curve serves as the plastic potential, and consequently the failure strain is directed perpendicular to the yield curve. This is the *normal, or associated, flow rule*. A characteristic property of plastic media is that the stress is independent of the absolute level of the rate of strain, but, however, does depend on the mode of strain (i.e., whether it is compression or tension and how much relative shear there is).

In the sea ice dynamics problem the normal flow rule has been used to solve the failure strain. How fast the strain takes place then depends on the other forces in the system. The flow rule is therefore formulated in terms of the strain rate, and the problem is closed for the absolute strain-rate level from the equation of motion. Consequently, the normal flow rule requires that:

$$\dot{\varepsilon}_1 = \lambda \frac{\partial F}{\partial \sigma_1}, \quad \dot{\varepsilon}_2 = \lambda \frac{\partial F}{\partial \sigma_2} \qquad (4.13)$$

where λ is a free parameter to be obtained as part of the solution. Strain rates are then obtained for the xy-coordinate system, assuming that the alignment of the principal axes of stress and strain rate is overlapping.

Example Take the yield curve as a circle $F(\sigma_1, \sigma_2) = \sigma_1^2 + \sigma_2^2 - \sigma_Y^2$. By definition, inside the yield curve ($F < 0$) the medium is rigid and fails at $\sigma_1^2 + \sigma_2^2 = \sigma_Y^2$. The associated flow rule gives $\dot{\varepsilon}_k = 2\lambda \sigma_k$, and λ derives from stress being equal to yield stress: $(\dot{\varepsilon}_1/2\lambda)^2 + (\dot{\varepsilon}_2/2\lambda)^2 = \sigma_Y^2$ or $\lambda = \sqrt{\dot{\varepsilon}_1^2 + \dot{\varepsilon}_2^2}/2\sigma_Y$ and further:

$$\sigma_k = \frac{\dot{\varepsilon}_k}{\sqrt{\dot{\varepsilon}_1^2 + \dot{\varepsilon}_2^2}} \sigma_Y$$

Clearly the stress is independent of the absolute level of the strain rate. For uniaxial compression $\dot{\varepsilon}_{xx} < 0$, $\dot{\varepsilon}_{xy} = \dot{\varepsilon}_{yx} = \dot{\varepsilon}_{yy} = 0$, and therefore we have $\dot{\varepsilon}_1 = 0$, $\dot{\varepsilon}_2 = \dot{\varepsilon}_{xx}$. The resulting stress is uniaxial, $\sigma_{xx} = -\sigma_Y$. ∎

Other flow rules exist in plasticity theory but for the drift ice problem they have not been applied. This is logical since the detailed form of the yield curve itself has not been established; in fact, different reasonable yield curves with the normal flow rule have produced equally good ice velocities within the limits of the observational validation of ice velocity (e.g., Zhang, 2000).

Plastic rheologies are considered best at representing the physical behaviour of compact drift ice. For a long time drift ice has been known to possess plastic properties. In particular the existence of yield strength is clear to people working in enclosed or semi-enclosed sea ice basins: compact ice is stationary until forcing exceeds a certain minimum level. Also a highly sublinear rheology is needed to produce narrow deformation zones, as found from the Beaufort Sea shear zone by Hibler et al. (1974a), among others. It was in the 1970s that the mathematical–physical basis for plastic drift ice models was constructed (Coon et al., 1974), a major achievement of the AIDJEX programme. Then the concept of a granular medium was introduced for drift ice, and new ideas came from understanding the similarities between the mechanics of drift ice and soil.

Further support for plasticity was obtained from local scale mechanical processes. Parmerter and Coon (1972) constructed a kinematic model of ridging, which is the main sink for kinetic energy in sea ice dynamics, and showed that ice

Sec. 4.3] Plastic laws 93

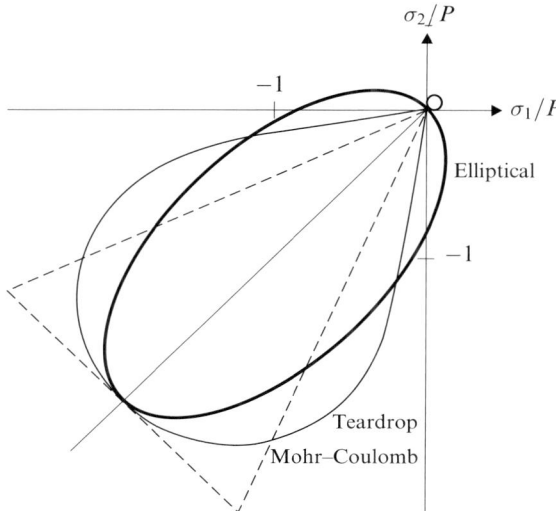

Figure 4.4. Plastic yield curves for drift ice: wedge or Mohr–Coulomb (Coon, 1974), teardrop (Rothrock, 1975a), and elliptic (Hibler, 1979).

stress appears to be independent of strain-rate magnitude, a strong characteristic feature of plasticity. They further examined the mechanical energy budget in ridging, and the results became the basis for the parameters of plastic drift ice rheologies (Rothrock, 1975a). This energy budget was later reworked with a discrete particle model (Hopkins and Hibler, 1991; Hopkins, 1994). The losses of kinetic energy in ridging are due to friction between ice blocks, creating potential energy, inelastic deformation of ice blocks, and breakage of the ice sheet. Potential energy can be observed in the topography of ridges, while the other losses would be very difficult to track. Therefore, total loss is often assumed to be proportional to potential energy change. This is arguably questionable, since it has become clear that potential energy loss is one order of magnitude less than frictional losses.

For the drift ice yield curve, the following additional requirements exist (based on the known qualitative properties listed in Section 4.1):

- Since drift ice has (almost) no tensile strength, the principal stresses must always be negative or zero σ_1, $\sigma_2 \leq 0$, and this means that the yield curve must be located in the quadrant III of the principal axes co-ordinate system.
- The yield curve is symmetric with respect to the line $\sigma_1 = \sigma_2$ for isotropy.
- The yield curve is elongated in the $\sigma_1 = \sigma_2$ axis to give a compressive strength higher than shear strength.

Figure 4.4 shows the common yield curves for drift ice that will be derived in the following subsections. The main parameter in drift ice plastic rheologies is compressive strength P. It depends on the ice state $P = P(J)$ and also varies due to the mode of failure (i.e., crushing, buckling, or rafting). In the yield curves of Figure 4.4, compressive strength is represented by the point farthest from the origin in the line $\sigma_1 = \sigma_2$. Other examples are (a) a square yield curve (bounded by the lines $\sigma_1 = 0$, $\sigma_2 = 0$, $\sigma_1 = -P/\sqrt{2}$, $\sigma_2 = -P/\sqrt{2}$) of Pritchard (1977), and (b) the cavitating fluid ($-P \leq \sigma_1 = \sigma_2 \leq 0$) of Flato and Hibler (1990).

The yield strength of ice increases with ice compactness and thickness. This means that the size of the yield curve increases but its shape is invariant. In compressive deformation ($\dot{\varepsilon}_I < 0$), compactness and thickness increase (i.e., drift ice is *strain hardening* under compression). Therefore for a given forcing, drift ice compression may proceed only to a certain limit, obeying common sense. In opening deformation ($\dot{\varepsilon}_I > 0$), the strength is zero and, if forced, the opening may continue without any limit.

Example In the one-dimensional case, let $P = P(h)$. Under compression, ice is packed thicker, and for strain-hardening ice $dP/dh > 0$. Compressing an ice bar of unit width and thickness h from the left boundary by a force F, it continues failing until stress equals yield stress at all points. At equilibrium, $P(h) = F/h$ gives the resulting thickness. ∎

4.3.2 Mohr–Coulomb rheology

The first plastic drift ice model was by Coon (1974) using the Mohr–Coulomb medium model, for which failure shear stress σ_S is proportional to compressive stress σ_N:

$$\sigma_S = \sigma_N \tan \vartheta \qquad (4.14)$$

where ϑ is the angle of friction between ice floes, taken as $\vartheta = 35°$ from the repose angle of ice ridges. A cut-off stress for σ_N is provided by the ice strength, which depends on the mode of failure. This results in a wedge-shaped yield curve (Figure 4.4). Coon (1974) adopted Drucker's postulate and consequently the normal flow rule.

According to Coon (1974), the cut-off stress in crushing failure is $\sigma_{cr} h$ when σ_{cr} is the crushing strength of ice sheet, and in buckling failure it is $\rho_w g l_0^2/(12h)$, where l_0 is the smallest floe size in the loading area; if l_0 is taken as the size of pieces that break in rafting (see Section 2.1), the cut-off stress is:

$$\frac{\pi^2}{192} \sqrt{\frac{Y \rho_w g h}{3(1-\nu^2)}}$$

For $h \sim 1\,\text{m}$, the local magnitude levels of crushing failure and buckling failure are, respectively, 1 MPa and 100 kPa for $h \sim 1\,\text{m}$ (i.e., crushing failure stress is one order of magnitude larger). They are both large but local; it is expected that, integrated over the continuum of length scales, the stress would be lower.

Example The Mohr–Coulomb model has also been used for sea ice ridges and hummocks (Figure 4.5). The shear strength of a pile of ice blocks is $\sigma_S = c + \sigma_N \tan \vartheta$, where c is cohesion. The values of cohesion and angle of

Figure 4.5. Field experiment into the strength of sea ice ridges in the Baltic Sea, winter 1987. The ridge keel is loaded by weights to examine the Mohr–Coulomb model.

friction depend on the degree of consolidation of the ridge. With no frozen bonds between ice blocks $c = 0$ and $\phi \approx 6\,\text{deg}$ (Schaefer and Ettema, 1986). ∎

4.3.3 AIDJEX elastic–plastic rheology

In the AIDJEX programme an elastic–plastic drift ice model was developed (Coon et al., 1974; Pritchard, 1975). A *linear elastic model* is used inside the yield curve, where stress is proportional to strain:

$$\boldsymbol{\sigma} = \boldsymbol{\sigma}(\boldsymbol{\varepsilon}; M_1, M_2) = M_1(\text{tr}\,\boldsymbol{\varepsilon})\mathbf{I} + 2M_2\boldsymbol{\varepsilon}' \qquad (4.15)$$

where M_1 is the bulk modulus and M_2 is the shear modulus. Note that here, inside the yield curve, stresses are compressive and tensile stresses are not allowed (M_1, $M_2 = 0$ for $\varepsilon_\mathrm{I} > 0$). The magnitude of mesoscale elastic constants is 10–100 MN/m, $M_1 \approx 2M_2$ (Pritchard, 1980b), and they allow the ice to deform up to yield stresses. The compressive elastic strength is $\sigma_c \approx 0.1\,\text{MN/m}$ (Pritchard, 1980b), and therefore mesoscale elastic strains remain less than about 10^{-3}. Note that in small-scale sea ice mechanics the elastic constants and compressive strength are one to two orders of magnitude larger, in three dimensions: $M_1 \approx 3\,\text{GN/m}^2$, $M_2 \approx 0.4\,\text{GN/m}^2$, $\sigma_c \approx 2\,\text{MN/m}^2$ (Mellor, 1986). The fast ice problem is a suitable area in which to apply the theory of elasticity.

The AIDJEX *plastic model* is based on the mechanics of ridging. Compressive strength is obtained by equating the plastic work done in ridging with the creation of potential energy and losses due to friction between ice blocks (Rothrock, 1975a). The rate of loss of energy in ridging is $R_p + R_f = \sigma_I \dot{\varepsilon}_I + \sigma_{II} \dot{\varepsilon}_{II}$, where R_p represents the creation of potential energy and R_f represents friction. The potential energy portion comes from the redistribution of ice thickness as (see Eqs 2.2 and 3.28):

$$R_p = \frac{dE_p}{dt} = c_p \int_0^\infty h^2 \psi \, dh = |\dot{\varepsilon}| \chi_r(\varphi) P_p \tag{4.16}$$

where $P_p = c_p \int_0^\infty h^2 \psi' \, dh$ represents potential energy produced per unit area and unit strain in pure convergence, $c_p = \frac{1}{2} \rho g (\rho_w - \rho)/\rho_w$, and ψ' gives thickness rearrangement in ridging (see Eq. 3.28), expressed as the sum of loss and gain as $\psi' = -\psi'_- + \psi'_+$. In the AIDJEX thickness distribution model, ridging transfers thin ice to ridged ice k times as thick. Therefore, $\psi'_-(h) = kh/(k-1)$ and $\psi'_+(kh) = kh/(k-1)$ for those thicknesses h that take part in building ridges.

To ascertain the frictional losses in ridge building, a Coulomb model is used where the frictional force equals the weight of the ice times the coefficient of sliding friction μ (Rothrock, 1975a). In building a ridge, the frictional force per unit length is:

$$F_f = \mu \rho g \frac{h_s^2}{2 \tan \varphi_s} + \mu_w (\rho_w - \rho) g \frac{h_k^2}{2 \tan \varphi_k} \tag{4.17}$$

where h_s, φ_s, h_k, and φ_k are the sail height, sail slope, keel depth, and keel slope of ridges (see Figure 2.16), and μ and μ_w are the coefficients of dry and wet friction, respectively. Assuming isostatic balance and reasoning that $\mu \approx \mu_w$, the sail term becomes small compared with the keel term and can be neglected. Then for ridging ice of thickness h_1:

$$F_f = \frac{\mu(\rho_w - \rho)g}{2 \tan \varphi_k} \left(\frac{\rho(k-1)}{\rho_w} \right)^2 h_1^2 = c_f h_1^2 \tag{4.18}$$

In the ridging process the loss of ice of thickness h_1 is $h_1 k/(k-1)$ (see Eq. 3.26), and consequently the frictional energy loss in ridging becomes:

$$R_f = |\dot{\varepsilon}| \chi_r(\varphi) P_f, \qquad P_f = c_f \int_0^\infty \frac{k}{k-1} h^2 \psi'_-(h) \, dh \tag{4.19}$$

The compressive strength of drift ice is then obtained as the sum of work done to increase the potential energy and overcome the friction. In pure compression, $|\dot{\varepsilon}| = |\dot{\varepsilon}_I|$, $\chi_r = 1$, and $R_p + R_f = \sigma_I \dot{\varepsilon}_I$, and since the loss and gain of ice must balance, the potential energy production can be written in terms of ψ'_-. We have:

$$P = P_p + P_f = \left(k c_p + \frac{k}{k-1} c_f \right) \int_0^\infty h^2 \psi'_-(h) \, dh \tag{4.20}$$

The ratio of frictional losses to potential energy is thus:

$$\Gamma = \mu \frac{k-1}{\tan \varphi_k} \frac{\rho}{\rho_w}$$

If the thickness of ice consumed in ridging is constant (h_1), then $P \propto h_1^2$. In the original work of Rothrock (1975a) the key parameters were chosen as $\mu = 0.1$ and $k = 5$, resulting in $\Gamma \approx 1$; but later it became clear from model tuning that $\Gamma \gg 1$ and both μ and k should be larger by a factor of about 3.

The value of compressive strength can be estimated from Eq. (4.20), and the result using the original parameters is $P \approx 5\,\text{kPa}$ for $h_1 = 1\,\text{m}$. This value is one order of magnitude too small, as should be clear from underestimating the frictional losses. Pritchard (1977) obtained $P = 40\,\text{kPa}$ based on fitting model simulations with kinematics data.

The shape of the yield curve still depends on the form of the ridging function $\chi_r(\varphi)$, which is unknown. Rothrock (1975a) suggested two possibilities for the yield curve: a teardrop shape (Figure 4.4) and a lens shape. Pritchard (1977) used the AIDJEX model and concluded that a wedge-shaped (Figure 4.4) or a square (bounded by the four lines $\sigma_{1,2} = 0$, σ_s) yield curve could better reproduce ice dynamics in the Beaufort Sea. A theoretical study on the yield curve of drift ice was made by Ukita and Moritz (1995).

4.3.4 Hibler's viscous–plastic rheology

Hibler's (1979) approach to the plastic flow solution involved a viscous–plastic rheology. In essence, no new physics were introduced, but the solution was very elegant. It satisfies the qualitative requirements for a proper rheology of drift ice and is an excellent tool for numerical modelling since it allows an explicit solution for stress as a function of strain rate. Most numerical sea ice models basically use this rheology, and its parameters have been tuned for different basins and scales (Figure 4.6).

The role of viscosity is to provide the stress field inside the yield curve. A major advantage is that viscous models depend on the strain rate and therefore are easily combined with strain rate-dependent flow rules. A linear viscous law can be taken as:

$$\boldsymbol{\sigma} = \zeta \dot{\varepsilon}_1 \mathbf{I} + 2\eta \dot{\varepsilon}' - P/2 \qquad (4.21)$$

where viscosities are constants. The existence of the pressure term $-P/2$ is connected to the yield curve; it guides the viscous stress to the plastic yield curve when the strain rate increases.

The yield curve is elliptic (Figure 4.4). The length of the major axis of the ellipse is given by compressive strength:

$$P = P^* h \exp[-C(1 - A)] \qquad (4.22)$$

where P^* is the compressive strength of compact ice of unit thickness, and C is the strength reduction constant for lead opening. The strength is directly proportional to ice thickness but highly sensitive to ice compactness. The inverse of C gives the e-folding scale for ice strength as a function of compactness. The length of the minor axis is P/e, and the aspect ratio of the ellipse is thus e. At the ends of the minor axis the gradient of the yield curve is parallel to the line $\sigma_1 = -\sigma_2$, which means that pure

Figure 4.6. Field experiment data are used for high quality tuning of sea ice mechanical phenomenology and for tuning rheological parameters. Large-scale stresses cannot be measured directly, but examined from measurements of ice state and ice kinematics. The photograph shows an automatic drifting station in the MIZEX-83 experiment, Greenland Sea.

shear strain will take place. Consequently, the length of the minor axis is determined by the shear strength of the ice.

To show the plastic flow solution in detail, the elliptic yield curve and the flow rule for the two principal strain rates are written as:

$$F(\sigma_1, \sigma_2) = \left(\frac{\sigma_1 + \sigma_2 + P/2}{P/2}\right)^2 + \left(\frac{\sigma_1 - \sigma_2}{P/2e}\right)^2 \qquad (4.23a)$$

$$\dot{\varepsilon}_1 = \lambda \frac{\partial F}{\partial \sigma_1} = \lambda \left(\frac{2(\sigma_1 + \sigma_2 + P/2)}{(P/2)^2} + \frac{2(\sigma_1 - \sigma_2)}{(P/2e)^2}\right) \qquad (4.23b)$$

$$\dot{\varepsilon}_2 = \lambda \frac{\partial F}{\partial \sigma_2} = \lambda \left(\frac{2(\sigma_1 + \sigma_2 + P/2)}{(P/2)^2} - \frac{2(\sigma_1 - \sigma_2)}{(P/2e)^2}\right) \qquad (4.23c)$$

Solving these equations for σ_1 and σ_2 and transforming to a fixed co-ordinate system (to do this we need to know that the principal stresses and principal strain rates are parallel) gives the rheological law:

$$\boldsymbol{\sigma} = \frac{P}{2}\left(\frac{\dot{\varepsilon}_I}{\Delta} - 1\right)\mathbf{I} + \frac{P}{2e^2\Delta}\dot{\boldsymbol{\varepsilon}}' \qquad (4.24)$$

Table 4.2. Parameters of the viscous–plastic ice rheology of Hibler (1979).

Parameter	Notation and standard	Range
Compressive strength	$P^* = 25\,\text{kPa}$	$10\text{--}100\,\text{kPa}$
Yield ellipse aspect ratio	$e = 2$	$1 < e \ll \infty$
Compaction hardening	$C = 20$	$C \gg 1$
Maximum creep	$\Delta_0 = 2 \times 10^{-9}\,\text{s}^{-1}$	$\Delta_0 < 10^{-7}\,\text{s}^{-1}$

where $\Delta = \sqrt{\dot{\varepsilon}_I^2 + (\dot{\varepsilon}_{II}/e)^2}$ is a strain-rate invariant corresponding to the strength of the ice. A great advantage of elliptic yield curves is that the stress–strain rate relationship can be written in the above closed form.

For the viscous law, which is valid inside the yield curve, viscosities can be expressed as $\zeta = \frac{1}{2}P/\Delta_0$, $\eta = \zeta/e^2$. Combined with the plastic law, the full viscous–plastic rheology reads as:

$$\boldsymbol{\sigma} = \frac{P}{2}\left(\frac{\dot{\varepsilon}_I}{\max(\Delta, \Delta_0)} - 1\right)\mathbf{I} + \frac{P}{2e^2 \max(\Delta, \Delta_0)}\dot{\boldsymbol{\varepsilon}}' \qquad (4.25)$$

where Δ_0 gives the maximum viscous creep rate, originally chosen as $\Delta_0 = 2 \times 10^{-9}\,\text{s}^{-1}$. As $\Delta \to \Delta_0$, the viscous and plastic stresses meet.

Normal levels for the plastic parameters of Hibler's (1979) rheology are $P^* = 25\,\text{kPa}$, $C = 20$, and $e = 2$ (Table 4.2). The strength constant P^* is the principal plastic rheology parameter. This has ranged from 5 kPa to 50 kPa. In the original paper (Hibler, 1979) it was 5 kPa; but it turned out that the low level was necessary because 8-day averaged winds were used. More recently, Arctic ice models have used a level of about 25 kPa. In the Baltic Sea, model outcome comparison with SAR ice kinematics has resulted in 10–40 kPa, with 27.5 kPa chosen as a working standard level (Zhang and Leppäranta, 1995; Leppäranta et al., 1998). Comparisons with sea surface elevation in open water and compact ice have shown that the strength could be as high as 100 kPa (Zhang and Leppäranta, 1995). Parameters e and C are usually fixed to the original values.

The pressure term is enough to force expansion by slow creep for a stationary ice field with free boundaries. However, this is not known to occur. This term is necessary to put the ellipse into the right quadrant in the principal axes system and should not be used alone. A more recent formulation overcomes this problem by letting $\boldsymbol{\sigma} \to 0$ (instead of $-\frac{1}{2}P\mathbf{I}$) as $|\dot{\boldsymbol{\varepsilon}}| \to 0$ (Hibler, 2001).

Example The cavitating fluid (Flato and Hibler, 1990) model can be derived as a special case of the Hibler (1979) rheology where $e \to \infty$. The rheology then becomes:

$$\boldsymbol{\sigma} = \frac{P}{2}\left(\frac{\dot{\varepsilon}_I}{\max[\Delta_0, |\dot{\varepsilon}_I|]} - 1\right)\mathbf{I} \qquad (4.26)$$

Figure 4.7. Ice pressure has captured one ship in the Baltic Sea. The strength of the ship is known but it is very difficult to predict local scale ice forcing from geophysical scale sea ice dynamics.
From Ramsay (1949).

The plastic flow is here given by[4] $\sigma = \frac{1}{2}(\text{sgn}(\dot{\varepsilon}_I) - 1)P$, while viscous flow is $\sigma = \zeta \nabla \cdot \boldsymbol{u} - \frac{1}{2}P$, where the bulk viscosity is $\zeta = \frac{1}{2}P/\Delta_0$. ∎

Recently an anisotropic version of the viscous–plastic model has been developed to take into account the geometric structure of lead patterns (Hibler and Schulson, 2000).

4.3.5 Scaling of ice strength

Scale questions have become a major issue in sea ice physics during the last 10 years. For cross-scale experiments, a particular campaign (SIMI, for Sea Ice Mechanics Initiative) was organized in the Beaufort Sea (e.g., Richter-Menge and Elder, 1998), and another took place in the Baltic Sea in 1997 (Haapala and Leppäranta, 1997b). In 2000, a dedicated symposium was organized around the scaling problem (Dempsey and Shen, 2001). Ice engineering has traditionally focused on the local scale (1–10 m), while in geophysical sea ice dynamics the scales of interest have been mesoscale to large scale (100–1,000 km). Between the engineering and geophysical scale there is the less known floe scale (100 m–1 km), through which the two main scales are interconnected (Figure 4.7).

---

[4] sgn is the sign function: $\text{sgn}(x) = 1$ for $x > 0$, 0 for $x = 0$, or -1 for $x < 0$.

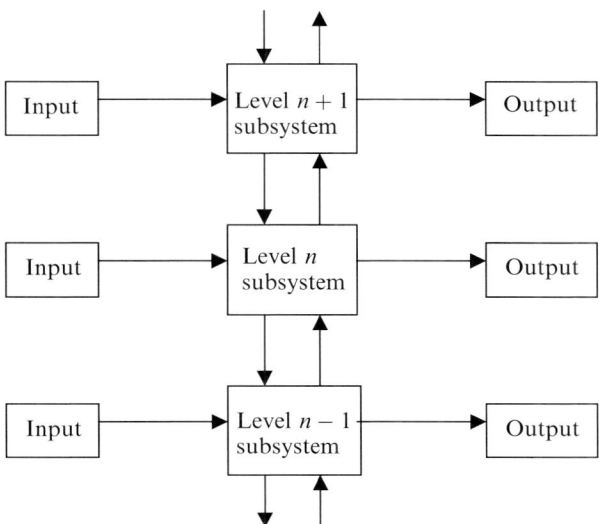

Figure 4.8. Structure of a hierarchical system.
Redrawn from Overland et al. (1995).

In the length range 10^1 m to 10^5 m, there are bands where different material properties play the main role (Overland et al., 1995; Weeks, 1998a): in the 10^1–10^2-m band these are fractures and thermal cracks; in the 10^2–10^3-m band these are individual ice floes; in the 10^3–10^4-m band these are floe assemblages and leads; and at larger scales the ice field becomes a continuum. Overland et al. (1995) proposed the concept of a hierarchical system to analyse sea ice over a wide range of scales (Figure 4.8). State variables at level $n+1$ vary smoothly and evolve slowly compared with those at level n; they serve as constraints, driving forces, or boundary conditions on level n. The effect of level $n-1$ on level n, via interactions and averaging, is known as the aggregate problem. The hierarchy theory states that levels $n+2$ or $n-2$ are on scales too large or too small to have a direct impact at level n.

A key quantity in the scaling problem is the ice stress or, rather, the ice strength. The local strength can be measured directly, while the mesoscale strength can only be derived from a mathematical model. It is known that, at its highest, the local strength is of the order of 1–10 MPa, while the mesoscale strength is 10–100 kPa in compact ice. So, there is a difference of two orders of magnitude. In a recent field study off the northern coast of Alaska, Richter-Menge et al. (2002) have shown clear connections between the small-scale and mesoscale internal stress fields in the ice.

Figure 4.9 presenting Sanderson's curve, gives a rough idea of how stress behaves as a function of the size of the loading area. Results from laboratory test to large-scale drift ice models are included. A decrease in strength of two orders of magnitude from local scale to mesoscale is clearly seen in groups C and D. The form of dependence appears as a power law $\sigma(L) \propto L^n$ with $n \approx -\frac{1}{2}$.

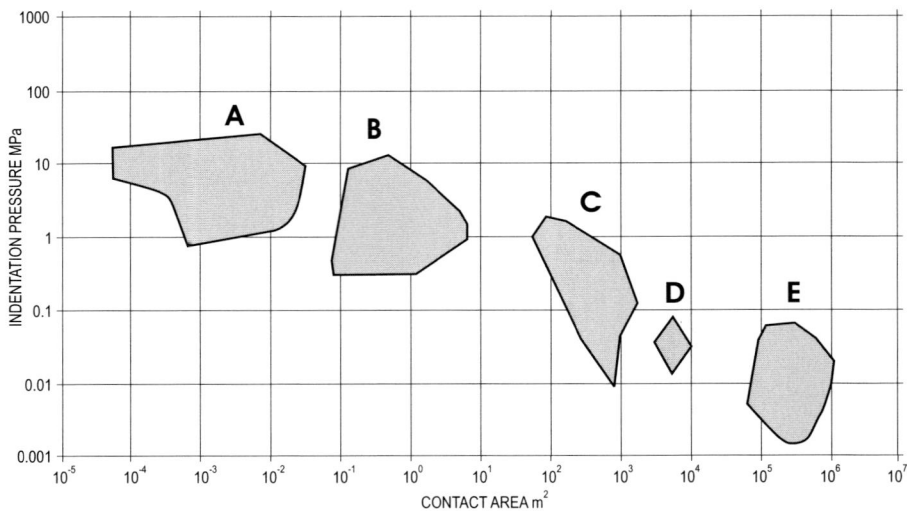

Figure 4.9. Sanderson's curve: the strength of sea ice vs. the loading area. A to C show local-scale tests, D shows Baltic Sea high-resolution ice drift models, and E shows Arctic Ocean meso-scale ice drift models.
From Sanderson (1988), but with Baltic Sea drift ice dynamics data (D) added.

The relationship between mesoscale and local scale stresses was examined using a probabilistic approach by Kheysin (1978). He proposed an exponential law for the spatial decay of correlation in the stress field and reported on experimental data showing that the correlation radius (e-folding length) of normal stresses is of the order of magnitude of 10 km. It is clear that the correlation length is not a constant but varies with the geophysical state of the ice. The exponential model differs in kind from the power law. It only gives small changes within the correlation radius and, thereafter, drops off faster than the power law.

Example: stress as a spatial diffusion process (Harr, 1977) Consider the transmission of stress as a diffusion process due to the random interactions between ice floes. Start with the local stress σ_{yy} at the origin and then progress along the y-axis:

$$\frac{\partial \sigma_{yy}}{\partial y} = D \frac{\partial^2 \sigma_{yy}}{\partial x^2}$$

where D is the diffusion coefficient, its dimension is length, and stress is considered dimensionless. Then $\sigma_{yy}(y)$ is a Gaussian process with zero mean and variance $2Dy$, and the probability density is:

$$p(x, y; D) = \frac{1}{\sqrt{4\pi Dy}} \exp\left(-\frac{x^2}{4Dy}\right)$$

The peak stress is $(4\pi Dy)^{-1/2}$, and the width over which the stress acts is $2(2Dy)^{1/2}$, defined as the length scale when the stress is within one standard deviation from the peak value. Thus, the peak stress × width = constant; the peak stress decreases as $y^{-1/2}$, while width increases as $y^{1/2}$. One would anticipate that $D \propto d$ (floe size) since diffusion proceeds as floes interact, implying that the smaller the floe size in ice mechanics the better it is at diffusing stress, scales in proportion to $d^{-1/2}$. ∎

In the geophysical regime, the main deformation processes involved in the convergence of compact ice are rafting and pressure ridge formation. They are quite different in kind and, consequently, the same similarity law is unlikely to be valid for both cases. The main internal resistance in rafting comes from the friction between rafting floes, while the accumulation of blocks in pressure ice formation has much larger friction, and a lot more potential energy has to be created. Leppäranta et al. (1998), in an effort to understand observed sea ice kinematics, showed that much lower resistance appeared in the rafting of thin ice (10–15 cm) than in the ridging of 30–40 cm thick ice. Using the Hibler (1979) viscous–plastic rheology, the compressive strength constant P^* was about 30 kPa for thick ice, but lower by nearly an order of magnitude for thin ice.

Hummocks and ridges normally form in ice thicker than 10–20 cm, while the thickest ice blocks found in ridges are around 1 m. Multi-year ice does not break down into pressure ice, but first-year ice and thinner ice in leads does. Consequently, our empirical knowledge of mesoscale or large-scale sea ice mechanics expects thicknesses of up to 1 m for the undeformed ice involved in deformation processes. Within this range at least, the internal stress in mesoscale dynamics, where hummocking and ridging cause the main loss of kinetic energy, follows the same rheological law. Compressive strength is proportional to thickness raised to the power n, $P \propto h^n$, with the exact value of n being an open question ($\frac{1}{2} \leq n \leq 2$). For a homogenous ice sheet, buckling failure would give $n = \frac{1}{2}$, crushing failure $n = 1$ (Coon, 1974), and in ridge formation $n = 2$ (Rothrock, 1975a).

4.4 GRANULAR FLOW MODELS

The idea of floe collisions as the basis for sea ice rheology was introduced in the 1970s (Ovsienko, 1976). The exact physical formulation was presented by Shen et al. (1986), who derived an analytical solution for rheology when spatial fluctuations in ice velocity give rise to floe collisions, which further transmit stresses within the ice field. Ice floes were taken as uniform size circular disks, diameter d, and thickness h. In a shear flow, random fluctuations in ice velocity give rise to floe collisions (Figure 4.10), and the frequency of collisions depends on the level of velocity fluctuations and ice compactness. The nature of the collisions is described by the restitution or inelasticity coefficient κ ($0 \leq \kappa \leq 1$, with $\kappa = 1$ for perfect elasticity).

The stress generated by the collisions is dictated by the momentum transfer rate within the system. To close the system the level of random velocity fluctuations needs to be determined. This is done by means of the kinetic energy equation, which provides the fluctuation level on the basis of energy dissipation in the deformation

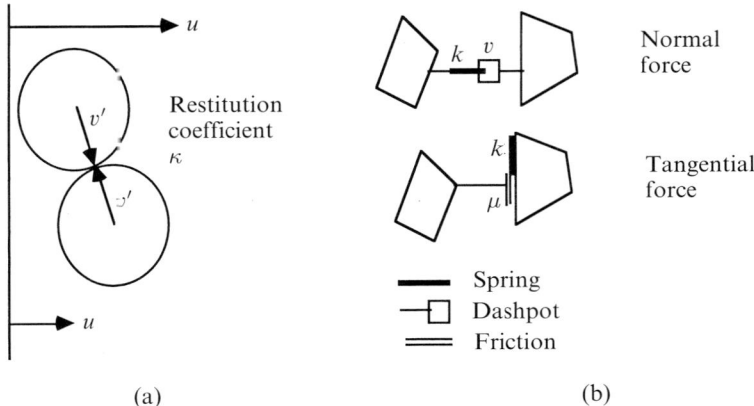

Figure 4.10. Floe–floe interaction: (a) the floe collision model, and (b) the discrete particle model.
(a) From Shen et al. (1986) and (b) from Hopkins (1994).

of ice. Thus, the fluctuation level is zero if and only if deformation is zero. These collisions lead (via averaging) to the continuum rheology (see Shen et al., 1986, 1987 for a detailed derivation):

$$\boldsymbol{\sigma} = m(1+\kappa)\frac{A^{3/2}}{\sqrt{A_0}-\sqrt{A}}(f_1\mathbf{I}+f_2\dot{\boldsymbol{\varepsilon}}') \qquad (4.27)$$

where $m = \frac{1}{4}\pi\rho h d^2$ is the mass of an ice floe, $A_0 = \pi/(2\sqrt{3})$ is equal to the maximum compactness of uniform circular floes, and f_1 and f_2 are viscosities:

$$f_1 = \frac{1}{4\pi}\frac{v'}{d}\dot{\varepsilon}_I - \frac{\sqrt{2}}{\pi^2}\left(\frac{v'}{d}\right)^2, \qquad f_2 = \frac{1}{6\pi}\frac{v'}{d} \qquad (4.28)$$

where v' is the velocity fluctuation level obtained from:

$$\left.\begin{array}{l}\dfrac{v'}{d} = \sqrt{k_1^2 - k_2} - k_1, \; k_1 = \left(\dfrac{4\sqrt{2}}{\pi^2(1-\kappa)} - \dfrac{2\sqrt{2}}{\pi}\right)\dot{\varepsilon}_I \\[10pt] k_2 = \dfrac{\dot{\varepsilon}_I^2}{4} + \dfrac{\dot{\varepsilon}_{II}^2}{8} - \dfrac{2}{3\pi(1-\kappa)}(3\dot{\varepsilon}_I^2 + \dot{\varepsilon}_{II}^2)\end{array}\right\} \qquad (4.29)$$

The derivation is based on the assumption $v'/d \gg |\dot{\varepsilon}|$ Note that v'/d is a first-degree homogenous function of strain-rate invariants,[5] and then from Eqs (4.27–4.29) it can be deduced that the stress is quadratic in the strain rate. Ice stress depends on floe characteristics (ρ, h, d), ice compactness, strain rate, and the restitution coefficient $\kappa \sim 0.9$. Note that quadratic dependence is of the form $\boldsymbol{\sigma} \propto \eta(\dot{\varepsilon}_I, \dot{\varepsilon}_{II})\dot{\boldsymbol{\varepsilon}}$, not $\boldsymbol{\sigma} \propto \dot{\boldsymbol{\varepsilon}}^2$ as given in the last term of the Reiner–Rivlin model (Eq. 4.5).

The collision model integrates kinematics within a drift ice particle into a large-

---

[5] Multiplying strain-rate invariants by c changes the fluctuation v'/D by the same factor c.

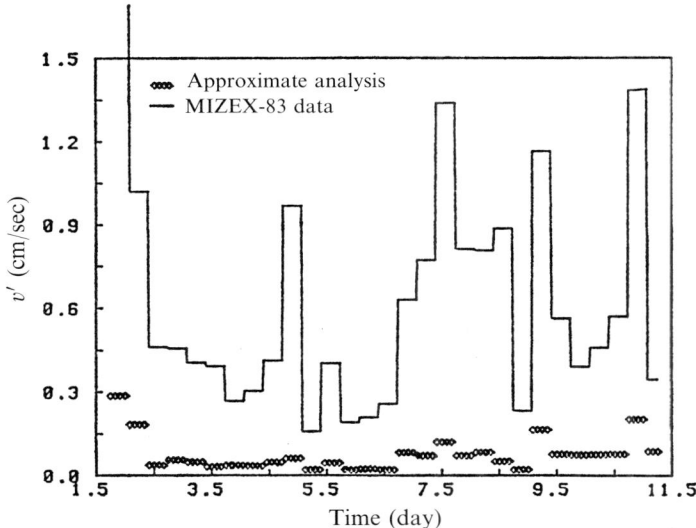

Figure 4.11. Velocity fluctuation level according to measurements and Monte Carlo simulations based on the measured strain rate. The field data are from MIZEX-83, Greenland Sea.
From Shen et al. (1987).

scale stress field. It is in principle correct. The stress level is very low for compactness less than 0.9, but there is a singularity as $A \to A_0$. In reality, when the stress level has reached the strength of individual floes or the plastic yield strength, floes break and override, and the collision model is no longer valid. This supports the physical picture that ice stress only becomes important when an ice field becomes compact; stresses then arise due to friction between floes and formation of hummocks and ridges. There is another singularity when $\kappa \to 1$ (i.e., when ice floes become fully elastic).

Inelastic floe collisions are incapable of transmitting large stresses. Lu et al. (1989) merged the floe collision model with a plastic model across the marginal ice zone's (MIZ) interior pack boundary, but the results did not reveal any new qualitative features about ice dynamics. However, collision rheology correctly represents the behaviour of drift ice at compactness levels up to ca 0.7–0.9, and its low stress levels explain why it is so easy for the ice pack to close up. Shen et al. (1986) examined observed kinematics during the Marginal Ice Zone Equipment (MIZEX-83) in the Greenland Sea and found a clear connection between predicted and observed velocity fluctuations (Figure 4.11).

Collision rheology gives a functional relationship between the strength of ice and ice compactness (Eq. 4.27): the factor is $B(A) = A^{3/2}/(\sqrt{A_0} - \sqrt{A})$. This becomes singular as $A \to A_0$ (≈ 0.907 for uniform circular floes), and $B(0.7) \approx 5.1$, $B(0.85) \approx 25.8$, $B(0.9) \approx 231.9$.

Example Harr (1977) Consider the transmission of forces across a line of unit width. At any point a force $F' > 0$ is or is not experienced, depending on whether there is, respectively, material contact or a void. For N points, the probability distribution of the number of $F' > 0$ points is binomial with a probability $p^{(N)}$. Increasing N, the Poisson distribution results (Feller, 1968), and the number of n forces $F' > 0$ has the probability:

$$p_n = \frac{\Gamma^n}{n!} e^{-\Gamma}$$

where Γ is the Poisson distribution parameter, $\Gamma = \lim_{N \to \infty} N p^{(N)}$. Since the probability of void must be equal to the relative amount of open water, we have $p_0 = 1 - A = e^{-\Gamma}$ and consequently $\Gamma = -\log(1 - A)$. This becomes singular as $A \to 1$; however, somewhere at $A < 1$ the ice floes break and pressure ice formation begins. When $A = 0.7$ or 0.99, Γ is 1.2 or 4.6, respectively, and the singularity is therefore very sharp. Mean and variance are equal to Γ, so the stress level is proportional to $-\log(1 - A)$. Although stress increases with compactness, its relative variability decreases and the coefficient of variation (standard deviation divided by the mean) is $\Gamma^{-1/2}$. If the mean stress level is $\langle \sigma \rangle$ and compactness is A, the range between the standard deviation and the mean is $\langle \sigma \rangle (1 \pm \Gamma^{1/2})$ and the probability of $\sigma = n \langle \sigma \rangle$ is $\Gamma^n e^{-\Gamma}/n!$ Overall, this reasoning does not hold when contacts between floes disappear, but the model illustrates how the stress field starts to change from a collisional system to a system in which stress is transmitted between ice floes in contact. The stress level from loose to dense contact fields increases several times. ∎

A more general approach to a floe–floe interaction system is provided by so-called "discrete particle models", where a mechanical model is constructed for the interactions and a simultaneous solution for a large number (thousands) of floes is obtained numerically (Hopkins and Hibler, 1991; Løset, 1993; Hopkins, 1994). This has the potential to improve the understanding of the geophysics of sea ice dynamics and provide possibilities to parameterize continuum models from individual floe–floe interaction processes.

Hopkins (1994) constructed a model for ice blocks breaking from a parent ice sheet and accumulating into an ice ridge. Floe–floe interactions were modelled by a viscoelastic normal force (Kelvin–Voigt medium, Section 4.1) and Coulomb friction was used for the tangential force (Figure 4.10). The viscous part of the normal force takes care of the inelasticity of floe contacts in much the same way as did the restitution coefficient in the collision model. The tangential force initially has an elastic part for static friction, but, once overcome, frictional sliding occurs.

The resulting ridges looked quite natural. Hopkins (1994) could also determine the mechanical energy budget to show that the potential energy created in ridge building is small compared with frictional losses. Figure 4.12 illustrates the result. The key parameter is the friction coefficient, and he used different dry (μ) and wet (μ_w) friction coefficients. Friction takes energy that is about one order of magnitude greater than that required to create potential energy. Comparable with potential

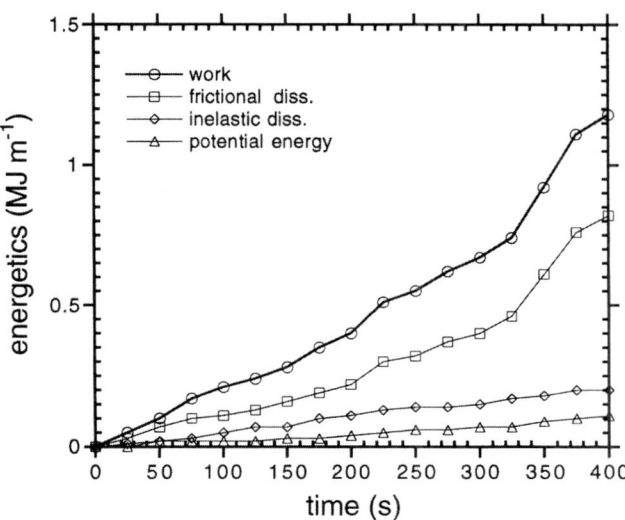

Figure 4.12. Energy budget in sea ice ridging as simulated by a discrete particle model. Ice thickness was 30 cm, the dry friction coefficient was 1.0, and the wet friction coefficient was 0.6.
From Hopkins (1994).

energy are the inelastic losses that occur in the deformation of ice blocks. Within the ranges $\mu \sim 0.4$–1.0 and $\mu_w \sim 0.3$–0.6 the energy budget was not much affected. Because potential energy and friction are correlated to some degree, the theory based on the potential energy is qualitatively good.

The results from discrete particle models have been good for local scale processes such as pressure ridge formation with its mechanical energy budget, which is key to understanding sea ice plastic rheology. Only recently have attempts been made to use them in full basin-scale sea ice dynamics problems. Comparison between continuum and discrete particle models for MIZ ice dynamics was made by Gutfraind and Savage (1997), giving similar results for both approaches. Rheem et al. (1997) used a mixed discrete particle–continuum approach for a study in the Sea of Okhotsk.

The rheology question is now complete. The core is the plastic model, which is based on the yield curve with associated flow rule. The size of the yield curve corresponds to the overall strength of the ice, while its shape specifies the behaviour of different modes of deformation. Open drift ice fields can be treated with the free drift (no stress) approach. Between open and compact states rheology changes from stress free in a viscous manner. Rheology gives the internal stress tensor as a function of ice state, strain, and strain rate. Then, the divergence of stress steps into the equation of motion as a forcing term, as will be shown in Chapter 5.

5

Equation of drift ice motion

5.1 DERIVATION OF THE EQUATION OF MOTION

The law of conservation of momentum, or the equation of motion, is derived for drift ice in this chapter from Newton's second law for a continuum. By integrating through the thickness of ice, the two-dimensional equations can be obtained. Atmospheric and oceanic forcing is treated in Section 5.2. Scale analysis and dimensional analysis are studied in Section 5.3 to derive the magnitudes of the terms of the momentum equation and the key dimensionless quantities. Section 5.4 concerns the dynamics of a single ice floe based on rigid body mechanics.

5.1.1 Fundamental equation

The motion of drift ice is two-dimensional (as described in Chapter 3). Its equation can now be derived using all three dimensions. For the sake of clarity three-dimensional vectors and tensors are underlined (e.g., $\underline{u} = (u, v, w)$, where w is the vertical velocity).

Figure 5.1 shows a photograph of ice motion past a lighthouse enclosed within drifting ice. The sea ice drift problem is schematically illustrated in Figure 5.2. The starting point is Newton's second law, $m\, d\underline{u}/dt = \underline{F}$, where $\underline{F}$ represents the forcing acting on the ice. First, normal *continuum mechanics* modifications are introduced (e.g., Hunter, 1976):

(i) Use mass per unit volume or density for ice continuum particles, or parcels.
(ii) The force due to the internal ice stress field is $\underline{\nabla} \cdot \underline{\Sigma}$, where $\underline{\nabla} = (\partial/\partial x, \partial/\partial y, \partial/\partial z)$ is included in $\underline{F}$.
(iii) Use an Eulerian frame to ascertain the advective acceleration terms.

Vasilii Vladimirovich Shuleikin (1895–1979). A Soviet oceanographer, sea ice physicist, and author of classic texts in the early development of the theory of sea ice drift. He was nominated Academician of the Soviet Academy of Sciences in 1946.

Reproduced with permission from the Russian State Museum of the Arctic and Antarctic, St Petersburg.

This results in the Cauchy equation of motion of a continuum:

$$\rho\left(\frac{\partial \underline{u}}{\partial t} + \underline{u} \cdot \underline{\nabla}\underline{u}\right) = \underline{\nabla} \cdot \underline{\Sigma} + \underline{F}_{\text{ext}} \tag{5.1}$$

where $\underline{F}_{\text{ext}}$ contains the external forces on the ice.

Second, *geophysical effects* are introduced:

(i) Coriolis acceleration is added to the inertial term (e.g., Pond and Pickard, 1983; Cushman-Roisin, 1994).
(ii) The Earth's gravity provides external body forcing, expressed as $\underline{\nabla}\Phi$, where Φ is the geopotential height of the sea surface. Here $\underline{\nabla}\Phi = g\underline{\nabla}'\xi$, where $\underline{\nabla}'$ is the three-dimensional gradient operator in the true horizontal–vertical co-ordinate system, and ξ is sea surface elevation.

Third, the *internal stress* of drift ice includes the stress due to interactions between ice floes, $\underline{\sigma}$, and the pressure p from the surrounding air and water, $\underline{\Sigma} = \underline{\sigma} - p\underline{I}$ (see Eq. 4.2). Now we have the three-dimensional momentum equation for sea ice as:

$$\rho\left(\frac{\partial \underline{u}}{\partial t} + \underline{u} \cdot \underline{\nabla}\underline{u} + 2\underline{\Omega} \times \underline{u}\right) = \underline{\nabla} \cdot (\underline{\sigma} - p\underline{I}) - \rho\underline{\nabla}\Phi \tag{5.2}$$

Sec. 5.1] **Derivation of the equation of motion** 111

Figure 5.1. Sea ice drifts past a lighthouse, northern Baltic Sea. The width of the lighthouse is about 5 m.

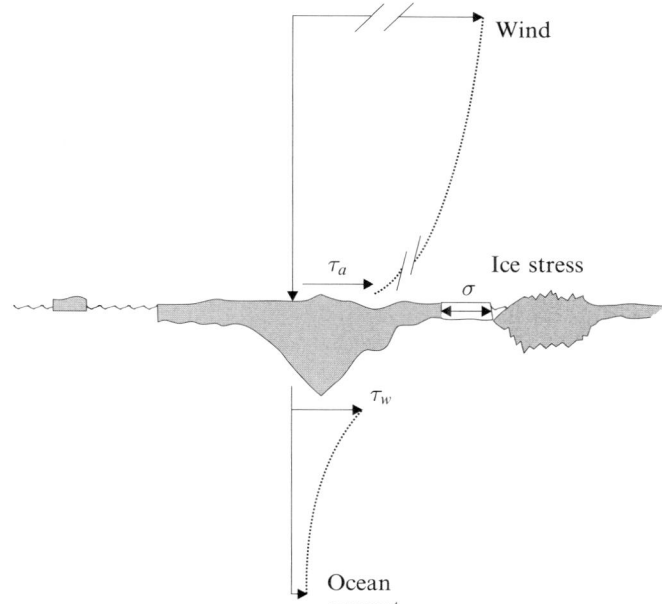

Figure 5.2. The ice drift problem. Ice is driven by atmospheric and oceanic flows and responds to forcing through its internal stress field.

where $\underline{\Omega} = (\Omega \cos \phi)\mathbf{j} + (\Omega \sin \phi)\mathbf{k}$ is the Coriolis vector, $\Omega = 7.292 \times 10^{-5}\,\text{s}^{-1}$ is the angular velocity of the Earth, and ϕ is the latitude.

Example For an incompressible, linear viscous fluid with constant viscosity η, the rheology reads $\underline{\boldsymbol{\sigma}} = 2\eta\underline{\dot{\boldsymbol{\varepsilon}}}'$ (see Section 4.1). Then $\underline{\nabla} \cdot \underline{\boldsymbol{\sigma}} = \eta \underline{\nabla}^2 \boldsymbol{u}$; that is, Eq. (5.2) becomes the Navier–Stokes equation on a rotating sphere:

$$\partial \boldsymbol{u}/\partial t + \boldsymbol{u} \cdot \underline{\nabla}\boldsymbol{u} + 2\underline{\Omega} \times \boldsymbol{u} = -\rho^{-1}\underline{\nabla}p + \nu\underline{\nabla}^2\boldsymbol{u} - \boldsymbol{g}$$

where ν is the kinematic viscosity of the fluid. ∎

As in ocean dynamics, the vertical component of the Coriolis acceleration arising from the eastward velocity component is very small compared with gravity acceleration. Also, in the horizontal Coriolis acceleration the part arising from vertical motion, $w2\Omega \cos \phi$, is very small compared with the part arising from horizontal motion. Consequently, Coriolis acceleration is reduced to $f\mathbf{k} \times \boldsymbol{u}$, where $f = 2\Omega \sin \phi$ is the *Coriolis parameter*.

Let us take the sea surface as the zero reference surface, approximating it locally as a Cartesian xy-plane. Although this is not a fixed surface on the Earth, it is easy to see (e.g., by scale analysis) that forces due to vertical motion of the sea level are negligible in the dynamics of sea ice.

5.1.2 Vertical integration

The equation of motion (5.2) is next integrated through the thickness of the ice. The horizontal velocity of sea ice has no vertical structure, as discussed in Chapter 3. Therefore, integration of the acceleration terms, the left-hand side of Eq. (5.2), is straightforward: they are simply multiplied by the ice thickness h. Similarly, integration of the geopotential term results in multiplication by h.

Integration of the divergence of internal ice stress is more complicated. We have:

$$\int_h \underline{\nabla} \cdot \underline{\boldsymbol{\Sigma}}\,dz = \int_h \underline{\nabla} \cdot \underline{\boldsymbol{\sigma}}\,dz - \int_h \underline{\nabla}p\,dz \qquad (5.3a)$$

where $\int_h$ stands for integration through the ice thickness. The actual integration boundaries are the upper and lower ice surfaces, and their altitudes with respect to the sea surface are the freeboard h' and the negative of the draft $-h''$, respectively. Therefore, in component form:

$$\int_{-h''}^{h'} \begin{bmatrix} \dfrac{\partial \Sigma_{xx}}{\partial x} + \dfrac{\partial \Sigma_{yx}}{\partial y} + \dfrac{\partial \Sigma_{zx}}{\partial z} \\[4pt] \dfrac{\partial \Sigma_{xy}}{\partial x} + \dfrac{\partial \Sigma_{yy}}{\partial y} + \dfrac{\partial \Sigma_{zy}}{\partial z} \\[4pt] \dfrac{\partial \Sigma_{xz}}{\partial x} + \dfrac{\partial \Sigma_{yz}}{\partial y} + \dfrac{\partial \Sigma_{zz}}{\partial z} \end{bmatrix} dz = \int_{-h''}^{h'} \begin{bmatrix} \dfrac{\partial \sigma_{xx}}{\partial x} + \dfrac{\partial \sigma_{yx}}{\partial y} + \dfrac{\partial \sigma_{zx}}{\partial z} \\[4pt] \dfrac{\partial \sigma_{xy}}{\partial x} + \dfrac{\partial \sigma_{yy}}{\partial y} + \dfrac{\partial \sigma_{zy}}{\partial z} \\[4pt] \dfrac{\partial \sigma_{xz}}{\partial x} + \dfrac{\partial \sigma_{yz}}{\partial y} + \dfrac{\partial \sigma_{zz}}{\partial z} \end{bmatrix} dz - \int_{-h''}^{h'} \begin{bmatrix} \dfrac{\partial p}{\partial x} \\[4pt] \dfrac{\partial p}{\partial y} \\[4pt] \dfrac{\partial p}{\partial z} \end{bmatrix} dz$$

$$(5.3b)$$

Derivation of the equation of motion

The components of the three-dimensional stresses $\underline{\Sigma}$ and $\underline{\sigma}$ are also underlined to differentiate them from integrated stresses, which are used here to represent the two-dimensional stress (see Eq. 4.3).

Equation of motion on the sea surface plane

The first and second components, or the x- and y-components, of Eq. (5.3b) are first considered. They are for the sea surface plane. Since this plane is the zero reference plane, the horizontal pressure gradient ∇p integrates to zero in the water and thus becomes equal to the air pressure gradient ∇p_a. Then:

$$\int_{-h''}^{h'} \begin{bmatrix} \frac{\partial \underline{\Sigma}_{xx}}{\partial x} + \frac{\partial \underline{\Sigma}_{yx}}{\partial y} + \frac{\partial \underline{\Sigma}_{zx}}{\partial z} \\ \frac{\partial \underline{\Sigma}_{xy}}{\partial x} + \frac{\partial \underline{\Sigma}_{yy}}{\partial y} + \frac{\partial \underline{\Sigma}_{zy}}{\partial z} \end{bmatrix} dz = \int_{-h''}^{h'} \begin{bmatrix} \frac{\partial \underline{\sigma}_{xx}}{\partial x} + \frac{\partial \underline{\sigma}_{yx}}{\partial y} \\ \frac{\partial \underline{\sigma}_{xy}}{\partial x} + \frac{\partial \underline{\sigma}_{yy}}{\partial y} \end{bmatrix} dz + \begin{bmatrix} \underline{\sigma}_{zx}(h') - \underline{\sigma}_{zx}(-h'') \\ \underline{\sigma}_{zy}(h') - \underline{\sigma}_{zy}(-h'') \end{bmatrix} - h\nabla p_a$$

(5.4)

Take the first term on the right-hand side. The two-dimensional internal ice stress is $\boldsymbol{\sigma} = \int_h \underline{\boldsymbol{\sigma}}_H \, dz$ (see Eq. 4.3). To get there the order of integration and derivation must be interchanged as was done for Eq. (4.6). Recalling this equation, we have for the x-component:

$$\int_{-h''}^{h'} \left(\frac{\partial \underline{\sigma}_{xx}}{\partial x} + \frac{\partial \underline{\sigma}_{yx}}{\partial y} \right) dz = \frac{\partial}{\partial x} \sigma_{xx} + \frac{\partial}{\partial y} \sigma_{yx}$$
$$- \left[\underline{\sigma}_{xx}(h') \left(1 - \frac{\rho}{\rho_w}\right) + \underline{\sigma}_{xx}(-h'') \frac{\rho}{\rho_w} \right] \frac{\partial h}{\partial x}$$
$$- \left[\underline{\sigma}_{yx}(h') \left(1 - \frac{\rho}{\rho_w}\right) + \underline{\sigma}_{yx}(-h'') \frac{\rho}{\rho_w} \right] \frac{\partial h}{\partial y} \quad (5.5)$$

Thus, the integral is split into the x-component of the divergence of the two-dimensional stress $\nabla \cdot \boldsymbol{\sigma}$ and the x-component of the ice thickness gradient correction, denoted below by $\Theta_{h,x}$. The y-component is treated similarly.

The second term on the right-hand side of Eq. (5.4) comes from the surface boundary condition that the ice shear stress must match the shear stresses of air and water, respectively, τ_a and τ_w on ice:

$$\begin{bmatrix} \underline{\sigma}_{zx}(h') - \underline{\sigma}_{zx}(-h'') \\ \underline{\sigma}_{zy}(h') - \underline{\sigma}_{zy}(-h'') \end{bmatrix} = \tau_a - (-\tau_w) \quad (5.6)$$

Consequently, the divergence of the total stress $\underline{\Sigma}$ is in the sea surface plane:

$$\left[\int_h \nabla \cdot \underline{\Sigma} \, dz \right]_H = \nabla \cdot \boldsymbol{\sigma} - \boldsymbol{\Theta}_h + \boldsymbol{\tau}_a + \boldsymbol{\tau}_w - h\nabla p_a \quad (5.7)$$

where $\boldsymbol{\Theta}_h = \Theta_{h,x}\mathbf{i} + \Theta_{h,y}\mathbf{j}$. To understand this, one needs $\underline{\sigma}(h')$ and $\underline{\sigma}(-h'')$ and, consequently, a full three-dimensional rheology of the ice sheet (see Nye, 1973).

But the thickness gradient is generally small enough for the correction term Θ_h to be neglected, as is usually implicitly done in sea ice dynamics (e.g., Rothrock, 1975b; Coon, 1980; Hibler, 1986; Leppäranta, 1998).

Denoting the co-ordinates of the true horizontal plane by $(x'y')$, the slope of the sea surface with respect to horizontal is:

$$\boldsymbol{\beta} = \mathbf{i}\frac{\partial \xi}{\partial x'} + \mathbf{j}\frac{\partial \xi}{\partial y'} \qquad (5.8)$$

The geopotential gradient on the sea surface plane thus integrates into $\rho h \nabla \Phi = \rho h g \boldsymbol{\beta}$, usually called *sea surface tilt*. Apart from shallow (depth less than the Ekman depth[1]) regions, this term can be expressed in terms of the surface geostrophic current $\boldsymbol{U}_{wg}$ by (e.g., Pond and Pickard, 1983):

$$g\boldsymbol{\beta} = -f\mathbf{k} \times \boldsymbol{U}_{wg} \qquad (5.9)$$

This is a convenient expression since geostrophic velocity is often used as the reference in the ice–water friction law. In shallow waters the geostrophic approximation given by Eq. (5.9) does not hold due to the influence of bottom friction, but the tilt expression must remain in the equation of motion. In fact, for deep basins the explicit tilt expression would also be preferable, but usually the non-geostrophic sea surface tilt is not known.

Finally, the general form of the equation of motion of sea ice on the sea surface plane is:

$$\rho h \left[\frac{\partial \boldsymbol{u}}{\partial t} + \boldsymbol{u} \cdot \nabla \boldsymbol{u} + f \mathbf{k} \times \boldsymbol{u} \right] = \nabla \cdot \boldsymbol{\sigma} + \boldsymbol{\tau}_a + \boldsymbol{\tau}_w - \rho h g \boldsymbol{\beta} - h \nabla p_a \qquad (5.10)$$

Figure 5.3 shows a schematic force diagram of ice drift. Usually, wind is the driving force, and this is balanced by the ice–ocean drag and internal friction of the ice. Coriolis acceleration is smaller than the three major forces and is always perpendicular to the ice motion. The other acceleration terms and sea surface tilt are smaller still and can influence in any direction.

Compared with vertically integrated ocean circulation models (see, e.g., Pond and Pickard, 1983) in the horizontal momentum equation: ice thickness corresponds to sea depth; ice–water friction corresponds to bottom friction; and the internal friction of ice corresponds to horizontal turbulent friction. In contrast to sea ice dynamics, in vertical integration the constancy of velocity does not hold in ocean circulation, resulting in a biased momentum balance due to the non-linearity of advection.

---

[1] The depth of the frictional influence of surface forcing in the ocean (e.g. Gill, 1982; Pond and Pickard, 1983); for sea ice it is typically 30–40 m.

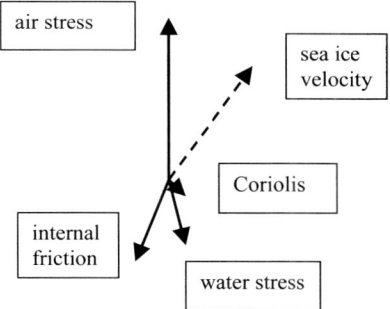

Figure 5.3. A typical diagram of major forces in drifting sea ice (northern hemisphere).

Vertical equation of motion

The vertical component of integrated total stress divergence is from Eq. (5.3):

$$\int_{-h''}^{h'} \left(\frac{\partial \Sigma_{xz}}{\partial x} + \frac{\partial \Sigma_{yz}}{\partial y} + \frac{\partial \Sigma_{zz}}{\partial z} \right) dz = \int_{-h''}^{h'} \left(\frac{\partial \sigma_{xz}}{\partial x} + \frac{\partial \sigma_{yz}}{\partial y} \right) dz + \left. (\sigma_{zz} - p) \right|_{-h''}^{h'} \quad (5.11)$$

The last term is equal to the hydrostatic pressure of water directly below sea ice, $\rho_w g h''$. By scaling analysis, it is seen that this pressure is much greater than the integral term on the right-hand side: since $(\sigma_{xz}, \sigma_{yz}) \sim \tau_a$, the integral is $\sim \tau_a h/L \sim 1$ Pa while $\rho_w g h'' \approx 10^4$ Pa.

The vertical geopotential gradient is $g_z \cong g$ since the sea surface is very close to horizontal. Thus, multiplied by ice density it integrates to $\rho g h$, which is comparable with the pressure at the ice bottom. It is easy to see (e.g., by scale analysis) that other terms in the vertical momentum equation are negligible. Consequently, the *vertical momentum equation* becomes the Archimedes law:

$$\rho_w g h'' - \rho g h = 0 \quad (5.12)$$

or $h''/h = \rho/\rho_w$. This equation corresponds to the vertical hydrostatic equation in ocean dynamics. Archimedes' law defines which portions of the ice sheet are above and below the sea surface; but, otherwise it is not needed in sea ice dynamics problems.

Boundary conditions

Consider a sea ice field Ω with a boundary curve Γ (Figure 5.4). The field is bounded by open water and a solid medium (land or landfast ice). The configuration changes with time, $\Omega = \Omega(t)$ and $\Gamma = \Gamma(t)$. The boundary conditions are:

$$\boldsymbol{\sigma} \cdot \mathbf{n} = 0 \quad \text{open water boundary} \quad (5.13a)$$

$$\boldsymbol{u} \cdot \mathbf{n} \leq 0 \quad \text{solid boundary} \quad (5.13b)$$

At the open water boundary the ice does not support normal stresses and the motion of ice changes the boundary configuration. At the land boundary, the ice is allowed

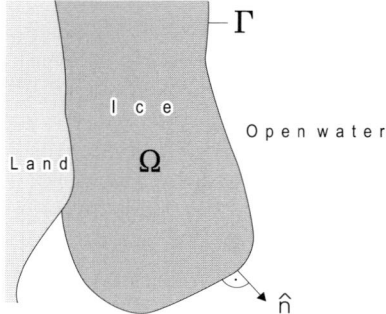

Figure 5.4. A sea ice field Ω with the boundary curve Γ consisting of open water and land sections.

to move away from the boundary into the drift ice basin, but it is not allowed to override the solid boundary medium. As soon as the ice has moved away (lost contact with the solid boundary), the drift ice boundary changes to the open water boundary.

The boundary conditions (5.13) are often replaced by a simplified form:

$$\text{Open water} \equiv \text{Ice with zero thickness} \qquad (5.14a)$$

$$\boldsymbol{u} = 0, \qquad \text{solid boundary} \qquad (5.14b)$$

Equation (5.14a) removes the question of the open water boundary, while Eq. (5.14b) is the usual no slip condition used for viscous flows.

5.1.3 Drift regimes

In general, sea ice dynamics phenomena can be divided into three categories: (i) *stationary ice*; (ii) *drift in the presence of internal friction*, or drift of interacting ice floes; and (iii) *free drift*, or drift of non-interacting ice floes. Cases (ii) and (iii) were named by Zubov (1945) and correspond to the fields of close drift ice and scattered ice, respectively.

The drift ice dynamics problem can be approached in three ways: free drift solution, analytical zonal models of drift in the presence of internal friction, and full numerical models. These are the topics of Chapters 6, 7, and 8. In the present chapter, the momentum equation is further discussed and analysed for external forcing, scales, and non-dimensional form.

5.1.4 Conservation of kinetic energy, divergence, and vorticity

The conservation law of kinetic energy (per unit area) $q = \frac{1}{2}\rho h |\boldsymbol{u}|^2$ is obtained from the momentum equation by scalar multiplication by ice velocity (Coon and Pritchard, 1979; Leppäranta, 1981b). This is quite straightforward, as done usually in fluid dynamics. For the internal friction term, the formula $\nabla \cdot (\boldsymbol{u} \cdot \boldsymbol{\sigma}) =$

$\boldsymbol{u} \cdot \nabla \cdot \boldsymbol{\sigma} + \boldsymbol{\sigma} : \nabla \boldsymbol{u} = \boldsymbol{u} \cdot \nabla \cdot \boldsymbol{\sigma} + \mathrm{tr}(\boldsymbol{\sigma} \cdot \dot{\boldsymbol{\varepsilon}})$ is utilized. The colon product ":" is defined by $\boldsymbol{B} : \boldsymbol{C} = B_{ij} C_{ij} \sum$ for two matrices $\boldsymbol{B}$ and $\boldsymbol{C}$. We have:

$$\frac{\partial q}{\partial t} + \boldsymbol{u} \cdot \nabla q = -\mathrm{tr}(\boldsymbol{\sigma} \cdot \dot{\boldsymbol{\varepsilon}}) + \nabla \cdot (\boldsymbol{u} \cdot \boldsymbol{\sigma}) + \boldsymbol{u} \cdot \boldsymbol{\tau}_a + \boldsymbol{u} \cdot \boldsymbol{\tau}_w - \rho h g \boldsymbol{u} \cdot \boldsymbol{\beta} - h \boldsymbol{u} \cdot \nabla p_a \quad (5.15)$$

The left-hand side gives the local change and advection of kinetic energy, while the right-hand side terms are, respectively, (i) work done by the stress field including frictional dissipation, (ii) work done by the surrounding ice, (iii) input of kinetic energy from the atmosphere, (iv) exchange of kinetic energy with the ocean, (v) input of kinetic energy from the sea surface slope, and (vi) input from the atmospheric pressure gradient.

To illustrate term (ii), the Gauss theorem says that for a region Ω with boundary $\partial \Omega$:

$$\int_\Omega \nabla \cdot (\boldsymbol{u} \cdot \boldsymbol{\sigma}) \, \mathrm{d}\Omega = \int_{\partial \Omega} (\boldsymbol{u} \cdot \boldsymbol{\sigma}) \, \mathrm{d}(\partial \Omega)$$

This term is understood as the transmission of stress across region boundaries.

Example For an ideal fluid, $\boldsymbol{\sigma} = -p \mathbf{I}$ and $-\mathrm{tr}(\boldsymbol{\sigma} \cdot \dot{\boldsymbol{\varepsilon}}) = p \nabla \cdot \boldsymbol{u}$ and $\nabla \cdot (\boldsymbol{u} \cdot \boldsymbol{\sigma}) = -\nabla \cdot (p \boldsymbol{u})$. There are no frictional losses, but the pressure field redistributes mechanical energy. For a linear viscous incompressible fluid, $-\mathrm{tr}(\boldsymbol{\sigma} \cdot \dot{\boldsymbol{\varepsilon}}) = -\eta \dot{\varepsilon}_{\mathrm{II}}^2$, which is the dissipation of kinetic energy due to viscous shear. ∎

The input of kinetic energy from the wind is $\boldsymbol{u} \cdot \boldsymbol{\tau}_a = \rho_a C_a |U_a| \boldsymbol{u} \cdot \boldsymbol{U}_a$. In purely wind-driven drift, ice velocity is approximately proportional to wind velocity, and therefore input is proportional to the cube of the wind speed. The ice–water exchange of kinetic energy is written as:

$$\boldsymbol{u} \cdot \boldsymbol{\tau}_w = \rho_w C_w |\boldsymbol{U}_w - \boldsymbol{u}| \boldsymbol{u} \cdot [\cos \theta_w (\boldsymbol{U}_w - \boldsymbol{u}) + \sin \theta_w \mathbf{k} \times (\boldsymbol{U}_w - \boldsymbol{u})] \quad (5.16)$$

If $\boldsymbol{U}_w = 0$, then $\boldsymbol{u} \cdot \boldsymbol{\tau}_w = -\rho_w C_w \cos \theta_w |\boldsymbol{u}|^3$ and the energy loss to the ocean boundary layer is proportional to the cube of ice speed.

The mechanical energy budget evaluated from observations is illustrated in Table 5.1. The results show that the principal source is the wind and that energy is used to overcome ice–water friction and the internal friction of the ice. The ice thickness was about 50 cm and, therefore, the overall level of kinetic energy was 1–10 J/m². This is considerably less than the average gain and loss over 1 hour and, therefore, the timescale of the kinetic energy of sea ice is very short.

The energy budget has not been widely used in sea ice dynamics investigations, despite the fact that it would shed light on internal stress transfer and dissipation mechanisms. However, apart from short-term elastic events, drift ice does not store recoverable mechanical energy.

Table 5.1. Kinetic energy budget of sea ice dynamics in the northern Baltic Sea. The values are in mJ/(m² s) and based on hourly data, acquired during a 2-week experiment.

	Mean	Standard deviation
Rate of change	0.3	1.6
Input from the wind	10.3	20.7
Input from currents	2.3	5.7
Input from sea surface tilt	0.1	0.5
Loss to the ocean boundary layer	−8.4	29.3
Loss to internal friction	−6.0	13.8

Source: Leppäranta (1981b).

The conservation laws of divergence $\dot{\varepsilon}_I$ and vorticity $\dot{\omega}$ are obtained from the momentum equation by applying divergence and curl operators (respectively, $\nabla \cdot$ and $\nabla \times$):

$$\rho h \left[\frac{D\dot{\varepsilon}_I}{Dt} + \nabla(\boldsymbol{u} \cdot \nabla) \cdot \boldsymbol{u} - f\dot{\omega} \right] = \nabla \cdot (\nabla \cdot \boldsymbol{\sigma} + \boldsymbol{\tau}_a + \boldsymbol{\tau}_w - h\nabla p_a) \quad (5.17a)$$

$$\rho h \left[\frac{D\dot{\omega}}{Dt} + (\dot{\omega} + f)\nabla \cdot \boldsymbol{u} \right] = \nabla \times (\nabla \cdot \boldsymbol{\sigma} + \boldsymbol{\tau}_a + \boldsymbol{\tau}_w) \quad (5.17b)$$

As in the case of kinetic energy, the inertial part or the left-hand side is small and the main balance is between internal friction and surface stresses.

Even though air and water are incompressible fluids, the Coriolis phenomenon causes divergence in surface stress. Therefore, the atmosphere and ocean are capable enough to produce divergence and vorticity in ice velocity. The physics are similar to the phenomenon known as "Ekman pumping" in the open ocean, where the divergence of surface stress sucks deeper water to the surface (e.g., Cushman-Roisin, 1994). In addition, the internal friction of ice and the inhomogeneity of the ice cover for surface roughness and thickness give rise to deformation and vorticity even in the presence of constant forcing.

5.2 ATMOSPHERIC AND OCEANIC DRAG FORCES

5.2.1 Planetary boundary layers

Planetary fluid boundary layers comprise two main parts (e.g., Tennekes and Lumley, 1972): a surface layer where the stress is approximately constant and an Ekman layer where velocity rotates due to the Coriolis effect and stress decreases to zero. The full boundary-layer solution of velocity is constructed by joining the surface layer and Ekman layer solutions together, normally assuming steady-state conditions. The thickness of the planetary boundary layer[2] is ~ 1 km in the atmosphere and ~ 50 m in the ocean.

Because of the important role of rotation, let us present the velocity solution of the planetary fluid boundary layer using complex variable techniques. As horizontal velocity is $U = U_1 + iU_2$, the boundary layer equations are (e.g., Cushman-Roisin, 1994):

$$ifU = ifU_g + \frac{1}{\rho}\frac{\partial \tau}{\partial z} \qquad (5.18a)$$

$$\frac{\partial p}{\partial z} = -g\rho \qquad (5.18b)$$

where U_g is geostrophic velocity and τ is vertical shear stress. This is expressed as:

$$\tau = \rho K \frac{\partial U}{\partial z} \qquad (5.18c)$$

where K is kinematic eddy viscosity. For the vertical direction the hydrostatic equation applies, and in barotropic flow the geostrophic velocity is independent of depth. At the top of the boundary layer, stress vanishes and, therefore, $U = U_g$, while, at the bottom of the boundary layer, velocity must be equal to the velocity of the boundary, $U = U_0$.

Sea ice lies between the atmospheric and oceanic boundary layers. These both follow the laws of the planetary boundary layer with the following main differences:

(i) The densities of air and sea water differ (respectively, $\rho_a \approx 1.3$ kg/m^3 and $\rho_w = 1,028$ kg/m^3, the density ratio being thus $\rho_a/\rho_w \approx 1.3 \times 10^{-3}$).
(ii) The velocities of air and water flows differ (respectively, $U_a \sim 10$ m/s and $U_w \sim 10$ cm/s, the velocity ratio being $U_a/U_w \sim 10^2$).
(iii) The height of boundary layers differ (respectively, $\delta_a \sim 1$ km and $\delta_w \sim 50$ m, the height ratio being $\delta_a/\delta_w \sim 20$).
(iv) $\delta_a \gg h_s$, $\delta_w \sim h_k$ (i.e., ridge sails are very small compared with the height of the atmospheric boundary layer, but ridge keels are of the order of the oceanic boundary layer depth).

In the presence of sea ice, its velocity (u) serves as the boundary condition for both media. Since $|U_a| \gg |u| \sim |U_w|$, for the atmospheric boundary layer we can take $U_0 \cong 0$; but, for the oceanic boundary layer the condition must be $U_0 = u$. Thus, in pure dynamics ice and ocean form a coupled system, but wind stress acts as an external force.

The atmospheric boundary layer over sea ice has been extensively examined (Rossby and Montgomery, 1935; Brown, 1980; Joffre, 1984; Andreas, 1998). The oceanic boundary layer beneath sea ice is much less known than the atmospheric layer above (McPhee, 1986; Shirasawa and Ingram, 1991). This is partly due to greater observational difficulties, but more so to the uniqueness of the ice–water

---

[2] Planetary fluid boundary layer thicknesses scale as $f^{-1}(\tau/\rho)^{1/2}$ (e.g., Tennekes and Lumley, 1972).

interaction. Atmospheric boundary layers have also been widely examined over land and open ocean and these results have been utilized in sea ice research as well.

Surface layer

In the surface layer, vertical shear stress is constant and velocity direction is constant (e.g., along the real axis). A velocity scale u^*, called the friction velocity, can be defined by $\tau = \rho u_*^2$; this serves as the velocity scale for the surface layer. The length scale comes from Prandtl's *mixing length hypothesis* that states that the size of turbulent eddies scales with distance from the boundary (i.e., the length scale is z: see Tennekes and Lumley, 1972). Therefore, $K \sim u_* z$ and, for the stress to be constant, the vertical velocity gradient is $\sim u_*/z$. In exact terms:

$$\frac{dU}{dz} = \frac{u_*}{\kappa z} \quad (5.19)$$

where $\kappa \cong 0.4$ is the von Karman constant. In neutral stratification, a logarithmic velocity profile results:

$$U(z) - U_0 = \frac{u_*}{\kappa} \log\left(\frac{z}{z_0}\right) \quad (5.20)$$

where z_0 is the *roughness length*. For undeformed sea ice, the roughness length is of the order of 0.1–1 cm, and characteristic friction velocities are 50 cm/s for the atmosphere and 2 cm/s for the ocean. For deformed sea ice, the effective aerodynamic roughness length is of the order of 1–10 cm, corresponding to hummocks and ridges that have geometrical extents of 1 m above the surface and 10 m below (see Section 2.4).

Measurements of atmospheric surface layers over sea ice have been widely made using profiling masts (Figure 5.5) and aircraft. The roughness length of the top surface of undeformed sea ice is of the order of 0.1 cm. Joffre (1982a) examined the surface layer over northern Baltic Sea ice. The roughness length ranged between 0.04 cm and 0.08 cm, and data-fitting resulted in:

$$z_0 = m u_*^2 \quad (5.21)$$

where $m = 1.5 \times 10^{-3}\, \text{s}^2/\text{m}$ for $u_* \geq 0.35\, \text{m/s}$, in good agreement with a compilation of various snow and ice data by Chamberlain (1983), resulting in an average of $m = 1.6 \times 10^{-3}\, \text{s}^2/\text{m}$. Note also that Eq. (5.21) indicates that sea ice roughness is grossly equivalent to the average roughness of the sea surface as determined by Charnock's formula (Charnock, 1955) with the same numerical coefficient m (e.g., Stull, 1988, p. 265) for non-gale wind situations. For instance, Garratt (1977) reviewed a wide range of sea surface roughness data that yielded $z_0 = 0.0144 u_*^2/g = 1.47 \times 10^{-3} u_*^2\, \text{s}^2/\text{m}$. Although roughness parameter z_0 is in general considered to represent only geometric surface characteristics, it can also depend on flow conditions such as hydrostatic stability (e.g., Joffre, 1982a).

There are many fewer data available of the roughness of the ice bottom (McPhee, 1986; Shirasawa and Ingram, 1991). Work during the Arctic Ice Dynamics Joint Experiment (AIDJEX) in the Beaufort Sea gave $z_0 \approx 20\, \text{cm}$,

Figure 5.5. A mast (height 10 m) for atmospheric surface layer measurements. Velocity, temperature, and humidity are measured at several altitudes to determine the fluxes using the so-called profile method.

which includes hummocks and ridges (McPhee, 1986). The local value is much less, however, down to 0.1 cm. Shirasawa (1986) and Shirasawa and Ingram (1997) obtained $z_0 \approx 0.5$–2.0 cm from near-bottom turbulence measurements in the Canadian Arctic.

Ekman layer

In the Ekman layer the eddy viscosity coefficient K is constant and Coriolis acceleration is important. Consequently, velocity-turning takes place. We have:

$$ifU = ifU_g + K\frac{d^2U}{dz^2} \tag{5.22}$$

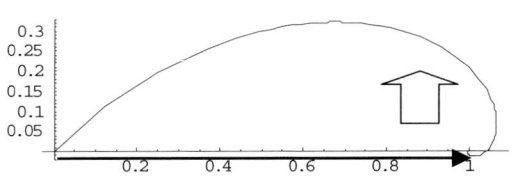

Wind velocity rotates counterclockwise down from the geostrophic wind to the surface. Surface stress is 45° to the left from the geostrophic wind.

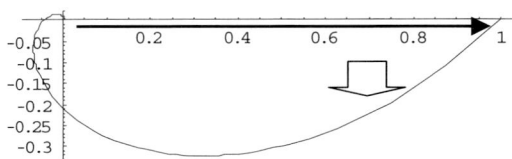

Surface flow is 45° to the right from the wind stress. Ocean current rotates clockwise down from the surface.

Figure 5.6. Theoretical form of the atmospheric and oceanic Ekman layers above and beneath sea ice (northern hemisphere). Note that oceanic velocities are two orders of magnitude less than atmospheric velocities.

The general solution is of the form $U = U_g + C_1 \exp(\lambda_1 z) + C_2 \exp(\lambda_2 z)$, where C_1 and C_2 are constants determined from boundary conditions, and $\lambda_{1,2} = \pm\sqrt{if/2K}$. The boundary conditions are the no-slip condition ($U = U_0$) at the surface and the stress-free geostrophic velocity at the top. A more simple condition can be used for the upper boundary, $U \to U_g$ as $z \to \infty$, resulting in:

$$U = U_g + (U_0 - U_g)\exp\{-[1 + \mathrm{sgn}(f)\mathrm{i}]\pi z/D\} \quad (5.23)$$

where $D = \pi\sqrt{2K/|f|}$ is the *Ekman depth*. At $z = D$, the latter term on the right-hand side has decreased to $e^{-\pi} \approx 4\%$ from the boundary level and rotated the angle of π clockwise.

Figure 5.6 illustrates the velocity distribution in the Ekman layers of the atmosphere and the ocean. For the atmosphere $U_0 \cong 0$. Starting from the top of the atmospheric boundary layer, velocity rotates to zero at the surface, counterclockwise in the northern hemisphere, and clockwise in the southern hemisphere. Surface stress now becomes $\tau = \rho K\, dU/dz = \rho K(U_g - U_0)[1 + \mathrm{sgn}(f)\mathrm{i}]\pi/D$. Since we also have $|\tau| = \rho u_*^2$, we get $u_*^2 = K|U_g - U_0|2\pi/D$.

There are three possibilities for the open ocean:

(a) If $|U_{wg}| \gg |u|$, then there is rotation in the oceanic boundary layer from the surface to the bottom, but with reversed directions compared with the atmosphere case.
(b) If $|U_{wg}| \ll |u|$, the Ekman spiral (e.g., Cushman-Roisin, 1994) results.
(c) If $U_{wg} \sim u$, the boundary layer modification is less sharp, but the faster moving medium always drives the slower one.

A key atmospheric parameter in sea ice dynamics research is the speed ratio U_{as}/U_{ag}

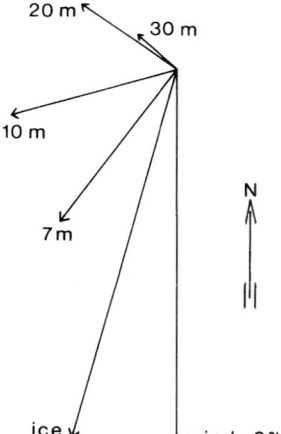

Figure 5.7. Averaged velocities of wind, sea ice, and currents (depths 7, 10, 30, and 40 m), 8–15 April 1975 in the Baltic Sea. For the velocity scale, the ice velocity averaged 3.1 cm/s (at maximum it was 35 cm/s).
From Lepparanta (1990).

between the surface wind U_{as} and the geostrophic wind U_{ag}. Its modulus is $R_g = |U_{as}|/|U_{ag}|$ and its argument is $\theta_a = \arg(U_{as}/U_{ag})$, the so-called cross-isobar angle. These parameters depend on the stability of stratification and their relevance comes from the fact that the wind is often known for its geostrophic value only from atmospheric pressure data, $ifU_{ag} = \rho^{-1}\nabla p_a$. In near-neutral conditions over sea areas, the ratio of surface wind speed to geostrophic wind speed is 0.5–0.7 and the cross-isobar angle is 10–30°. During AIDJEX the average speed ratio and turning angle was found to be 0.59 and 25.9° in the Beaufort Sea (Albright, 1980; Brown, 1980). Joffre (1985) investigated the diurnal cycle of the ratio R_g for two different campaigns lasting about 3 weeks in the Baltic Sea. The resulting R_g displayed a maximum of 0.9 at about 23:00 h (local time) and a minimum of 0.55 at about 08:00 h. Joffre (1982b) also found that the cross-isobar angle varied between 15° and 30°.

Ice is an excellent platform for oceanic boundary layer investigations. It gives the pure shear stress forcing case with no disturbances from surface waves and provides a stable site for measurements. Ekman himself made observations in winter 1901 on the ice of Oslo Fjord and was the first to record the Ekman profile in the ocean. It took a long time for the Ekman theory to be validated in open ocean conditions. Figure 5.7 shows the Ekman spiral recorded beneath drifting sea ice in the Baltic Sea (based on 1-week velocity records).

Stratification

Sverdrup (1928) emphasized the role of stratification in the planetary boundary layers. He concluded that in the Siberian shelf in summer there is a 25–40-m, low-density upper layer that moves almost uniformly with the ice and with a very weak momentum transfer to the deeper water. For the wind stress he also concluded that the strong atmospheric stability in winter reduces wind drag.

Rossby and Montgomery (1935) examined the air–ice and ice–water stresses using the data of Sverdrup (1928) and Brennecke (1921), and obtained a floe-scale air drag coefficient of 2.5×10^{-3}, close to modern values. Their roughness parameter was 3.68 cm, which corresponds to a physical roughness of the order of 1 metre (explained by the presence of pressure ridges). Rossby and Montgomery also underlined the importance of stability in low wind speeds in agreement with Sverdrup. For the ocean boundary layer they agreed with Sverdrup that the presence of a stable shear layer in summer significantly reduces ice–water friction. However, they criticized Sverdrup's assumption for the bulk drift of the upper ocean with ice and, instead, showed that there is always a shear layer beneath the ice, which corresponds well to present understanding.

In general, in non-neutral conditions Eq. (5.19) is replaced by (e.g., Andreas, 1998):

$$\frac{dU}{dz} = \frac{u_*}{\kappa z} \phi_m\left(\frac{z}{L_{MO}}\right) \quad (5.24)$$

where ϕ_m is the "universal function" for momentum transfer and L_{MO} is the *Monin–Obukhov length* (e.g., Tennekes and Lumley, 1972), which describes the stability of the stratification:

$$L_{MO} = \frac{u_*^2}{\kappa b \theta_*} \quad (5.25)$$

where $b = g/T_v$ is a buoyancy parameter, $\theta_* = -Q/(\rho_F c_p u_*)$ is a temperature scale, T_v is virtual temperature,[3] Q is heat flux, ρ_F is density of the fluid, and c_p is specific heat at constant pressure. In neutral conditions, $Q = 0$; therefore, $L_{MO} = \infty$, $z/L_{MO} = 0$, and $\phi_m(0) = 1$. In stable conditions, heat flux is negative (toward the boundary) and then $z/L_{MO} > 0$. In unstable conditions the opposite is true.

Heat flux and buoyancy depend on temperature and humidity (atmosphere) or salinity (oceans); therefore, additional equations along the lines of Eq. (5.24) are needed. The three quantities are coupled, and the corresponding boundary-layer equations must be simultaneously solved. This question goes beyond the scope of this book; however, details can be found in Andreas (1998) and McPhee (1986). The main point here is the influence of atmospheric and oceanic stratification on drag parameters, the subject of Section 5.2.2.

To extend the Ekman layer for stratified flows, the Rossby number similarity theory is used with the roughness Rossby number $Ro_* = u_*/(fz_0)$ as the scaling parameter (McPhee, 1986; Andreas, 1998). Omstedt (1998) used a second-order turbulence model to examine the boundary layer beneath sea ice. This approach includes the equations of turbulent energy and its dissipation and, therefore, avoids the need for drag law parameters. However, this model is one-dimensional

---

[3] The temperature, which dry air would need to have the same density as moist air (i.e., $T_v = T(1 + 0.6078q)$, where q is specific humidity, see Gill, 1982).

and its applicability is consequently limited to local boundary-layer problems only (Leppäranta and Omstedt, 1990; Omstedt et al., 1996).

5.2.2 Drag force formulae

In general fluid dynamics, based on dimensional analysis, the drag force of the fluid on a solid plate is $\tau = \rho_0 C U^2$, where ρ_0 is the density of the fluid, U is the velocity of the fluid relative to the plate, and C is the (dimensionless) drag coefficient (e.g., Li and Lam, 1964). The drag coefficient depends on the Reynolds number $Re = UL/\nu$, ν being the kinematic viscosity of the fluid, and surface roughness. In laminar flow, $C \propto Re^{-1}$ and therefore $\tau \propto U$. But in turbulent flow the Reynolds number becomes very large and the drag coefficient is then independent of it, $\tau \propto U^2$.

For a planetary boundary layer, geostrophic flow serves as the natural, undisturbed reference velocity. Surface geostrophic velocities are usually good approximations for velocities immediately above boundary layers. As the boundary-layer flow is turbulent, the drag coefficient depends on surface roughness and stratification. In planetary boundary layers the Coriolis effect causes the flow to turn, and therefore a second stress law parameter is required: the *turning angle*, or *Ekman angle*, between the geostrophic flow and the surface stress (see Figure 5.6). This angle also depends on surface roughness and stratification.

By using the quadratic drag law, the necessary parameters of which are the drag coefficient and boundary-layer turning angle (e.g., Brown, 1980; McPhee, 1982), the atmospheric and oceanic drag laws can then be written as:

$$\tau_a = \rho_a C_a |U_{ag}| \exp(i\theta_a) U_{ag} \quad (5.26a)$$

$$\tau_w = \rho_w C_w |U_{wg} - u| \exp(i\theta_w)(U_{wg} - u) \quad (5.26b)$$

where ρ_a and ρ_w are air and water densities, C_a and C_w are air and water drag coefficients, and θ_a and θ_w are the boundary-layer turning angles in air and water. Note that to ascertain wind stress exactly there should be $(U_a - u)$ instead of U_a, but since $U_a \gg u$ the approximation used is good. The factors $\exp(i\theta_a)$ and $\exp(i\theta_w)$ simply rotate the velocity vectors by the angles θ_a and θ_w. In the northern hemisphere the angles are positive, turning counterclockwise, and vice versa in the southern hemisphere. Drag coefficients consist of skin friction and form drag due to ridges, $C = C^S + C^F$. For the air–ice interface they arguably are of similar magnitude (Arya, 1975), as seems to be the case for the ice–water interface.

Note that air and water stresses are written in vector form as:

$$\tau_a = \rho_a C_a U_{ag}(\cos\theta_a + \sin\theta_a \mathbf{k}\times) U_{ag} \quad (5.27a)$$

$$\tau_w = \rho_w C_w |U_{wg} - \mathbf{u}|(\cos\theta_w + \sin\theta_w \mathbf{k}\times)(U_{wg} - \mathbf{u}) \quad (5.27b)$$

Geostrophic drag parameters have been widely used. For the atmosphere, drag coefficient measurements are largely based on surface wind (altitude of 10 m, WMO recommendation for wind measurements) where the turning angle is zero. The drag parameters change for geostrophic wind on the basis of a boundary-layer model. When using atmospheric model output for the wind field, the surface wind is calculable. Similarly, in coupled ice–ocean modelling, oceanic top-layer velocity is used as the reference, with drag parameters depending on the thickness of this top layer. The ratio between the drag coefficients for geostrophic wind and the surface wind should be $C_{aw}/C_{ag} \approx 2$–4. In the AIDJEX experiment this ratio was 2.9 (Albright, 1980).

Neutral stratification

In neutral stratification the drag coefficient depends on surface roughness. Banke et al. (1980) modelled the air–ice surface drag coefficient for the surface wind as:

$$C_a = 1.10 + 0.072\gamma \quad (5.28)$$

where γ is the root-mean-square (r.m.s.) elevation, defined by Banke et al. (1980) as the variance of the surface elevation for wavelengths less than 13 m (the best correlation was found when the variance was taken over short wavelengths only). Their data covered r.m.s. elevations from 4 cm to 14 cm corresponding to drag coefficients from 1.2×10^{-3} to 2×10^{-3} or roughness lengths z_0 from 0.01 cm to 0.2 cm. Results by Andreas and Claffey (1995), however, suggest that snowdrifts tend to deform into streamlined shapes and thereby reduce the drag coefficient; therefore, Eq. (5.28) oversimplifies the air–ice coupling effects.

Due to pressure differences on both sides of ridges an additional *form drag* results. Attempts have been made to relate the atmospheric form drag coefficient to ridge statistics. In Arya's (1975) model the form drag coefficient is proportional to the product of ridge density μ and mean sail height $\langle h_s \rangle$, later confirmed by Banke et al. (1980) as:

$$C_a^F = \tfrac{1}{2} C_F \mu \langle h_s \rangle \quad (5.29)$$

where C_F is the form drag coefficient for a single ridge sail, $C_F = 0.012 + 0.012\phi_s$ (ϕ_s being the sail slope angle in degrees). Thus, $C_F \approx 0.4$ for $\phi_s = 30°$. For $h_s \sim 1$ m and $\mu \sim 5 \text{ km}^{-1}$, $C_a^F \sim 1.0 \times 10^{-3}$ (i.e., close to the skin friction drag coefficient). This result has been found to be valid in the Baltic Sea as well (Joffre, 1983). Including the form drag effect, Joffre (1983, 1984) found that overall roughness is approximately between 20% and 70% larger than local skin roughness.

Table 5.2 shows a collection of local surface wind drag coefficients for several regions obtained from mast measurements over undeformed ice surfaces. Variations are from 1.4×10^{-3} to 1.9×10^{-3}, and a representative drag coefficient is 1.5×10^{-3}. The data in Rossby and Montgomery (1935) over the Siberian Shelf was based on soundings, therefore integrating over a larger area, and resulted in $C_a = 2.5 \times 10^{-3}$. Snowdrifts may add about 50% (Andreas and Claffey, 1995), and for heavily ridged ice the drag coefficient may be as high as 4×10^{-3}.

The geostrophic drag coefficient is about 1.2×10^{-3} for both polar regions, corresponding to a surface wind drag coefficient of 3.5×10^{-3} (Brown, 1980; Martinson and Wamser, 1990). The geostrophic wind integrates roughness over a larger area, including the form drag due to hummocks and ridges. It therefore represents the real atmospheric force that is driving the ice.

Table 5.2. The drag coefficient for surface wind (measured at the standard altitude of 10 m) for neutral atmospheric stratification.

Region	Drag coefficient	Method	Reference
Gulf of St Lawrence	1.4×10^{-3}	Mast	Smith (1972)
Beaufort Sea	1.6×10^{-3}	Sonic anemometer	Banke et al. (1980)
Weddell Sea	1.9×10^{-3}	Mast	Andreas and Claffey (1995)
Baltic Sea	1.5×10^{-3}	Mast	Joffre (1982a)

Table 5.3. The parameters of the quadratic water–ice drag laws for neutral oceanic stratification.

Region	Drag coefficient	Turning angle (deg)	Level	Reference
Barrow Strait	5.4×10^{-3}	0	1 m	Shirasawa and Ingram (1997)
Beaufort Sea	5.0×10^{-3}	25	GSC	McPhee (1982)
Baltic Sea	3.5×10^{-3}	17	GSC	Leppäranta and Omstedt (1990)
Weddell Sea	1.6×10^{-3}	15	GSC	Martinson and Wamser (1990)

GSC = Geostrophic current.

In the open ocean, the drag coefficient of surface wind depends on wind speed. According to Smith (1980), it is 1.1×10^{-3} for weak winds and $(0.61 + 0.063 U_a) \times 10^{-3}$ for wind speeds ranging from 6 m/s to 22 m/s. Therefore, air–ice drag is normally larger than air–water drag. For strong winds (about 15 m/s) these drags are equal over undeformed ice, but, over deformed ice, the drag may be twice as much.

Table 5.3 shows a collection of oceanic drag parameters for several regions. Analysing ice drift based on the Ekman theory, Shuleikin (1938) assumed $\theta_w \approx 30°$, a constant. There is a natural variation of ±50% in drag coefficients due to stability and roughness. Nansen (1902) stated that floating ice increases the velocity of the oceanic boundary layer because no energy is needed for wave formation and the surface of ice is rougher than open water (absorbing more energy from the wind) as long as the internal ice resistance to motion is not too strong. The reference depth of 1 m has been used in several turbulence measurements beneath sea ice. The study of Shirasawa and Ingram (1997) gives the local drag coefficient of smooth ice, while other studies refer to geostrophic flow and include the form drag caused by hummocks and ridges. A specific case is *dead water* (Ekman, 1904), where a large amount of ice momentum is used to generate internal waves. This necessitates the existence of a shallow, stable surface layer, on the bottom of which internal waves form. The representative drag coefficient may then be five times the normal ice–ocean drag coefficient (Waters and Bruno, 1995).

Most research on sea ice boundary layers has been made in Arctic seas. The representative geostrophic drag coefficients and turning angles of $C_a = 1.2 \times 10^{-3}$,

$\theta_a = 25°$, $C_w = 5 \cdot 10^{-3}$, and $\theta_w = 25°$, found here are the result of AIDJEX field studies in the Beaufort Sea (Brown, 1980; McPhee, 1982). In the Baltic Sea the drag coefficients are 25–50% smaller than in the Arctic due to the roughness of the ice. But, in the Weddell Sea the water drag coefficient seems to be as low as 1.6×10^{-3} while the air drag coefficient is close to the Arctic value (Martinson and Wamser, 1990). This may be because the bottom of Antarctic sea ice is smooth and frictional oceanic boundary layers are thinner than those in Arctic or even those in the Baltic Sea.

As for geostrophic reference velocities, in Arctic seas $C_a/C_w \approx 0.24$; this happens to be about the ratio of the scales in the top and bottom topography of deformed ice. Because the geometries of top and bottom ice surfaces are well correlated, the ratio C_a/C_w should not vary much between different regions. In the Baltic Sea, $C_a/C_w \approx 0.19$ (Leppäranta and Omstedt, 1990). But, in the Antarctic the ratio C_a/C_w turns out to be larger, since the ice bottom is relatively smoother; Martinson and Wamser (1990) data give $C_a/C_w \approx 0.8$.

In general, the drag coefficient and turning angle may be defined for an arbitrary velocity reference level, with their values then depending on the chosen level. If the velocity profile and the stratification are known, then transformation of the drag law parameters between different levels is straightforward (e.g., in measurement campaigns, instrumentation altitudes can fix the reference level). In shallow waters, where the depth is much less than the Ekman depth, the oceanic boundary layer covers the whole water body, and then the drag coefficient depends on the depth of the water and the turning angle can be ignored.

Taking the logarithmic profile for the surface layer, in neutral conditions the roughness length and drag coefficient are related by $C = [\kappa/\log(z/z_0)]^2$.

Non-neutral stratification

In a stratified fluid the drag parameters also depend on the stability of the stratification. According to the Monin–Obukhov similarity theory, they are functions of the Monin–Obukhov length. Figure 5.8 presents the geostrophic drag parameters for the atmosphere (Andreas, 1998). The neutral values are $C_a = 1.2 \times 10^{-3}$ and $\theta_a = 15°$. In very stable conditions $C_a \approx 10^{-4}$ and $\theta_a \approx 35°$, while in unstable conditions $C_a \approx 1.5 \cdot 10^{-3}$ and $\theta_a \to 0$.

Joffre (1982a) examined the surface wind drag coefficient over a wide range of stability conditions ($-1 < z/L_{MO} < 10$) in the Baltic Sea. He found that C_a is strongly modulated by thermal stability (shown also in Stull, 1988, p. 268), decreasing from about 2×10^{-3} under unstable conditions to close to zero for strong stratification. His formulae read:

$$\frac{1}{\sqrt{C_a}} = 70.51 \frac{z}{L_{MO}} + 24.56 \quad \text{if} \quad \frac{z}{L_{MO}} > 0 \quad (5.30a)$$

$$\frac{1}{\sqrt{C_a}} = 2.98 \frac{z}{L_{MO}} + 25.55 \quad \text{if} \quad \frac{z}{L_{MO}} < 0 \quad (5.30b)$$

The fit was especially good under stable conditions.

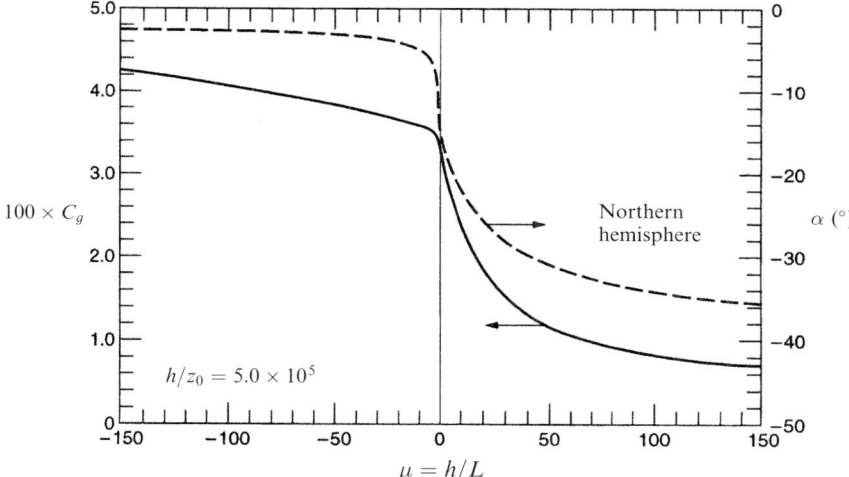

Figure 5.8. Geostrophic drag coefficient $C_{ag}(=C_g^2)$ and turning angle as functions of boundary layer height h scaled using the Monin–Obukhov length L.
Reproduced from Andreas (1998), with permission of the University of Helsinki.

Overland (1985) reviewed the air–sea drag coefficient for first-year ice from aircraft measurements and reconciled the range of observed drag coefficients of $10^3 C_a \approx 1.2$–3.7 for the surface wind (referenced at an altitude of 10 m) for all sea ice types, based on ice roughness and seasonal meteorology. For sea ice dynamics investigations it is necessary to define an effective drag coefficient that relates regional stress to regional wind, because sea ice is heterogeneous. Regional stress is influenced by the distribution of surface roughness, the buoyancy flux from quasi-periodic leads, and external atmospheric conditions, principally the inversion height. For wind speeds greater than 5 m/s and air temperatures below freezing, the effective drag coefficient is $10^3 C_a \approx 2.5$–3.0 for nearly continuous pack ice, such as first-year ice in seasonal ice zones and the central Arctic Basin. The range of values of $10^3 C_a$ is 3.0–3.7 for unstable boundary layers typical of off-ice winds in the marginal ice zone (MIZ) or even greater if the ice has been broken up by a recent storm. The effective drag coefficient when there is just a small concentration of sea ice is generally greater than the oceanic value.

Linear drag law

Quadratic water stress introduces non-linearity to the ice drift problem. Despite not being very robust, linear models can be useful in several cases. To preserve the approximate linearity between ice and wind velocities, it is convenient to linearize wind stress too. The linear forms are expressed as:

$$\tau_a = \rho_a C_{a1} \exp(i\theta_a) U_a \qquad (5.31a)$$

$$\tau_w = \rho_w C_{w_1} \exp(i\theta_w)(U_w - u) \qquad (5.31b)$$

where the linear drag law coefficients may be estimated from quadratic ones using fixed scaling speeds (subscript 0): $C_{a1} = C_a |U_{a0}|$ and $C_{w1} = C_w |(U_w - u)_0|$. Another way is to take the Ekman spiral down to the surface and from $\tau = \rho_0 K \, dU/dz|_{z=0}$ we have $C_1 = (Kf)^{1/2}$. The turning angle becomes $\mathrm{sgn}(f)\pi/4$ in the Ekman model.

5.3 SCALE ANALYSIS AND DIMENSION ANALYSIS

5.3.1 Magnitudes

The ice parameters in the momentum equation are thickness, velocity, and stress, all of which depend on time and space. The magnitudes of their *scales* are now rather well known from the long-term database built up over the years by manned and automatic drifting station programmes (Figure 5.9 and Table 5.4).

Very thin ice is the term used for ice 0.1 m thick and deformed ice that for 10 m thick ice. Velocity is zero for stationary ice, and strength is zero for open drift ice. The timescale for velocity changes is typically $1 \, \mathrm{d} \approx 10^5 \, \mathrm{s}$ and in stationary flows it is

Figure 5.9. The Soviet Union North Pole [Severnyj Poljus] drifting station programme was commenced in 1937, continuing regularly until 1991. It was recently (2003) restarted. The photograph shows the first station SP-1, under the leadership of Ivan Papanin, who wrote the book *Life on an Ice Floe* about his experience on SP-1.
Reproduced with permission from the Russian State Museum of the Arctic and Antarctic, St Petersburg.

Sec. 5.3]   Scale analysis and dimension analysis   131

Table 5.4. Typical and extreme levels from the long-term database.

	Typical	Low	High
Thickness, H	1 m	0.1 m	10 m
Velocity, U	0.1 m/s	0	1 m/s
Strength, P	10 kPa	0	100 kPa
Time, T	10^5 s	10^3 s	∞
Length, L	100 km	10 km	1,000 km

∞. For compact ice the length scale is 100 km in mesoscale dynamics. It has been known to go down to 10 km in shear and marginal ice zones, while in the general ice circulation of the polar oceans it can reach lengths of 1,000 km. Thin ice can show extreme behaviour (Figure 5.10): due to wind forcing, the whole ice cover in the Bay of Bothnia, Baltic Sea was driven in 6 days into the north-east corner (into an area 20% of its initial value).

Magnitude analysis of the momentum equation is performed using Table 5.4. The derivatives are estimated as $\partial \mathbf{u}/\partial t \sim U/T$, etc. The internal friction of ice is the most open term because there are still unsolved phenomenological problems; thereafter the key parameters are the air–ice and ice–water drag coefficients, and the

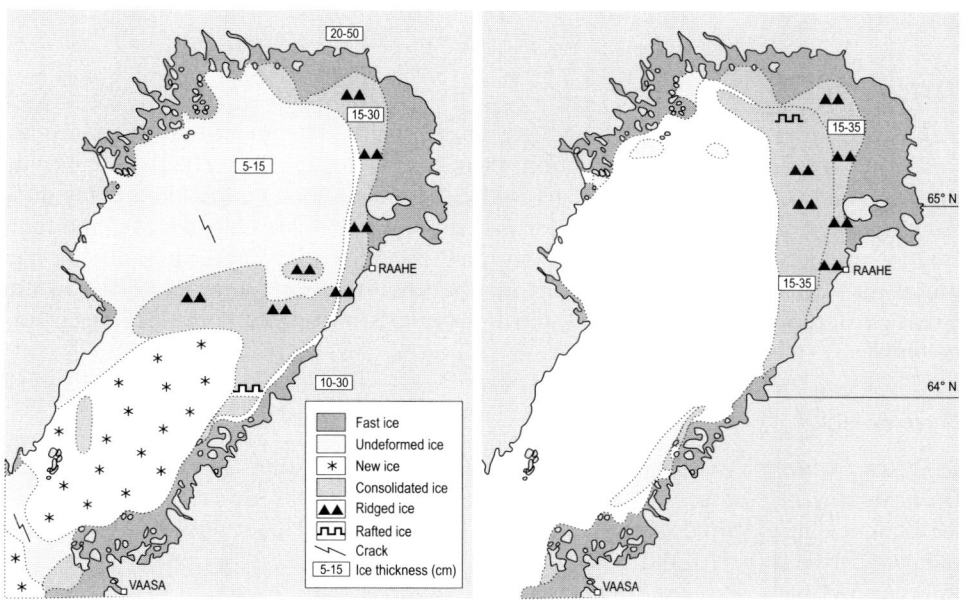

Figure 5.10. Ice situation in the Bay of Bothnia, Baltic Sea on 20 and 26 February 1992. Strong southerly to westerly winds drove the ice to the north-east corner of the basin (the basin length is 320 km).

Table 5.5. Scaling of the terms of the momentum equation of drift ice. Representative elementary scales are: $H = 1\,\text{m}$, $U = 10\,\text{cm/s}$, $P = 10\,\text{kPa}$, $U_{ag} = 15\,\text{m/s}$, $U_{wg} = 5\,\text{cm/s}$, $T = 1$ day, and $L = 100\,\text{km}$. The "Value" column gives the base-ten logarithm of the scale in pascals (i.e., -3 corresponds to $10^{-3}\,\text{Pa}$).

Term	Scale	Value ($^{10}\log$ [in Pa])	Comments
Local acceleration	$\rho H U / T$	-3	-1 for rapid changes ($T = 10^3\,\text{s}$)
Advective acceleration	$\rho H U^2 / L$	-4	Long-term effects may be significant
Coriolis term	$\rho H f U$	-2	Mostly < -1
Internal friction	PH/L	$-1(-\infty)$	Compact ice, $A > 0.9$ ($A < 0.8$)
Air stress	$\rho_a C_a U_{ag}^2$	-1	Mostly significant
Water stress	$\rho_w C_w |U - U_{wg}|^2$	-1	Mostly significant
Sea surface tilt	$\rho H f U_{wg}$	-2	Mostly < -2
Air pressure	$H \rho_a f U_{ag}$	-3	Minor term

Coriolis parameter. Atmospheric and oceanic boundary-layer angles do not influence the magnitude of drag forces.

As for air and water velocity, typical scales in geostrophic flows are $U_{ag} \sim 15\,\text{m/s}$ and $U_{wg} \sim 5\,\text{cm/s}$. The air pressure effect is related to geostrophic wind since $f\mathbf{k} \times \mathbf{U}_{ag} = -\rho_a^{-1} \nabla p_a$ (i.e., $h\nabla p_a \sim H\rho_a f U_{ag} \sim 10^{-3}\,\text{Pa}$). The governing background scales are the atmospheric cyclones and ocean circulation systems (100–1,000 km) and oceanic eddies (10–100 km).

Typical scales

Table 5.5 shows the outcome based on typical scales. The governing terms are air and water stresses and internal friction, their level being $10^{-1}\,\text{Pa}$. The significance of internal friction is limited to compact ice fields, while in open ice fields the free drift situation prevails. Coriolis acceleration and sea surface tilt are an order of magnitude smaller, and local and advective acceleration are smaller still. The air pressure term is minor and will be excluded from further analysis. It is not commonly listed in the equation of motion of drift ice, but may if necessary be absorbed into the sea surface tilt term.

Scale ranges

In the case of *thin-ice cover* (0.1 m), compared with the standard situation (Table 5.5) the air and water stresses are unchanged, but the other terms are proportional to the ice thickness and drop down by one order of magnitude; this results in a governing balance between the air and water stresses. In the case of *thick-ice cover* (10 m), the internal friction is dominant in compact ice, while the Coriolis and pressure gradient terms become comparable with the air and water stresses. Absence of ice strength automatically implies the free drift condition, while high ice strength makes the internal friction dominant.

In the case of *stationary ice*, internal friction balances any external force acting on the ice. Due to the plasticity of compact ice, a stationary situation is true of any forcing lower than yield stress. In the case of *rapidly moving ice* (1 m/s), the acceleration terms increase, but the question is which force drives the ice. If it is air stress, then air and water stresses increase by two orders of magnitude and form the dominant balance; if it is water stress, the geostrophic current velocity is also ~ 1 m/s and the magnitude of water stress is still 10^{-1} Pa. It is unlikely that internal friction or sea surface tilt would cause such high ice velocities.

Acceleration terms are not important in the standard case. Local acceleration disappears in the steady state ($T = \infty$) and rises to 0.1 Pa when the timescale is shortened to 10^3 s; this is the inertial timescale of drift ice. The ratio of advective to local acceleration is UT/L, equal to 10^{-1} in the standard case. In the case of a short length scale and high ice velocity the advective term could become significant.

The *length scale* becomes important in internal friction and advection. In the case of short length scales they both increase and, in compact drift ice, internal friction becomes the governing term. In compact ice a short length scale increases internal friction, which suppresses the role of advection. Consequently, in sea ice dynamics advective acceleration may only become important in a rapid free drift with a short length scale (e.g., under low wind conditions when the ice is just tracking the ocean surface layer currents).

In vertically integrated shallow-water models, the dominant terms are the wind and bottom stresses (analogous to wind stress and water stress here). But, instead of internal friction in the ice case, the Coriolis term and acceleration are important in the shallow-water case. This difference is due to the large difference between the thicknesses of sea ice and the oceanic boundary layer and between the rheologies of ice and water.

5.3.2 Dimensionless form

For a more quantitative approach to the scaling problem, the momentum equation is examined in dimensionless form. Let the scaling factors be H, U, P, T, and L as in Section 5.3.1 for the thickness, velocity and stress of ice, and for time and horizontal length scales. We denote dimensionless quantities with asterisks (e.g., $u^* = u/U$). Assuming $U \neq 0$, a convenient scaling of the momentum equation is made by the ice–water stress magnitude $\rho_w C_w U^2$:

$$\frac{\rho}{\rho_w C_w} \left(\frac{H}{UT} h^* \frac{\partial u^*}{\partial t^*} + \frac{H}{L} h^* u^* \cdot \nabla^* u^* + \frac{fH}{U} h^* \mathbf{k} \times u^* \right)$$

$$= \frac{1}{C_w} \frac{PH}{\rho_w U^2 L} \nabla^* \cdot \boldsymbol{\sigma}^* + \frac{\rho_a C_a}{\rho_w C_w} (\cos\theta_a + \sin\theta_a \mathbf{k} \times) U_a^* U_a^*$$

$$+ (\cos\theta_w + \sin\theta_w \mathbf{k} \times)|U_w^* - u^*|(U_w^* - u^*) - \frac{\rho}{\rho_w C_w} \frac{gH}{U^2} h^* \boldsymbol{\beta} \quad (5.32)$$

134 Equation of drift ice motion [Ch. 5

The following fundamental dimensionless quantities arise to characterize the importance of the different terms:

- Local acceleration: *Strouhal number* $\quad Sr = \dfrac{\rho}{\rho_w C_w} \dfrac{H}{UT}$

- Advective acceleration: *draft aspect ratio* $\quad \delta_A = \dfrac{\rho}{\rho_w C_w} \dfrac{H}{L}$

- Coriolis acceleration: *Rossby number* $\quad Ro = \dfrac{\rho}{\rho_w C_w} \dfrac{fH}{U}$

- Internal friction: *friction number* $\quad X = \dfrac{1}{\rho_w C_w U^2} \dfrac{PH}{L}$

- Wind stress: *Nansen number* $\quad Na = \sqrt{\dfrac{\rho_a C_a}{\rho_w C_w}}$

- Pressure gradient: *Froude number* $\quad Fr = \dfrac{\rho}{\rho_w C_w} \dfrac{gH}{U^2}$

All these dimensionless quantities have the ice–water drag coefficient C_w in the denominator, emphasizing the very close link between drift ice dynamics and the dynamics of the ocean boundary layer. The ocean has a firm grip on the ice, and many of the characteristics of sea ice drift are oceanic: ice transfers momentum from the wind to the ocean and receives part of it back as modified by the ocean. However, it is also clear that ice itself has an active role to play in the interaction between the atmosphere and the ocean. Consequently, we have $0 \ll C_w \ll \infty$. Scaling also included the velocity U, which can be assumed non-zero; the zero case would be simply a stationary ice cover held together by a plastic yield limit.

When $C_w \to \infty$, ice follows the ocean surface current exactly and it only affects the air drag force on the ocean. When $C_w \to 0$, the dynamics would be governed by the balance $\rho h f \mathbf{k} \times \mathbf{u} = \nabla \cdot \boldsymbol{\sigma} + \boldsymbol{\tau}_a$. With no internal friction, the result would be the Ekman transport equation $\mathbf{u} = -\mathbf{k} \times \boldsymbol{\tau}_a / \rho h f$ (i.e., ice motion is perpendicular to the wind, and $|\mathbf{u}| \propto |\mathbf{U}_a|^2$ and $|\mathbf{u}| \sim 2\,\text{m/s}$ for $|\mathbf{U}_a| \sim 10\,\text{m/s}$). With strong internal friction, the influence of Coriolis acceleration reduces and ice adjusts to the wind field by rearranging its state. This situation is sometimes close to reality: wind-driven drift is strongly slowed by internal friction.

Using the dimensionless quantities defined above, Eq. (5.32) is written:

$$Sr\, h^* \frac{\partial \mathbf{u}^*}{\partial t^*} + \delta_A h^* \mathbf{u}^* \cdot \nabla^* \mathbf{u}^* + Ro\, h^* \mathbf{k} \times \mathbf{u}^* = X \nabla^* \cdot \boldsymbol{\sigma}^* + Na^2 (\cos\theta_a + \sin\theta_a \mathbf{k} \times) U_a^* \mathbf{U}_a^*$$

$$+ (\cos\theta_w + \sin\theta_w \mathbf{k}\times)|\mathbf{U}_w^* - \mathbf{u}^*|(\mathbf{U}_w^* - \mathbf{u}^*)$$

$$- Fr\, h^* \boldsymbol{\beta} \qquad (5.33)$$

In ice–water stress just the normalized ice–ocean velocity difference remains and the magnitude is one. Wind velocity is naturally scaled by the Nansen number down to

$Na\,U_a^*$ for a representative wind-driven ice velocity. If other terms are small, $|U_w^* - u^*| \sim Na\,U_a^*$ and the direction between $U_w^* - u^*$ and U_a^* will depend on boundary-layer angles. The ratio $\Gamma_{iw} = u^*/U_w^*$ tells us whether the ice is driving the oceanic boundary layer ($\Gamma_{iw} > 1$) or vice versa ($\Gamma_{iw} < 1$). Wind and the geostrophic water current are outer boundary velocities between which the ice and water velocities lay. Consequently, the ratio $\Gamma_{aw} = Na\,U_a/U_{wg}$ tells us whether the ice motion is dominantly driven by wind ($\Gamma_{aw} > 1$) or by oceanic circulation ($\Gamma_{aw} < 1$).

Internal friction looks more complicated, with the coefficient $X = P^*H/(\rho_w C_w U^2 L)$. The numerator is the strength integral across the ice thickness and the denominator is the surface stress times the fetch (i.e., X compares the strength and forcing). A natural length scale for drift ice becomes:

$$L = \frac{P^*H}{F} \tag{5.34}$$

where F is forcing per unit area. In particular, in the wind-driven drift of compact ice, a natural ice length scale is $L = P^*H/\tau_a$. For $P^* \approx 10\,\mathrm{kPa}$, $H \approx 1\,\mathrm{m}$, and $\tau_a \approx 0.1\,\mathrm{N/m^2}$ we have $L \approx 100\,\mathrm{km}$.

The inertial timescale is included in the Strouhal number (Sr); this equals $T_I = \rho H/(\rho_w C_w U)$. However, when $H \sim 1\,\mathrm{m}$ and $U \sim 10\,\mathrm{cm/s}$, $T_I \sim 2 \times 10^3\,\mathrm{s}$ (a short timescale can be seen in Figure 3.9). Figure 5.11 shows in detail a case in which sea ice drift followed a sudden large change in the direction of strong wind. This is very short in comparison with the timescale of ice advection and deformation (typically 1–100 days, see Section 3.4). Since the timescales of forcing are usually much longer than the inertial timescale of ice, a *quasi-steady* approach can often be taken. This means solving the steady-state form of the momentum equation first, allowing the advection and deformation of ice with the obtained velocity field, and thereafter returning to the steady-state momentum equation with a new ice state field, etc.

The advective timescale is $T_D = L/U$, and advection is proportional to $\delta_A = \rho H/(\rho_w C_w L)$. Thus, increasing the value of velocity for intensive dynamics does not help advection because ice–water friction increases at the same rate; only over a very short length scale ($H/L \sim \rho_w C_w/\rho \sim 10^{-3}$) may momentum advection be important.

The Coriolis term has a Rossby number coefficient $Ro = \rho f H/(\rho_w C_w U)$, which includes the Coriolis period $f^{-1} \sim 10^4\,\mathrm{s}$ (a fundamental timescale in geophysical fluid dynamics). This Rossby number can be written as $Ro = T_I f$ (i.e., the Coriolis term becomes important when the inertial timescale is comparable with the Coriolis period—a very high ice thickness would be needed for this).

As a consequence, in drift ice dynamics the internal timescales generally satisfy:

$$T_I \ll f^{-1} \ll T_D \tag{5.35}$$

The coefficient of sea surface tilt is the Froude number $Fr = \rho g H/(\rho_w C_w U^2) \sim 10^5$. Therefore, a very steep slope (10^{-5}) would be needed for the pressure gradient to become important. In free drift the important length scales of ice motion are those in

136 Equation of drift ice motion [Ch. 5

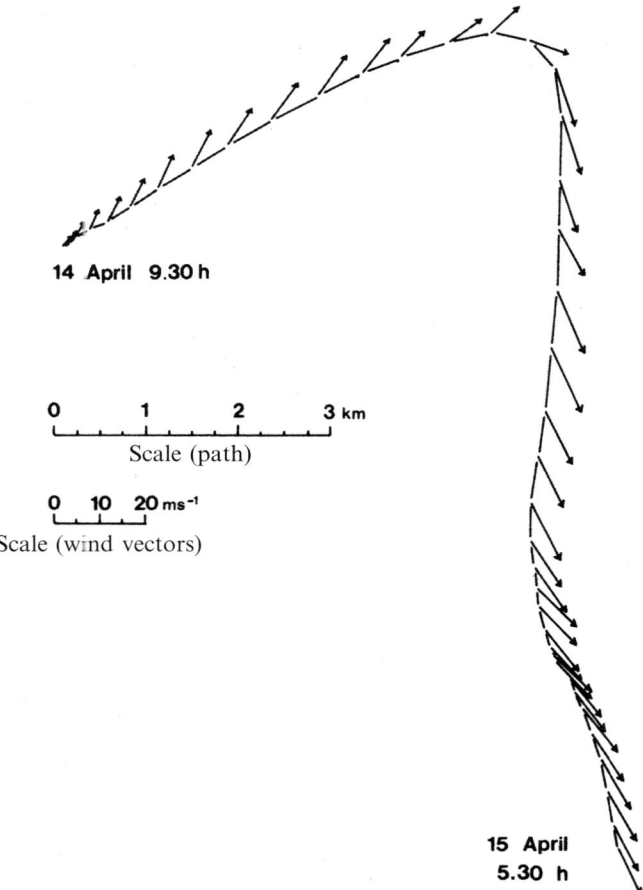

Figure 5.11. Observed path of a drifting station and wind in the Baltic Sea, April 1979. The time interval between trajectory points and wind vectors is 30 min.
From Leppäranta (1981b).

the wind and currents, but when internal friction is important it introduces a length scale (Eq. 5.34) related to forcing and strength.

Stationary ice

In the case of a stationary ice field the momentum equation is reduced to the static case $\boldsymbol{u} \equiv 0$. A convenient scaling is obtained using the internal friction of ice, PH/L. Then, we have:

$$\nabla^* \cdot \boldsymbol{\sigma}^* + \frac{L}{PH}\boldsymbol{F} = 0 \qquad (5.36)$$

where $\boldsymbol{F} = \boldsymbol{\tau}_a + \boldsymbol{\tau}_w - \rho gh\boldsymbol{\beta}$ is the forcing. Now, both surface stresses $\boldsymbol{\tau}_a$ and $\boldsymbol{\tau}_w$ are independent of the ice (apart from surface roughness), but the pressure gradient is proportional to the ice thickness. The key dimensionless number is $X_0 = PH/FL$. When $X_0 > 1$ the forcing is below the strength of the ice and the stationarity condition holds; but when $X_0 < 1$ the opposite is true.

5.3.3 Basin scales

The theory of drift ice dynamics has been applied to a wide range of basins. At the upper bound, the length scale of polar oceans is up to 3,000 km. Large Arctic and subarctic seas (e.g., the Barents Sea and Okhotsk Sea) are at the 1,000 km size. Further down there are small mediterranean and marginal seas with size 100–300 km (e.g., the Baltic Sea and the Gulf of St Lawrence). There are also large lakes in this size band that have drift ice (e.g., the Great Lakes in North America, see Wake and Rumer, 1983, and Lakes Ladoga and Vänern in Europe). Small bays of freezing seas and lakes, 10–30 km in size, exist with occasional drift ice conditions. Even ice flows in large rivers can be modelled using the present drift ice dynamics theory (e.g., ice flows in the Niagara River in North America, see Shen et al., 1990).

Two widely investigated areas for sea ice dynamics are the Arctic Ocean and the Baltic Sea. In the former the ice thickness and the size of the basins are one order of magnitude larger than in the latter. This scaling applies to the ice morphology as well (see Table 2.1). The stability ratio h/L is also about the same.

The Baltic Sea belongs to the seasonal sea ice zone. The size of its basins is 100–300 km, the thickness of undeformed ice can reach 1 m, and ice ridges are typically 5–15 m thick. The extent of the fast ice zone is largely influenced by the formation of grounded ridges, in the Baltic extending to sea depths of 5–15 m. The motion of ice ranges between zero and the free drift state, depending on the forcing and thickness of the ice. Compact, thick-ice fields may remain stationary. When drifting the geometry of the fast ice boundary has a strong aligning influence on ice motion. In fact, the whole ice field belongs to the coastal boundary layer. Compact ice fields show plastic behaviour with a yield strength of 10–100 kPa and significant shear strength, while slight opening in the ice fields transforms the flow to a viscous regime.

Table 5.6 shows the characteristics of sea ice dynamics in the polar seas and in the Baltic Sea. The principal reason for the difference in values is ice thickness (forcing and characteristic ice velocity are about the same in all these areas). The

Table 5.6. Comparison between the characteristics of sea ice drift in Arctic, Antarctic, and Baltic Sea waters. The velocity scale is 10 cm/s in all cases.

		Arctic	Antarctic	Baltic
Ice thickness	H	2–5 m	0.5–2 m	0.1–1 m
Free drift	Na	2%	3%	2%
Internal length scale	PH/F	250 km	100 km	50 km
Inertial timescale	$H/C_w U$	2 h	1.5 h	0.5 h
Adjustment timescale	L/U	30 d	10 d	5 d

138 Equation of drift ice motion [Ch. 5

ice–water drag coefficient is relatively low in Antarctica (apparent in the free drift level and inertial timescale).

5.4 DYNAMICS OF A SINGLE ICE FLOE

A single ice floe (say, Ω) has three degrees of freedom in its movement: translational velocity U and angular motion $\dot{\omega}$ around the vertical axis. For a point at distance r from the mass centre of the floe, the velocity is (see Figure 3.1):

$$u = U + \dot{\omega} \mathbf{k} \times r \tag{5.37}$$

By definition of the mass centre $\int_\Omega hr \, d\Omega = 0$. The motion of the floe must satisfy the conservation of linear and angular momentum (e.g., Landau and Lifschitz, 1976):

$$\rho\langle h\rangle \left(\frac{dU}{dt} + f\mathbf{k} \times U\right) = S_f^{-1} \int_\Omega (\tau_a + \tau_w - \rho h \nabla \Phi) \, d\Omega \tag{5.38a}$$

$$\rho\langle h\rangle r_I^2 \frac{d\dot{\omega}}{dt} = S_f^{-1} \mathbf{k} \cdot \int_\Omega r \times (\tau_a + \tau_w - \rho h \nabla \Phi) \, d\Omega \tag{5.38b}$$

where $\langle h \rangle$ is mean thickness, S_f is the floe area and r_I is the radius of gyration defined by:

$$\int_\Omega \rho h \, d\Omega \, r_I^2 = \int_\Omega \rho h r^2 \, d\Omega \tag{5.39}$$

Note that the right-hand side is the moment of inertia of the floe about the vertical axis. Translational motion is determined from the mean floe thickness, but for rotation the radius of gyration also needs to be known. The point of action of inertial and body forces is the mass centre of the floe, while surface forces act on the geometric centre, which is located at the distance $r^* = S_f^{-1} \int_\Omega r \, d\Omega$ from the mass centre. The quantity r^* measures floe inhomogeneity; for $h = $ constant we have $r^* = 0$. In general:

$$\rho\langle h\rangle \left(\frac{dU}{dt} + f\mathbf{k} \times U\right) = \tau_a - \rho\langle h\rangle \nabla \Phi + S_f^{-1} \int_\Omega \tau_w \, d\Omega \tag{5.40a}$$

$$\rho\langle h\rangle r_I^2 \frac{d\dot{\omega}}{dt} = \mathbf{k} \cdot (r^* \times \tau_a) + S_f^{-1} \mathbf{k} \cdot \int_\Omega r \times \tau_w \, d\Omega \tag{5.40b}$$

The integrals of ice–water stress cannot be directly evaluated since they depend quadratically on the ice–water velocity difference. Because of this non-linear relation, a rotating ice floe affects resistance to translation. Linearizing the ice–water stress $\tau_w = \rho_w C_{w1}[\cos\theta_w + \sin\theta_w \mathbf{k}\times](U_w - u)$, Eqs (5.40) become:

$$\rho\langle h\rangle \left(\frac{dU}{dt} + f\mathbf{k} \times U\right) = \tau_a + \tau_{w1} - \rho\langle h\rangle \nabla \Phi - \rho_w C_{w1}\dot{\omega}[\cos\theta_w + \sin\theta_w \mathbf{k}\times]r^* \tag{5.41a}$$

$$\rho\langle h\rangle r_I^2 \frac{d\dot{\omega}}{dt} = \mathbf{k} \cdot [r^* \times (\tau_a + \tau_{w1})] - \rho_w C_{w1} \dot{\omega} \cos\theta_w \langle r^2\rangle \tag{5.41b}$$

Dynamics of a single ice floe

where $\tau_{w1} = \rho_w C_{w1}[\cos\theta_w + \sin\theta_w \mathbf{k}\times](\mathbf{U}_w - \mathbf{U})$, and $\langle r^2 \rangle$ is the mean square radius of the floe. For a homogeneous floe, $\mathbf{r}^* = 0$; the floe rotation effect disappears from translational motion but remains as a damping term in rotation. Since then $\langle r^2 \rangle = r_I^2$, the timescale of this damping is:

$$t_d = \frac{\rho h}{\rho_w C_{w1} \cos\theta_w} \qquad (5.42)$$

The timescale is proportional to ice thickness and equals about 1 hour for $h = 1$ m. Thus, if a rotation is initiated by floe shear or collision, its lifetime will be short. For an extreme case, a very low drag coefficient would allow relatively long relaxation times.

The derivation and analysis of the equation of motion of drift ice is now complete. Together with the ice conservation law (Section 3.4) and ice rheology (Chapter 4), it forms the closed system used to solve the dynamics of drift ice with a given definition of state (Section 2.5). In the following three chapters, the solution to the drift ice dynamics problem is presented using three different approaches.

6

Free drift

6.1 STEADY-STATE SOLUTION

This chapter presents the free drift solution of sea ice dynamics. It is based on the momentum equation only. There is no internal stress, and the ice conservation law is not normally solved since it would eventually lead to an unrealistic ice-state field. The solution describes the velocity at one point or along the path of one ice floe, easily obtained for both steady-state and unsteady situations. The free drift case is analogous to the pure Ekman drift case in ocean dynamics.

Free drift is defined as the drift of sea ice in the absence of internal friction of ice. The term was introduced by McPhee (1980), although the situation itself has been known from the times of Fridtjof Nansen and Vagn Walfrid Ekman and the *Fram* expedition (1893–1896). Free drift is a good approximation for individual, separate ice floes and for ice fields with low compactness, less than about 0.8 (e.g., McPhee, 1980; Leppäranta and Omstedt, 1990). One may follow the motion in either a Lagrangian frame, in which there is no advective acceleration, or in a Eulerian frame, in which advective acceleration remains. However, it is a small term and the advantage of ignoring it is that no spatial derivatives are left in the momentum equation (see Eq. 5.10) in free drift.

In free drift theory, utilization of a complex number technique is particularly useful (see Section 3.3 for notation, etc.). Ignoring advective acceleration, the steady-state free drift obeys the algebraic equation:

$$\rho_a C_a \exp(i\theta_a)|U_a|U_a + \rho_w C_w \exp(i\theta_w)|U_w - u|(U_w - u) + i\rho h f(U_{wg} - u) = 0 \quad (6.1)$$

Note that the Coriolis parameter and turning angles are positive in the northern hemisphere and negative in the southern hemisphere. Normally the geostrophic current is chosen as the reference in ice–water stress, $U_w = U_{wg}$.

There are four velocities in the momentum equation. Let their representative scales be: U_a for wind velocity, U for ice velocity, U_w for water velocity, and U_{wg} for

Vagn Walfrid Ekman (1874–1954) was born in Sweden. He was a student of Bjerknes and Nansen, and presented in his doctoral thesis (1902) his theory of ocean currents, including the theoretical explanation of Nansen's observations of the drift of the vessel *Fram* in the Arctic Ocean. He later became a professor at the University of Lund (Sweden).
Reproduced from the V.W. Ekman Family Photo Album, with courtesy of Dr Artur Svansson.

geostrophic water velocity. The ratio $\Gamma_{iw} = U/U_w$ tells us whether ice is driving the ocean ($\Gamma_{iw} > 1$) or vice versa ($\Gamma_{iw} < 1$), and the ratio $Na\,U_a/U_w$ tells us whether wind or the current is the dominant force. The solution is determined by boundary-layer parameters and the Coriolis parameter (i.e., Nansen number, Rossby number, see Section 5.3.2, and Ekman angles).

Geostrophic sea ice drift

Let us assume there is no wind. In such a case, we can clearly see that, irrespective of drag law parameters and ice thickness, the only solution is $u = U_{wg}$. Sea ice follows the geostrophic flow of the ocean surface layer. Thus the pressure gradient provides the correct limiting behaviour and continuity with the surface layer and, though small, is therefore important in coupled ice–ocean modelling.

Note that in the case of sea ice the balance between the Coriolis acceleration and sea surface tilt β reads $i\rho h f u = -\rho g h \beta$ (see Eq. 5.9), which gives $u = igf^{-1}\beta$, exactly the same solution as for the surface geostrophic current in the ocean (e.g., for $\beta \sim 10^{-6}$ we have $|u| \sim 0.1$ m/s).

6.1.1 Classical case

Free drift of thin ice

Next, wind is added. When ice thickness is small, the Coriolis and pressure gradient terms are small and the solution is easily obtained as:

$$u = Na \exp(-i\theta_0) U_a + U_{wg} \qquad (6.2)$$

where the Nansen number $Na = \sqrt{\rho_a C_a / \rho_w C_w}$ equals the wind factor, and $\theta_0 = \theta_w - \theta_a$ is the deviation angle, or Ekman angle, between the wind-driven ice drift and wind.

The ratio C_a/C_w is key to determining the free drift speed. The roughnesses of the upper and lower surfaces are to some degree correlated, and therefore the ratio C_a/C_w is not very sensitive to ice type. A representative value for Arctic Ocean ice cover is $Na \approx 1.7\%$ for geostrophic references and $Na \approx 2.5\%$ for the surface wind, while in Antarctica these values are much greater (e.g., surface wind $Na \approx 3.5\%$) due to the low ice–water drag coefficient (see Section 5.2).

Positive rotation is counterclockwise. The angle θ_w is positive in the northern hemisphere and as $\theta_a = 0$ for the surface wind, we have $\theta_0 = \theta_w > 0$ and the minus sign in Eq. (6.2) turns rotation clockwise (i.e., ice drift is to the right of the wind direction). In the southern hemisphere $\theta_w < 0$ and rotation is opposite. In the case of geostrophic wind, $\theta_w \approx \theta_a$.

The deviation angle is $\theta_0 \approx 0$ for geostrophic wind or $25°$ ($-25°$) for surface wind in the northern (southern) hemisphere.

Example: Zubov's (1945) isobaric drift rule For the geostrophic wind U_{ag}, $\theta_0 \approx 0$. If $U_{wg} = 0$, then $u = Na\, U_{ag}$ (i.e., ice drifts along the isobars of atmospheric pressure at a speed 1.7% times the geostrophic wind speed). This law was earlier used as the rule of thumb, easily applied when atmospheric pressure charts are available. ∎

General solution

Using the geostrophic current as the reference, the solution can be written in the form:

$$u = \alpha \exp(-i\theta) U_a + U_{wg} \qquad (6.3)$$

where α and θ are the general (scalar) wind factor and the deviation angle; they depend on boundary-layer parameters and the Rossby number $Ro = \rho H f/(\rho_w C_w U)$, where H and U are the thickness and velocity scales, respectively. The solution is obtained from Eq. (6.1) as follows. First, rewrite the equation as:

$$\exp(i\theta_a)|U|U = \exp(i\theta_w)|u|u + iRo\, Uu \qquad (6.4)$$

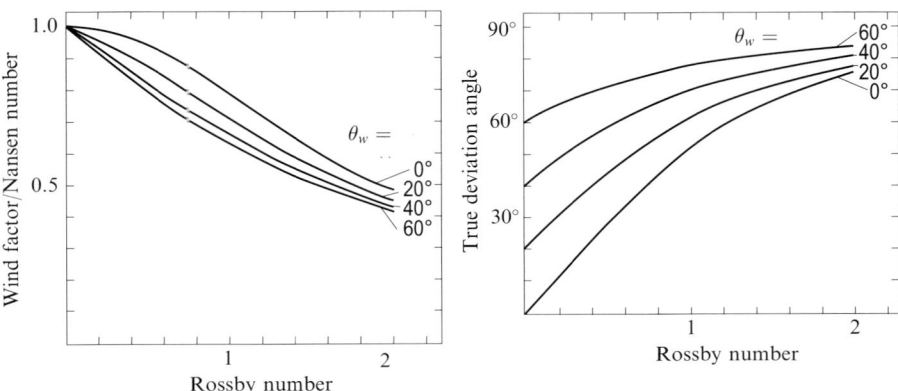

Figure 6.1. The free drift solution: wind factor α (*left*) and deviation angle θ (*right*) as a function of the Rossby number $Ro = (\rho H f)/(\rho_w C_w Na\, U_a)$ in the northern hemisphere (the signs of Ro, θ_w, and θ are reversed for the southern hemisphere).

where the velocity scale is $U = Na|U_a|$. Dividing Eq. (6.4) by $|U|^2$ and taking the modulus and argument of both sides gives us:

$$\alpha^4 + 2\sin\theta_w\, RoNa\, \alpha^3 + Ro^2 Na^2\, \alpha^2 - Na^4 = 0 \tag{6.5a}$$

$$\theta = \theta_a - \arctan[\tan\theta_w + RoNa/(\alpha\cos\theta_w)] \tag{6.5b}$$

Note that Ro and θ_w are negative in the southern hemisphere, following the sign of Coriolis acceleration. The first equation is a fourth-order polynomial of α, which in principle can be solved in closed form; however, this is quite cumbersome and does not allow a simple general expression for α. A numerical iteration scheme provides us with an easier way. Figure 6.1 shows the solution in graphical form. It is directly seen that $\alpha < Na$, $|\theta| > |\theta_0|$, and:

$$\alpha \to Na \quad \text{and} \quad \theta \to \theta_0 \quad \text{as } Ro \to 0$$

For $|Ro| \ll 1$, the first correction term to Na is $-\frac{1}{2}\sin\theta_w\, Ro$. But the accuracy of the thin-ice solution ($Ro \to 0$) is good for $|Ro| < 0.2$: α is within 10% below Na and θ is within $10°$ from θ_0. At the other extreme, $|Ro| > 2$, the solution is asymptotically:

$$\alpha = Na\sqrt{\sqrt{\frac{Ro^4}{4}+1} - \frac{Ro^2}{2}} \approx \frac{Na}{Ro}$$

$$\theta \approx \theta_a + \arctan\left(\tan\theta_w + \frac{Ro^2}{\cos\theta_w}\right)$$

In the Arctic, the usual surface wind parameters are 2% for the wind factor and $30°$ for the deviation angle (Nansen, 1902; Ekman, 1902), as later confirmed with drift buoy data (Thorndike and Colony, 1982). High wind factors have been found in Antarctica. Brennecke (1921) obtained averages of $\alpha \approx 2.8\%$ and $\theta = -34°$, and

Martinson and Wamser (1990) reported $\alpha \approx 3\%$ for surface wind in the Weddell Sea. The ratio of the drag coefficients C_a/C_w must therefore be larger in the Weddell Sea than in the Arctic; this is probably due to the hydrodynamically smoother lower surface of ridges and hummocks in the south. Rossby and Montgomery (1935) attributed this to hydrographic conditions in the water beneath the ice. The wind factor in the Baltic Sea is 2–2.5% (Leppäranta, 1981b, 1990), while in the Okhotsk Sea it is more than 3% (Fukutomi, 1952).

Example Kinematics data can be utilized to obtain information about the air–ice and ice–water drag parameters (Leppäranta, 1981b). The wind factor and deviation angle approach Na and θ_0 as $Ro \to 0$; thus, for a drifter on a fixed ice floe, asymptotically $|u - U_{wg}|/U_a \to Na \propto \sqrt{C_a/C_w}$ and $\theta \to \theta_0$ as $Ro \to 0$. Often C_a is much better known than C_w, and particularly in the case of surface wind $\theta \to \theta_w$ as $Ro \to 0$. ∎

The theory behind our general solution is best applicable in summer conditions, when there are free paths between floes and open-water areas do not freeze (Figure 6.2). Shuleikin (1938) solved the wind-driven ice–ocean flow problem

Figure 6.2. "In summer in the Arctic Ocean" by Gunnar Brusewitz, 1980. Gunnar Brusewitz (1924–) is an author and artist from Sweden, best known for his drawings of birds. This picture is from the Ymer-80 expedition in the European sector of the Arctic Ocean.
Reproduced from Gunnar Brusewitz, *Arktis sommar*, 1980, with permission from Wahlström & Widstrand, Stockholm.

using the Ekman theory. He used the balance between the air and water stresses and the Coriolis term, resulting in Eq. (6.3) with $U_{wg} = 0$ and exact expressions for the wind factor and deviation angle.

At low wind speeds the straight free drift model is questionable, since the stratification of air and water may play a major role. When the wind speed is less than 2 m/s there is no longer any clear connection between ice motion and wind (Rossby and Montgomery, 1935), since the local wind is overcome by the wind field over a larger area and by ocean currents as well. At $|U_a| \approx 3\text{–}4$ m/s, there is a rapid decrease in the wind factor since then the stable layer disappears from beneath the ice.

In shallow waters (depth considerably less than the Ekman depth), turning angle and Coriolis acceleration can be ignored, as the solution is in principle easily obtained from $\rho_w C_w |u - U_w|(u - U_w) = \tau_a - \rho h g \beta$, where β is sea surface tilt. Since the whole water column is in the frictional boundary layer, the water drag force becomes depth-dependent.

6.1.2 One-dimensional channel flow

One-dimensional flow models are useful in sea ice dynamics. In free drift it is the external friction coefficients that have the dominant role, while for drift in the presence of internal friction much of the resistance comes from compression. Consequently, first-order physics can be taken as non-rotational. The mathematics becomes much more simple and the solutions easy to interpret. The one-dimensional free drift equation comes from Eq. (6.1) as:

$$\tau_a + \rho_w C_w |U_w - u|(U_w - u) - \rho g h \beta = 0 \tag{6.6}$$

where $\tau_a = \rho_a C_a |U_a| U_a$ is wind stress, U_w is current velocity beneath the boundary layer, and β is the slope of the water surface. All variables are now real. If $\tau_a - \rho g h \beta > 0$, then $u > U_w$ and the solution is $u = U_w + [Na^2|U_a|U_a - gh''\beta/C_w]^{1/2}$ (h'' is the draft); otherwise, $u = U_w - [-(Na^2|U_a|U_a - gh''\beta/C_w)]^{1/2}$.

Consider an open channel of finite depth H and bottom slope $-\beta$ (Figure 6.3). The coupled momentum equations for ice and water are:

$$\tau_a + \tau_w + \rho g h \beta = 0 \tag{6.7a}$$

$$\left. \begin{aligned} \frac{d}{dz}\left(K_w \frac{dU_w}{dz}\right) + g\beta &= 0 \\ U_w(0) &= 0 \\ K_w \frac{dU_w}{dz}(H) &= \frac{\tau_w}{\rho_w} \end{aligned} \right\} \tag{6.7b}$$

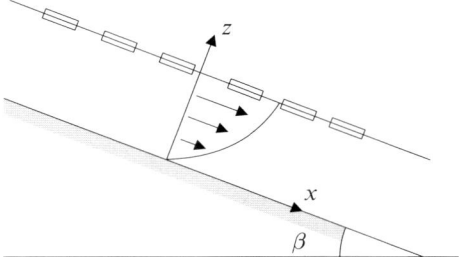

Figure 6.3. Schematic picture of open channel flow.

where K_w is the eddy viscosity coefficient of water, and the vertical co-ordinate z is taken up from the channel bottom. With $K_w = $ constant, this system can be directly integrated, and the solution for velocity is:

$$U_w = \frac{g\beta}{K_w}\left(H + \frac{\rho}{\rho_w}h - \frac{z}{2}\right)z + \frac{\tau_a}{\rho_w K_w}z \tag{6.8}$$

and by continuity $u = U_w(H)$. Fixing the total volume of water (solid + liquid), we may write $H_0 = H + h'' = $ constant, where H is the depth of liquid water and $h'' = (\rho/\rho_w)h$ is the draft, which corresponds to the volume of water stored in the ice. The surface velocity is then (cf. Figure 6.3):

$$U_w(H) = \frac{g\beta}{2K_w}(H_0^2 - h''^2) + \frac{\tau_a}{\rho_w K_w}(H_0 - h'') \tag{6.9}$$

This shows that the more ice is formed and consequently the less liquid water there is, the smaller the surface velocity will be. The reason is that when the thickness of ice increases, the depth of liquid water decreases and the bottom friction becomes relatively more important.

Example In classical hydraulics the Chezy formula is applied for flow calculations (e.g., Li and Lam, 1964). The flow is driven by gravity alone and is assumed uniform, and the velocity is:

$$U_w = \sqrt{\frac{8gm\beta}{\mu}}$$

where m is the hydraulic radius and μ is the friction coefficient, which can be determined from Manning's formula[1] $\sqrt{8g/\mu} = 0.262\sqrt{g}m^{1/6}n^{-1}$ where n is the Manning coefficient. For a channel with a half-circle cross-section $m = \frac{1}{2}H$, while a rectangular cross-section gives $m = \frac{1}{2}H(1 + \delta)$, where δ is the channel aspect ratio. The Manning coefficient is $n \approx 0.020\,\text{m}^{1/6}$ for earth material. Thus, $U_w \propto \sqrt{\beta}$ and $U_w \propto H^{2/3}$, while in Eq. (6.9) $U_w \propto \beta$ and $U_w \propto H^2$. The difference is due to Eq. (6.9) having the eddy viscosity, which in general is not constant; rather, it depends on

---

[1] Classical formula in river hydraulics (see, e.g., Li and Lam, 1964).

the characteristics, geometry, and flow. For $\beta \sim 10^{-4}$ and $m = 5\,\mathrm{m}$ we have $U_w \sim 1.2\,\mathrm{m/s}$. Assuming that ice drifts with the Chezy flow, the result agrees with the surface velocity from Eq. (6.9) when the eddy viscosity is $K_w = 0.041\,\mathrm{m^2/s}$, a realistic number (e.g., Cushman-Roisin, 1994). ∎

In a channel with a closed end, a surface slope develops under wind forcing. In the steady state the slope balances the surface stress, so that the vertical mean velocity of water is zero. The coupled ice–water model is thus:

$$\tau_a - \rho_w C_w |u|u - \rho g h \beta = 0 \qquad (6.10\mathrm{a})$$

$$\rho_w C_w |u|u - \rho_w g H \beta = 0 \qquad (6.10\mathrm{b})$$

where for simplicity τ_a, $u > 0$ has been assumed. The solution is $u = Na\,U_a\sqrt{H/(H + h'')}$. If $h''/H \sim 0.1$, then the wind factor is 0.95 times the deep-sea value. The free drift assumption, however, breaks down when the ice starts to pack at the end of the channel. This case will be examined in Chapter 7.

6.1.3 Shallow waters

In shallow water, where the sea depth is less than the Ekman depth ($\sim 30\,\mathrm{m}$), the Earth's rotation can be ignored for ice and water. There is no geostrophic water current to stand for the drag reference, and the bottom friction of the water body must be taken into account. Then, for a coupled ice–ocean steady-state model, with complex variables:

$$\rho_a C_a |U_a| U_a + \rho_w C_w |U_w - u|(U_w - u) - \rho g h \beta = 0 \qquad (6.11\mathrm{a})$$

$$-\rho_w C_w |U_w - u|(U_w - u) - \rho_w C_b |U_w| U_w - \rho_w g H \beta = 0 \qquad (6.11\mathrm{b})$$

where U_w is the water velocity at mid-depth and C_b is the bottom drag coefficient. Both ice–water and bottom drag coefficients depend on depth.

Assume an infinite ocean with constant wind. In such a case there is no build-up of sea surface slope, $\beta = 0$, and the solution is:

$$u = \sqrt{\frac{\rho_a C_a}{\rho_w C_w}} U_a + U_w, \; U_w = \sqrt{\frac{\rho_a C_a}{\rho_w C_b}} U_a \qquad (6.12)$$

Assuming that $C_w \sim C_b$, we have $|u|/|U_w| \sim 2$. As sea depth decreases, both water and bottom drag coefficients increase with $|u|/|U_w| \approx$ constant and $u, U_w \to 0$.

We can then define a semi-infinite ocean as the negative *real* half in the complex plane and take the wind blowing toward the shore. At steady state, the mean water velocity U_w must be zero, and the solution is the same as that in the channel model, $u = Na\,U_a\sqrt{H/(H + h'')}$.

6.1.4 Linear model

Quadratic water stress introduces non-linearity to the free drift system. However, this non-linearity is not very strong and a linear model is sufficiently good for several cases. To preserve the approximate linearity between ice and wind velocities, it is convenient to linearize the wind stress as well. The linear forms are expressed as:

$$\tau_a = \rho_a C_{a1} \exp(i\theta_a) U_a \tag{6.13a}$$

$$\tau_w = \rho_w C_{w1} \exp(i\theta_w)(U_w - u) \tag{6.13b}$$

where linear drag law coefficients may be estimated from quadratic ones using fixed scaling speeds: $C_{a1} = C_a |U_{a0}|$ and $C_{w_1} = C_w |(U_w - u)_0|$; this is essentially the same drag law that results from the classical Ekman solution, with $C_a |U_a| = \sqrt{K_a f}$ and $C_w |U_w| = \sqrt{K_w f}$ and the boundary-layer angles changed to θ_a and θ_w from 45°. Defining the Nansen and Rossby numbers for the linear drag law as:

$$Na_1 = \frac{\rho_a C_{a1}}{\rho_w C_{w1}}, \qquad Ro = \frac{\rho f h}{\rho_w C_{w1}} \tag{6.14}$$

the resulting steady-state solution is:

$$u = a_1 U_a + U_w, \qquad a_1 = \frac{\rho_a C_{a1} \exp(i\theta_a)}{\rho_w C_{w1} \exp(i\theta_w) + i f \rho h} = \frac{\exp(i\theta_a)}{\exp(i\theta_w) + i Ro} Na_1 \tag{6.15}$$

where the coefficient a_1 includes the linear wind factor $\alpha_1 = |a_1|$ and the linear deviation angle $\theta_1 = \arg a_1$. It depends on boundary-layer parameters and ice thickness, but in contrast to the quadratic solution it is independent of wind speed.

Example: Ekman solution In his PhD thesis, Ekman (1902) was the first to devise linear friction laws. He obtained $C_{*1} = \sqrt{K_* f}$ and $\theta_a = \theta_w = 45°$, and he neglected the geostrophic current. The solution was therefore:

$$u = \frac{\rho_a \sqrt{K_a}(1+i)}{\rho_w \sqrt{K_w}(1+i) + i\sqrt{2} f \rho h} U_a + U_w$$

For $h \to 0$, the thin-ice solution becomes $a = \rho_a \sqrt{K_a} / (\rho_w \sqrt{K_a})$. ∎

6.2 NON-STEADY-STATE SOLUTION

The small amount of inertia of drift ice was clearly illustrated in historical data; consequently, not much attention was paid to it. Shuleikin (1938) analysed the free drift case and noted inertial oscillations due to the Coriolis term and the short response time of about 1 h. Zubov (1945) also reported the quick (a few hours) response of ice to wind. An example of the rapid response of ice drift to changes in wind is shown in Figure 5.11.

6.2.1 One-dimensional flow with quadratic surface stresses

The full, non-linear, free drift equation does not allow for a general analytical solution. The one-dimensional case can be easily solved for constant forcing. The equation reads:

$$\rho h \frac{du}{dt} = \rho_a C_a \operatorname{sgn}(U_a) U_a^2 + \rho_w C_w \operatorname{sgn}(U_w - u)(U_w - u)^2 \qquad (6.16)$$

For the relaxation case, take $U_a = 0$, $U_w = 0$, and $u(t=0) = u_0$. The Strouhal number (see Section 5.3.2) can be taken as $Sr = \rho h/(\rho_w C_w |u_0| T)$, where T is the timescale. The exact solution is $u(t) = u_0/(1 + t/T)$; for $h \sim 1$ m and $u_0 \sim 0.1$ m/s the timescale is $\sim \frac{1}{2}$ h. For a linear drag law with $C_1 = C_w|u_0|$, the solution would be $u(t) = u_0 \exp(-t/T)$.

For constant wind and $u(t=0) = 0$, the natural velocity scale is $U = Na\, U_a$ and, consequently, $Sr = \rho h/(\rho_w C_w |U| T)$, and the solution with quadratic surface stresses is $u(t) = U \tanh(t/T)$. It is also interesting to examine this solution using fully dimensional parameters:

$$u(t) = Na\, U_a \tanh\left(\frac{\sqrt{\rho_a C_a \rho_w C_w}}{\rho h} U_a t \right) \qquad (6.17)$$

Thus, $T = \rho h/\sqrt{(\rho_a C_a \rho_w C_w)} U_a$; for $U_a \sim 5$ m/s and $h \sim 1$ m, $T \sim \frac{1}{2}$ h. Thus, in the spin-up, or accelerating, case the response time, being proportional to $1/\sqrt{C_a C_w}$, decreases with both air and water drag coefficients, while the steady-state velocity is proportional to $\sqrt{C_a/C_w}$. Since surface and bottom roughnesses are positively correlated, rougher ice responds faster; however, steady-state velocity is not very sensitive to roughness. Finally, note that if $U_w = $ constant, the above quadratic solution will be valid for $u - U_w$.

6.2.2 Linear model

The two-dimensional non-steady problem can be analytically examined using the linear model. This model possesses the strong property of superposition, which allows us to construct general solutions from simple elementary solutions. The linear model is written:

$$\rho h \frac{du}{dt} = [\rho_w C_{w_1} \exp(i\theta_w) + i\rho h f](U_{wg} - u) + \tau_a \qquad (6.18)$$

The steady-state system is described by Nansen and Rossby numbers (Eqs 6.14), and for the unsteady part the Strouhal number (see section 5.3.2) is taken, expressed with the linear drag coefficient $C_{w1} = C_w U$, as:

$$Sr = \frac{\rho h}{\rho_w C_{w1} T} \qquad (6.19)$$

and therefore $T = \rho h/(\rho_w C_{w1})$ is the natural timescale of the system. Equation (6.18) is a linear non-homogeneous equation, and the solution is directly integrated from:

$$u(t) = u(0) \exp(-\lambda_i t) + \int_0^t \exp[-(\lambda_i + if)(t-s)]\{\lambda_i U_{wg} + \tau_a/(\rho h)\} ds \qquad (6.20)$$

where $\lambda_i = \exp(i\theta_w) T^{-1}$. Furthermore, $\operatorname{Re} \lambda_i$ is the inverse response time and

Im $\lambda_i + f$ is the oscillation frequency of the system. If they are both non-zero, a damping oscillatory solution results for constant forcing.

For constant forcing we have:

$$u(t) = u(0)\exp(-\lambda_i t) + [1 - \exp(-\lambda_i t)](U_{wg} + a_1 U_a) \tag{6.21}$$

where a_1 is the steady-state, complex wind factor given in Eq. (6.15). Thus, the ice adjusts from the initial velocity to the steady-state velocity, and the adjustment timescale is $T = |\lambda_i|^{-1}$. For $U \sim 0.1\,\text{m/s}$ and $h \sim 1\,\text{m}$, we have $T \sim \frac{1}{2}\,\text{h}$. Note that the oscillation frequency is different from the Coriolis frequency; this bias is an artefact left over from using a fully passive ocean in this model and will be removed when using a fully coupled ice–ocean model in Section 6.3.

Tidal ice flow

The periodic forcing case can be applied to the tidal problem. Tidal currents can be taken independently of the ice appearing as a periodic forcing in Eq. (6.20). Oceanic velocity is fixed by $U_w = U_0 \exp(i\omega t)$ where U_0 and ω are tidal velocity amplitude and frequency. Let us take a simple approach by ignoring wind and rotational effects ($\theta_w = 0$ and $f = 0$). The transient term can also be left out, and then:

$$u(t;\omega) = \int_{-\infty}^{t} \exp\left(-\frac{t-s}{T}\right) \frac{U_0 \exp(i\omega s)}{T} \, ds \tag{6.22}$$

The solution becomes:

$$u(t;\omega) = \Gamma(\omega) U_0 \exp(i\omega t), \qquad \text{where } \Gamma(\omega) = \frac{1 - i\omega T}{1 + (\omega T)^2} \tag{6.23}$$

Here $\Gamma(\omega)$ is the transfer function; this tells us that the frequency of ice velocity follows the tidal current, while phase is delayed by ωT rad and amplitude is damped by $1/\sqrt{1 + (\omega T)^2}$. Since $T \sim \frac{1}{2}\text{h}$ for $h \sim 1\,\text{m}$, in semi-diurnal tides the phase lag is $\sim 0.5\,\text{h}$ and amplitude is 97% of the amplitude of the tidal current. Allowing for Coriolis acceleration of the ice, it is easy to see that the product ωT in the transfer function changes into $(\omega + f)T$. For semi-diurnal tides $|\omega| \approx |f|$ in high polar regions. If $\omega = -f$, the tidal and inertial currents rotate in the same direction and then $\Gamma(\omega) = 1$ (i.e., the ice follows the tidal water flow exactly). Otherwise ($\omega = f$), for $T = \frac{1}{2}\text{h}$ the phase lag becomes 1 h and the amplitude damping factor is 0.89. A half-day cycle for divergence–convergence of an ice drift had already been reported by Nansen (1902) and Sverdrup (1928). They suggested it to be a tidal phenomenon. However, in polar oceans the semi-diurnal tidal period and inertial period are very close in duration.

The tidal period is much longer than the inertial response time of drift ice, and consequently the transfer function Γ is not much different from unity. Therefore, a quasi-steady model would also be applicable. Because of the linearity of the system, ice follows the tide in a free drift tidal flow and the wind-driven drift component can be superposed on top.

6.2.3 Drift of a single floe

Drifting stations have sometimes been installed on large ice floes or islands (Figure 6.4), and observations about their movements as a rigid body exist. In particular, the path of inhomogeneous floes is affected by the geometry of the floe via rotational motion.

Let us now return to the normal vector formulation of two-dimensional quantities and equations. A single ice floe drifts with three degrees of freedom: translational velocity u and angular velocity $\dot{\omega}$ around the vertical axis obtained from the conservation laws of momentum and angular momentum (see Section 5.4). These conservation laws were given by Eqs (5.41) for linear ice–water stress. For homogeneous floes the inertial timescale of rotation is obtained from the Strouhal number as $T \sim \rho h/(\rho_w C_{w1})$ (Eq. 5.42), equal to the inertial timescale of translational motion. High ice thickness or a low ice–water drag coefficient would allow relatively long relaxation times.

Inhomogeneous floes possess a non-zero distance $r^* \neq 0$ between the mass centre and the geometric centre. Surface stresses act on the geometric centre, while inertial

Figure 6.4. Fletcher's ice island, or ice island T-3, in the Beaufort Sea, a widely used base for ice drift studies in the 1950s.
From Rodahl (1954).

and body forces act on the mass centre, and the resulting couple influences the orientation of ice floes with respect to the direction of forcing. Rotation causes an additional ice–water stress term for translational motion, but its magnitude is normally small; for example, for $\dot\omega \sim 1$ c.p.d. (cycles per day), $|r^*| \sim 100$ m, and $|U_w - u| \sim 10$ cm/s the relation between translational and rotational stress terms is $\sim 10^{-1}$.

In the inhomogeneous case, the rotation equation obtains a forcing term. Let us first write $\mathbf{k} \cdot [r^* \times (\tau_a + \tau_{w1})] = |r^*|(\Delta\tau_x \cos\omega + \Delta\tau_y \sin\omega)$, where $\Delta\tau = \tau_a + \tau_{w1}$, τ_{w1} is the ice–water stress from translational motion and ω is the direction of r^*. Let us then assume that $\Delta\tau \approx$ constant and align the x-axis with it. Then:

$$\rho\langle h\rangle r_I^2 \frac{d^2\omega}{dt^2} + \rho_w C_{w1} \cos\theta_w \langle r^2\rangle \frac{d\omega}{dt} + \Delta\tau |r^*| \sin\omega = 0 \qquad (6.24)$$

When ω is small, $\sin\omega \approx \omega$ and we have a linear second-order equation that represents pure damping or oscillatory damping whether the discriminant of this equation is positive or negative, respectively. The pure damping case takes place if:

$$\frac{\langle r^2\rangle^2}{\langle h\rangle r_I^2 |r^*|} > \frac{4\rho\Delta\tau}{(\rho_w C_{w1} \cos\theta_w)^2} \qquad (6.25)$$

The left-hand side is $\sim d/(\delta|r^*|)$, where d and δ are the size and aspect ratio of the floe. The right-hand side is $\sim 4\rho U^2/\tau$. Simple scale considerations tell us that Eq. (6.25) is usually true. Taking $d \sim 1$ km, $\delta \sim 10^{-3}$, and $|r^*| \sim 100$ m, the left-hand side becomes $\sim 10^4$; and with $U \sim 10$ cm/s and $\tau \sim 10^{-1}$ Pa, the right-hand side becomes $\sim 4 \times 10^2$. This means that floe rotation normally obeys the pure damping solution: once the forcing changes, the floe orientation moves steadily to the new equilibrium position, and the angular velocity is:

$$\dot\omega \approx \frac{\Delta\tau |r^*| \sin\theta_w}{\rho_w C_{w1} \cos\theta_w \langle r^2\rangle} \qquad (6.26)$$

A damping oscillator solution could result for strongly inhomogeneous floes ($|r^*| \sim d$) or for low ice–water friction C_{w1}.

Example (Leppäranta, 1981b) One large ice floe was mapped in a Baltic Sea field experiment with the following resulting measures: characteristic diameter 3.3 km, thickness 0.5 m, radius of gyration 1.3 km, and radius of inhomogenuity $|r^*| = 0.24$ km. Then $\langle r^2\rangle^2/(\langle h\rangle r_I^2 |r^*|) \sim 1.5 \times 10^5$, where the approximation $\langle r^2\rangle \sim d^2/2$ has been used. It is very difficult to break the inequality expressed in Eq. (6.25), and a pure damping solution results. The angular velocity is of the order of 10^{-6} s^{-1}. Interestingly, floe geometry can cause similar vorticities in much the same way as external forcing does. ∎

The ratio of rotational kinetic energy to translational kinetic energy is $(\dot\omega r_I/|u|)^2$. Usually, $\dot\omega r_I < 1$ cm/s, and so the kinetic energy is nearly all translational. In practice, the surface stress resultant $\Delta\tau$ varies in timescales of 0.1–1 day. In such periods floe rotations are within a few degrees and appear as aperiodic adjustments to the changing forcing.

6.3 LINEAR, COUPLED ICE–OCEAN MODEL

In the non-steady case a prescribed oceanic velocity can be a limiting assumption; therefore, a coupled model should be used. Consequently, a vertically integrated mixed-layer model is employed for the ocean. Now, U_w represents the mean velocity of the mixed layer and H is the thickness of the mixed layer. The resulting equation for the mixed layer velocity is:

$$\rho_w H \frac{dU_w}{dt} = -[\rho_w C_{w1} \exp(i\theta_w) + i\rho_w H f] U_w + \rho_w C_{w1} \exp(i\theta_w) u \qquad (6.27)$$

which includes on the right-hand side ice–ocean coupling (first and third terms) and Coriolis acceleration.

In April 1975 an almost ideal free drift situation took place in the northern Baltic Sea (Figure 6.5). There was a drift ice field 100 km across without any contact with the fast ice boundary. The correlation between the velocities of ice and wind was about 0.9, and the spectra showed good correspondence with each other (Figure 6.6) Due to the shallowness of the basin (the mean depth is 43 m), inertial oscillation is weak.

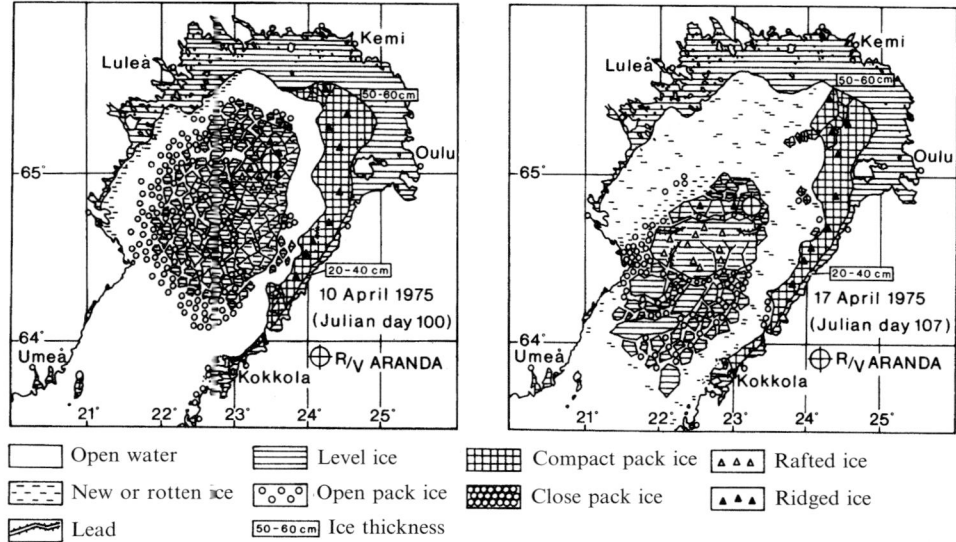

Figure 6.5. Drift of an ice patch, 100 km across and free from any contact with solid boundaries (April 1975, Bay of Bothnia, Baltic Sea). R/V *Aranda* was the base.
From Leppäranta (1990).

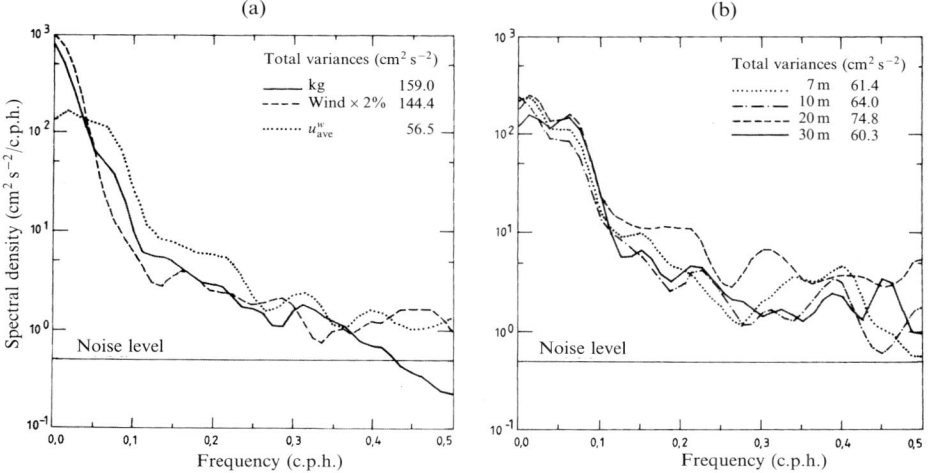

Figure 6.6. Spectra of wind, ice drift, and mixed layer velocities in the near-free drift situation shown in Figure 6.5. On the right plot the spectra for current velocities at 7 m, 10 m, 20 m, and 30 m depths are shown, while u_{ave}^w on the left plot is the mean mixed layer velocity estimated from them.
From Leppäranta and Omstedt (1990).

6.3.1 General solution

The coupled model (Eqs 6.18 and 6.27) is a pair of linear first-order equations with constant coefficients that can be solved by standard techniques. The coupled system includes three internal frequencies: the Coriolis frequency f, the ice response λ_i, and the mixed layer response:

$$\lambda_w = \frac{C_{w1} \exp(i\theta_w)}{H} \qquad (6.28)$$

The corresponding timescales have different magnitudes:

$$|\lambda_i|^{-1} \sim 10^3 \text{ s} \ll |\lambda_w|^{-1} \approx |f|^{-1} \sim 10^4 \text{ s}$$

Additionally, the forcing may have its own timescales.

The coupled model is easily rewritten, with velocity terms multiplied by inverse timescales, in the form:

$$\frac{du}{dt} = -(\lambda_i + if)u + \lambda_i U_w + \lambda_a U_a \qquad (6.29a)$$

$$\frac{dU_w}{dt} = -(\lambda_w + if)U_w + \lambda_w u \qquad (6.29b)$$

where

$$\lambda_a = \frac{\rho_a C_{a1}}{\rho h} \exp(i\theta_a) = Na_1 \lambda_i \exp[i(\theta_a - \theta_w)] \qquad (6.30)$$

The general solution is obtained by the elimination method (consult a basic analysis textbook for this, such as Adams, 1995). Briefly, U_w is eliminated from Eqs (6.29) and a second-order equation is obtained for u. The roots of the characteristic polynomial are $r_1 = -if$, $r_2 = -(\lambda_i + \lambda_w + if)$, and the general solution is then:

$$u = A\exp(r_1 t) + B\exp(r_2 t) + u_p \qquad (6.31a)$$

where the terms A and B describe the general solution to the homogeneous case, and $u_p = u_p(t)$ is a particular integral of the full equation. The solution for U_w then results from Eq. (5.29a) as:

$$U_w = A\exp(r_1 t) - \frac{\rho h}{\rho_w H} B\exp(r_2 t) + \frac{1}{\lambda_i}\frac{du_p}{dt} + \left(1 + \frac{f}{\lambda_i}\right)u_p + \frac{\lambda_a}{\lambda_i} U_a \qquad (6.31b)$$

Example Estimation of water drag parameters from velocity data using the momentum integral method (Hunkins, 1975). Vertical integration of ocean boundary-layer equations gives:

$$\frac{dU_w}{dt} + ifU_w = -\frac{\tau_w + \tau_b}{\rho_w H}$$

where τ_b is stress at the bottom of the boundary layer. Stresses τ_w and τ_b influence the evolution of momentum. If current measurements are available across the boundary layer, the sum of these stresses can be evaluated. If the sea depth is more than the Ekman depth, one may assume $\tau_b \approx 0$ and obtain the ice–water stress τ_w. Once we know the ice velocity, the drag parameters can be obtained from $\tau_w = \rho_w C_w \exp(i\theta_w)|U_{w0} - u|(U_{w0} - u)$, where U_{w0} is the chosen reference velocity. ∎

6.3.2 Inertial oscillations

Inertial oscillation is one of the fundamental frequencies in the dynamics of the upper ocean (e.g., Gill, 1982). The period of this oscillation equals $T_f = 2\pi/|f| = \pi/(\Omega|\sin\phi|)$, clockwise in the northern hemisphere and counterclockwise in the southern hemisphere. At the poles the period is 12 hours and then it increases toward the equator; for $|\phi| > 74°$ it is still below 12.5 h. Consequently, in high polar regions, inertial frequency is not easily distinguishable from the semi-diurnal tidal frequency with the same sign for rotation. In the subarctic seas of the seasonal sea ice zone (e.g., Sea of Okhotsk and Gulf of St Lawrence) the latitude goes down to 45° where $T_f \approx 17$ h (Figure 6.7). The inertial signal has been found from drifting sea ice (Hunkins, 1967; McPhee, 1978; Ono, 1978) when near-free drift conditions exist. However, the internal friction of ice seems to depress the signal. In shallow seas the loss of stratification in winter also increases the influence of bottom friction on the damping of the inertial oscillation.

In the absence of wind, Eqs (6.29a, b) show that the ice and ocean velocities must be equal (apart from possible transient terms from the initial conditions). Then, they can be simply added together for the solution (McPhee, 1978):

$$\frac{dM}{dt} = -iMf \qquad (6.32)$$

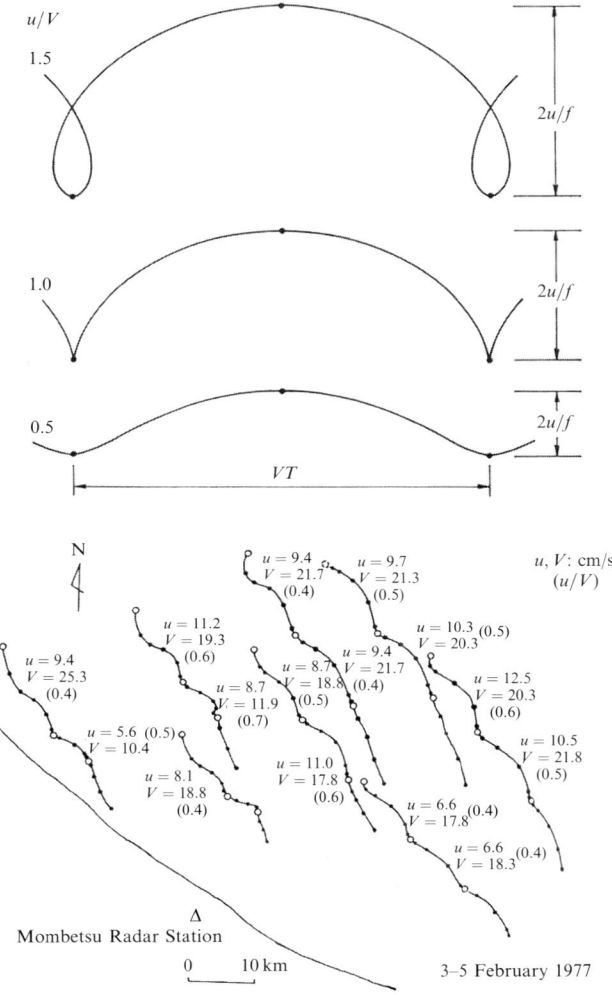

Figure 6.7. Inertial motion of sea ice superposed on alongshore translational ice motion. Upper graph: theoretical, u/V is relative magnitude of inertial motion. (*Lower graph*) Observations in the Sea of Okhotsk.
Reproduced from Ono (1978), with permission from Hokkaido University.

where $M = \rho_w H U_w + \rho h u$ is the total mass transport of ice and water. The solution is $M = M_0 \exp(-ift)$, where M_0 is the initial transport, similarly to the mass transport in the dynamics of the ice-free upper ocean. Inertial motion draws a clockwise (counterclockwise) circle in the northern (southern) hemisphere, and the radius of this circle is $|U_{iw}|/f$, where $U_{iw} = M/(\rho_w H + \rho h)$ is the ice–ocean joint velocity that is equal to the average weighted by the ice and mixed layer masses; for $|U_{iw}| \sim 0.1$ m/s the radius is ~ 1 km.

Now, let us consider the coupled model (Eqs 6.29) in the homogeneous case. With initial conditions $u = u_0$ and $U = U_0$ at $t = 0$, the solution is:

$$u = U_{iw}\exp(-ift) + (1 - \rho h/M)(u_0 - U_0)\exp[-(\lambda_i + \lambda_w + if)t] \quad (6.33a)$$

$$U_w = U_{iw}\exp(-ift) + \rho h/M(U_0 - u_0)\exp[-(\lambda_i + \lambda_w + if)t] \quad (6.33b)$$

The ice and ocean are in a state of persistent inertial (Coriolis) oscillation, and transient terms are needed to adjust the equalization of ice and ocean velocities. The adjustment timescale is $\mathrm{Re}(\lambda_i + \lambda_w)^{-1} \sim 1\,\mathrm{h}$, primarily determined by the ice response.

If constant wind stress is included in Eq. (6.29), the particular integral u_p can be evaluated as the steady-state solution, and then we have:

$$u_p = \left(\frac{1 - i\lambda_i b/f}{1 + b + if/\lambda_i}\right) Na_1 \exp[i(\theta_a - \theta_w)]U_a \quad (6.34)$$

where $b = \rho h/(\rho_w H)$. The full solution can now be obtained from Eqs (6.31), and the initial conditions will determine the constants A and B. With $u = U_w = 0$ at $t = 0$, the solution is (ignoring transient terms):

$$u = u_p + U_{iw}\exp(-ift) \quad \text{and} \quad U_w = U_{w1} + U_{iw}\exp(-ift)$$

where $U_{iw} = M_w = \rho_w H - i\tau_a/[(M_w + m)f]$ (i.e., ice and ocean have individual steady-state components and together undergo inertial oscillation of the same amplitude, frequency, and phase).

The amplitude of inertial oscillation compared with steady-state velocity is small, $\sim M_w f/|(\rho_w C_{w1})| \sim 1\%$. The uncoupled steady state of velocity is $\tau_a/(\rho_w C_{w1}\exp(i\theta_w) + imf)$, with a slightly larger magnitude and slightly smaller deviation angle.

A friction term for the mixed layer, due to bottom friction or internal friction, could be added to the coupled model in the form $-\mu U_w$, $\mu \sim \rho_w C_b |U_{w0}|$; this would damp inertial motion out; in the solution both roots of the characteristic polynomial would have non-zero real components taking care of damping. Frictional dissipation is small in deep water, $[\mu/(\rho_w H)]^{-1}$ being of the order of days, and inertial oscillations are clearly observable. But as soon as the depth is less than Ekman depth, mixed layer friction corresponds to bottom friction, $\mu/(\rho_w H) \sim \lambda_w$, λ_w^{-1} being of the order of hours, and damping takes place rapidly, as has been reported in observations (Leppäranta 1990). Using a similar linear friction law for the internal friction of ice, rapid damping would also result; this is reflected in the fact that the inertial signal in ice drift is much clearer in summer than in winter.

6.3.3 Periodic forcing

Wind forcing on sea ice drift is purely external in that it is independent of ice velocity. To examine the spectral response of ice drift to a periodic wind, let us choose:

$$U_a = U_{a0}\exp(i\omega t) \quad (6.35)$$

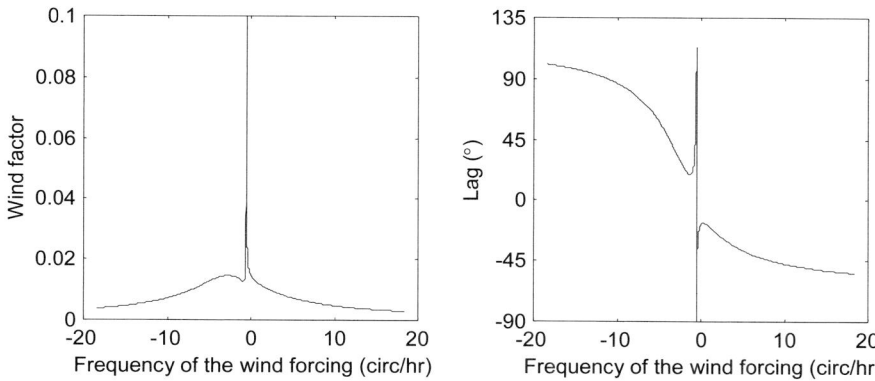

Figure 6.8. The geostrophic wind factor (*left*) and lag (*right*) in a wind-driven non-steady ice drift as functions of the frequency of wind forcing.

where U_{a0} is the amplitude and ω the frequency of wind velocity. The particular integral can be directly obtained by the integrating factor method, and the result is:

$$u_p = a_1 U_{a0} \exp(i\omega t) \tag{6.36}$$

where the transfer function $a_1 = a_1(\omega)$ is:

$$a_1 = \lambda_a \frac{\lambda_w + i(\omega + f)}{i(\lambda_i + \lambda_w)(\omega + f) - (\omega + f)^2} \tag{6.37}$$

That is, ice follows wind with a wind factor of $|a_1|$ and deviation angle $\arg a_1$, which is in this case the lag in a cycle as well. The solution is illustrated by the gain and phase shift of ice motion with respect to the wind (Figure 6.8). As $\omega \to 0$, the steady-state solution results; and as $\omega \to \infty$, the wind factor slowly decreases with increasing $|\omega|$, except for peaking at the clockwise Coriolis frequency $\omega = -f$. There is no observational evidence for this peak, but it also occurs in advanced turbulence modelling of the ocean beneath the ice (Omstedt et al., 1996). As $\omega \to \infty$, the transfer function a_1 asymptotically becomes $-i\lambda_a/\omega$, which means that $|a_1|$ asymptotically vanishes as $|\lambda_a|/\omega$, and $\arg a_1$ asymptotically approaches $\arg \lambda_a - \pi/2$ or the ice follows the surface wind ($\arg \lambda_a = 0$) one-quarter of a cycle behind.

6.3.4 Free drift velocity spectrum

Observed sea ice velocity spectra (Figure 3.10) are governed by wind forcing at frequencies below one cycle per day (c.p.d.). At higher frequencies, the inertial peak is commonly observed (as are tidal frequencies in places), but otherwise the high-frequency spectra are red noise. However, available observations only provide coverage up to about 5 c.p.d. because of measurement noise problems. The free drift solution with periodic forcing allows us to construct the theoretical frequency spectrum of sea ice velocity.

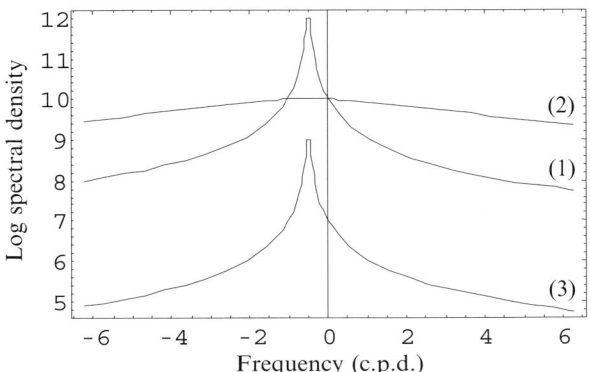

Figure 6.9. Sea ice velocity spectra normalized by forcing: (1) ice–ocean coupled model, (2) ice-only model, and (3) locked ice–ocean model.

For a general forcing $F/(\rho h)$ the spectrum is $p_F = p_F(\omega)$. The resulting spectrum of sea ice velocity is obtained from the coupled model as:

$$p_i(\omega) = \frac{(\lambda_i^2 + f^2 + \omega^2)}{[(\lambda_i + \lambda_w)^2 + (\omega + f)^2](\omega + f)^2} p_F(\omega) \qquad (6.38)$$

This has the following properties:

(i) $p(0) = \dfrac{(\lambda_i^2 + f^2)}{[(\lambda_i + \lambda_w)^2 + f^2]f^2} p_F(0)$.

(ii) There is a delta peak at the inertial frequency $\omega = -f$.

(iii) Asymptotically $p(\omega) \to p_F(\omega)/\omega^2$ as $\omega \to \infty$.

The form normalized by dividing by the forcing spectrum is shown in Figure 6.9. This fully free drift spectrum is then compared with two specific cases: (a) ice-only, where the ocean is passive; and (b) locked ice–ocean, where the ice and mixed layer flow together.

In the ice-only case (Eq. 6.18), ocean velocity is external and the ice velocity spectrum becomes:

$$p_i(\omega) = \frac{p_F(\omega)}{\lambda_i^2 + (\omega + f)^2} \qquad (6.39)$$

This equals $p_F(0)/(\lambda_i^2 + f^2)$ at $\omega = 0$, peaks at $p_F(-f)/\lambda_i^2$ at $\omega = -f$, and asymptotically reaches $p_F(\omega)/\omega^2$ as $\omega \to \infty$ (Figure 6.9). The ice response time dominates the form of the spectrum. For the locked ice–ocean system (McPhee, 1978), adding the forcing $F' = F\rho h/M$ to Eq. (6.32) produces the joint ice and ocean velocity spectrum:

$$p_w(\omega) = \frac{p_{F'}(\omega)}{(\omega + \lambda)^2} \qquad (6.40)$$

which equals $p_{F'}(0)/f^2$ at $\omega = 0$, has a delta peak at $\omega = -f$, and asymptotically

$p_w(\omega) \to p_{F'}(\omega)/\omega^2$ as $\omega \to \infty$ (Figure 6.9). Since $F'/F = \rho h/M_w \sim 0.05$, the locked system shows a much lower level than the fully coupled system.

A particular ice model with inertial embedding was presented by Heil and Hibler (2002). In this approach, sea ice velocity is extracted from the joint ice and ocean boundary layer velocity by assuming a drag coefficient and a turning angle. The stationary solution is correct, and therefore low frequencies reproduce very well in the ice velocity spectrum, as does the inertial peak. However, at higher frequencies the spectrum falls off too quickly. Using the present linear model, inertial embedding is carried out by replacing the ice mass ρh in the acceleration term of Eq. (6.18) by $\rho h - i\rho_w C_{w1}/f$. If we work through the analytic model with this modified ice inertia, the main features of the Heil and Hibler (2002) model would became apparent.

Quadratic ice water stress

The linear free drift model is just a linear filter for forcing where frequencies are unchanged. This is also approximately true of the free drift model with quadratic ice–water stress. For periodicities above the inertial timescale, the quasi-steady approach is accurate, giving $u = U_w + \alpha \exp(-i\theta)U_a$ (Eq. 6.3), and therefore ocean current and wind velocity frequencies appear unchanged in the ice motion. The parameters α and θ depend on wind speed, and thus the transformation is not exactly linear.

In high frequencies, when the ice responds according to the quadratic ice–water stress, $|U\exp(i\omega)|U\exp(i\omega) = U^2 \exp(i\omega)$ and the frequency hardly changes when ice and ocean speed levels are not close to each other. A study using a coupled free drift, second-order turbulence model forced by the wind did not indicate any frequency change in ice motion (Omstedt et al., 1996). Figure 6.10 compares the analytical solution (given in Eq. 6.36 for ice) and the turbulence model outcome. The results only differ at very high frequencies, basically because the drag formulation in terms of the mean mixed layer velocity does not follow rapid changes well. But strong non-linearities could arise if the ice was forced by wind and ocean currents equally from both sides and at frequencies above the inertial scale (not a likely scenario, though).

Momentum advection transfers energy to higher frequencies—as is usually the case in fluid dynamics—but it is a minor term in sea ice drift. Instead, the following takes place: ice drift transfers kinetic energy to the ocean, where frequency transfer takes place toward higher frequencies, before then returning to the ice drift.

6.4 SPATIAL ASPECTS OF FREE DRIFT

The free drift solution has severe limitations. As it is only valid for compactnesses less than 0.8, ice deformation may invalidate the free drift state by driving compactness over the critical limit (e.g., when two ice floes 10 km apart approach each other at 2.5 cm/s, the compactness between them would increase from 0.6 to 0.8 in 24 h).

An even more severe limitation is that the ice drift problem cannot be closed with a free drift model. The ice would eventually pile up on shore. A similar case is

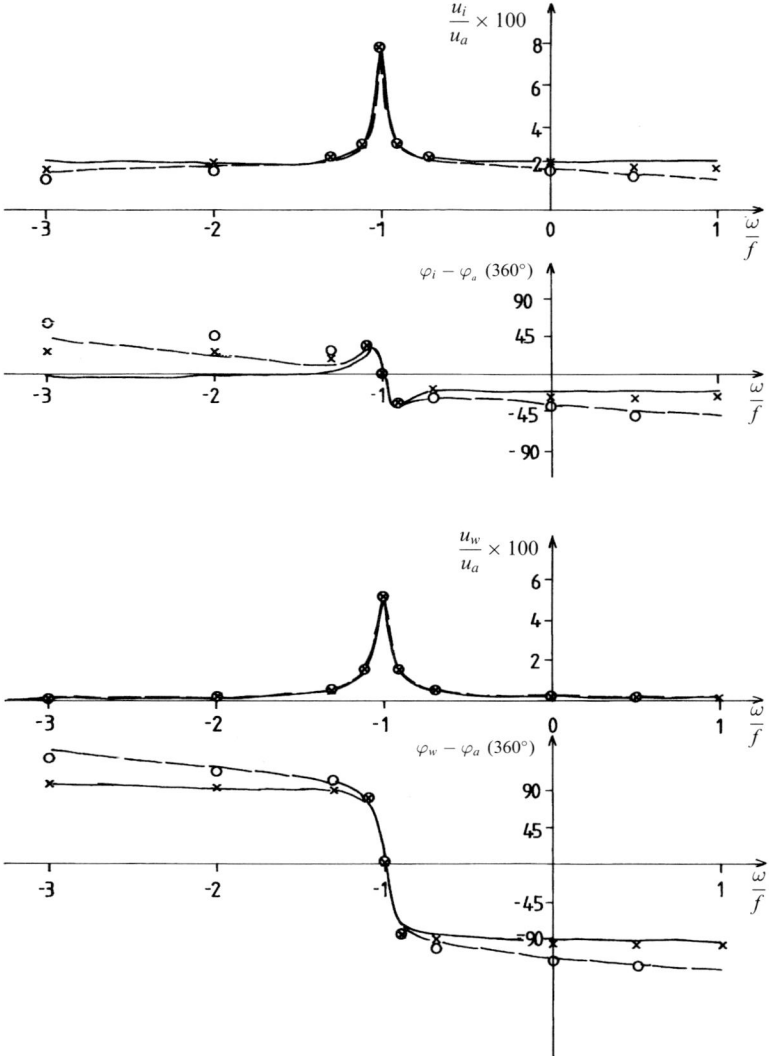

Figure 6.10. Speed and direction of ice drift (u_i, φ_i) and mixed layer (u_w, φ_w) scaled with wind (u_a, φ_a), as functions of forcing frequency ω scaled with Coriolis frequency f. Analytical two-body model: solid line for $h = 0.5$ m, dashed line for $h = 3$ m; ice model coupled with second-order turbulence model: x for $h = 0.5$ m and 0 for $h = 3$ m. The angles G_a, G_i, and G_w are positive anticlockwise.

the Ekman surface current model: it must be combined with the pressure gradient and mass conservation law to obtain the circulation over the whole study basin. However, the internal friction force combined with the ice conservation law enables the ice problem to be closed.

Because of its limitations in treating ice deformation properly, the ice conservation law is not usually analysed in free drift theory.

6.4.1 Advection

Let us give a simple example to illustrate the role of advective acceleration in sea ice drift. Take the steady-state, one-dimensional free drift equation:

$$\rho h u \frac{du}{dx} = \tau_{ax} - \rho_w C_w u^2 \tag{6.41}$$

where $u > 0$ has been assumed. Writing $u\,du/dx = \frac{1}{2}\,du^2/dx$, we have a linear equation for u^2. The general solution then becomes:

$$u^2 = u_0^2 \exp(-2\lambda x) - \int_0^x \frac{2\tau_{ax}(x')}{\rho h} \exp[-2\lambda(x - x')]\,dx' \tag{6.42}$$

where $\lambda = \rho_w C_w/(\rho h)$. The advective length scale is $(2\lambda)^{-1} \sim 100$ m for $h \sim 1$ m, much smaller than the forcing length scale. Consequently, the advective term is very small.

Example Take the wind stress as $\tau_{ax} = \tau_0 x/L$. The solution is:

$$u^2 = u_0^2 \exp(-2\lambda x) - \frac{\tau_0}{\rho_w C_w L 2\lambda}[1 - \exp(-2\lambda x)] + \frac{\tau_0 x}{\rho_w C_w L}$$

Without advection the solution would be $u^2 = \tau_0 x/(\rho_w C_w L)$. At distances $x \gg \lambda$, advection causes a spatial lag of $1/(2\lambda)$ to velocity distribution. ∎

6.4.2 Divergence and vorticity

In this subsection, velocities are taken as normal vectors (i.e., $\boldsymbol{u} = u\mathbf{i} + v\mathbf{j}$, etc.). The free drift solution (Eq. 6.3) can then be written as $\boldsymbol{u} = (\alpha_1 + \alpha_2 \mathbf{k}\times)\boldsymbol{U}_a + \boldsymbol{U}_w$, where α_1 and α_2 are scalars. Assuming α_1 and α_2 to be constant, we have:

$$\nabla \cdot \boldsymbol{u} = \alpha_1 \nabla \cdot \boldsymbol{U}_a + \alpha_2 \nabla_z \times \boldsymbol{U}_a + \nabla \cdot \boldsymbol{U}_w \tag{6.43a}$$

$$\nabla_z \times \boldsymbol{u} = \alpha_1 \nabla_z \times \boldsymbol{U}_a + \alpha_2 \nabla \cdot \boldsymbol{U}_a + \nabla_z \times \boldsymbol{U}_w \tag{6.43b}$$

where ∇_z stands for the vertical component of vorticity. For surface wind the deviation angle between ice drift and wind is about 30°, $\alpha_1 \approx 2\%$, and $\alpha_2 \approx 1\%$. For geostrophic wind, Zubov's (1945) rule would give ice drift parallel to the isobars of atmospheric pressure, (i.e., incompressible drift). However, this rule is exactly valid as $h \to 0$, and with finite ice thickness the cross-isobar parameter α_2 would be nonzero.

If the horizontal wind and water flow are incompressible, the vorticity of wind alone may drive the divergence of ice drift. This is similar to the Ekman pumping phenomenon in atmosphere–ocean dynamics. For a wind vorticity of $10^{-4}\,\text{s}^{-1}$, a typical ice drift divergence level of $10^{-6}\,\text{s}^{-1}$ would result. Counterclockwise wind vorticity gives opening and clockwise vorticity gives closing of the ice (in the northern hemisphere). The vorticity of ice is driven by both wind and current vorticities.

In addition to this, variations in ice thickness or roughness may affect free drift parameters sufficiently to produce divergence and vorticity under constant wind (e.g., $\nabla \cdot (\alpha_1 \boldsymbol{U}_a) = \boldsymbol{U}_a \cdot \nabla \alpha_1$ for $\boldsymbol{U}_a =$ constant). The free drift solution shows that we may have $\nabla \alpha_1 \sim 0.001/10\,\mathrm{km}$ and therefore $\nabla \cdot (\alpha_1 \boldsymbol{U}_a) \sim 10^{-6}\,\mathrm{s}^{-1}$ for $|\boldsymbol{U}_a| \sim 10\,\mathrm{m/s}$.

Example: floe clustering (Leppäranta and Hibler, 1984)   Free drift velocity depends on ice thickness, with thicker floes drifting slower and more to the right of the wind. Leppäranta and Hibler (1984) used this mechanism to show that faster floes catch up slower ones and form clusters, one possible mechanism for the banding together of ice floes in the marginal ice zone (MIZ). A similar reasoning was used by Zubov (1945) for the drift of scattered ice, illustrating how dynamics depends on ice compactness. As floes have individual velocities, this may lead to collisions and consequently grouping of floes. In case of low ice concentration, ice strips converge with small floes (moving faster) on the wind side. When wind ceases, these strips diverge due to inertia.   ∎

The conservation law of kinetic energy (Eq. 5.15) is easily understood in free drift: energy is transferred to the ocean after a short time lag, and energy storage within the ice is very low. The rate of this energy transfer is mainly determined by drag coefficients.

In this chapter the free drift solution was presented for sea ice drift: the steady-state case resulted in an algebraic equation, while the non-steady-state case was solved using a linearized form of the momentum equation. The free drift solution is fairly realistic when ice compactness is less than 0.8. But this is a severe limitation, since ice compactness changes easily and levels about 1 are usual. In Chapter 7 the internal friction of ice is added into the free drift model, and its influence on sea ice drift characteristics is examined by analytical modelling.

7

Drift in the presence of internal friction

7.1 THE ROLE OF INTERNAL FRICTION

In this chapter the drift of ice in the presence of internal friction is examined using analytical methods. The models are simple, consisting of one-dimensional channel flow models and 1.5-dimensional zonal flow models, which basically only allow steady-state solutions with analytical methods. Section 7.4 introduces the mathematical modelling of tests in ice tanks where a properly scaled drift ice-type medium is used.

In compact ice fields there is large internal resistance to motion, and it is even possible for stationary ice conditions to prevail under considerable forcing. A fundamental problem exists in that the system of sea ice dynamics equations cannot be properly closed under conditions of free drift because it is ice stress that controls the mechanical accumulation of ice material. The ice cannot go ashore or pile up to arbitrary thicknesses, as would result in free drift in a basin with land boundaries or anywhere that has variable forcing fields.

The situation can be compared with that of ocean dynamics. Ekman's theory of wind-driven currents provides a good estimate for local surface-layer currents and transport, but for a whole basin it does not give a realistic circulation. To achieve this goal, the hydrostatic pressure gradient is needed to control the piling up of water and to keep all the water in the basin. Thus, hydrostatic pressure has a similar role in ocean dynamics to that of ice stress in ice dynamics. But since sea ice as a material is essentially different from liquid seawater and since the mechanical energy used for deformation is irrecoverable in sea ice dynamics, the ice and water circulation systems are qualitatively different.

The inclusion of internal friction necessitates a formulation of drift ice rheology (see Chapter 4). This tells us how ice resists different kinds of deformation, and thus in general how spatial interactions of ice floes step into the picture. The presence of internal friction in drift ice has the following main consequences:

Malcolm Mellor (1933–1991), a classic figure in snow and ice science, in particular in sea ice mechanics. He was born in England, studied at Melbourne University, Australia, and from 1961 he worked in the U.S. Army Cold Regions Research and Engineering Laboratory, in Hanover, New Hampshire.

Reproduced with permission from the U.S. Army Cold Regions Research and Engineering Laboratory, Hanover, New Hampshire.

(i) The overall level of drift speed is lowered since part of the input of mechanical energy goes into ice deformation.
(ii) Ice is motionless if external forcing does not reach the yield limit.
(iii) Spatial variations of forcing show up smoother in the ice velocity, because ice is more rigid than the atmosphere and ocean.
(iv) Narrow deformation zones occur, in particular at land (or fast ice) boundaries.
(v) The internal stress field in the ice is capable of transferring mechanical energy over long distances in the ice cover.

7.1.1 Examples

First, let us give a few examples to illustrate the influence of internal friction on the dynamics of sea ice.

Example: drift speed vs. ice compactness Figure 7.1 shows some observations of ice drift speed in different ice compactness conditions (Shirokov, 1977). When compactness reaches about 0.8 and increases further, the mean wind factor decreases and the variance of the wind factor increases sharply (i.e., the wind factor becomes sensitive

Sec. 7.1]   The role of internal friction   167

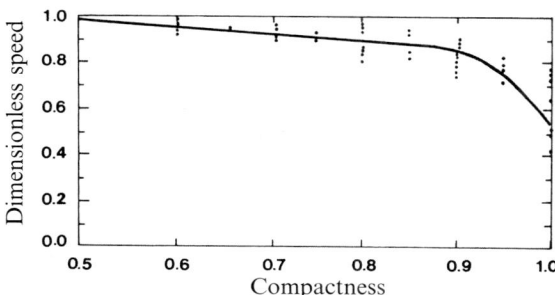

Figure 7.1. Empirical data of ice drift speed scaled with the theoretical free drift speed as a function of ice compactness.
Redrawn from Shirokov (1977).

to small compactness changes and becomes less predictable by free drift theory). The reason can only be in the internal friction of the ice. ∎

Example: sea level elevation and ice fields in a closed bay In a bay closed at one end, at dynamic equilibrium the wind stress τ_0 on the surface and sea surface slope β are balanced; that is:

$$\tau_0 - \rho_w g H \beta = 0$$

where H is the depth of the bay. In open sea $\tau_0 = \tau_a$, while in an ice-covered sea $\tau_0 = -\tau_w = \tau_a + \rho g h \beta$ in free drift conditions. Therefore, sea surface slopes in open water and ice conditions are, respectively, $\tau_0/(\rho_w g H)$ and $\tau_0/(\rho_w g H + \rho g h)$. But, since the ice is formed from water in the bay, $\rho_w H + \rho h =$ constant for all realizable (H, h) pairs and, consequently, sea surface slope is the same. Here it is assumed that the drag coefficients over ice and open water are equal, which is close to the truth in the Baltic Sea for rough (high-wind) sea surfaces. Sea level records from the Bay of Bothnia, Baltic Sea show (Figure 7.2), however, that sea surface slope is smaller under ice conditions than under ice-free conditions with the same wind forcing (Lisitzin, 1957). Therefore, wind forcing is reduced through compact ice fields due to internal friction of the ice, and the reduced surface stress is about one-third that of the open-water case. The reduction in sea surface slope was later reproduced by a numerical ice–ocean model (Zhang and Leppäranta, 1995). ∎

Example: ice–ocean differential motion In the free drift of ice the ice velocity gradient equals the ocean velocity gradient plus a linear transformation of the wind velocity gradient (see Eq. 6.3, with α and θ constants). In the Marginal Ice Zone Experiment (MIZEX-83), Greenland Sea ice kinematics and ocean current data were collected at four sites in a 10-km area (Leppäranta and Hibler, 1987). Differential motion was at a much smaller level in the ice than in the oceanic surface layer and variance spectra were similar in form but an order of magnitude less in the ice (Figure 7.3). The obvious explanation for this is the internal friction of the ice. ∎

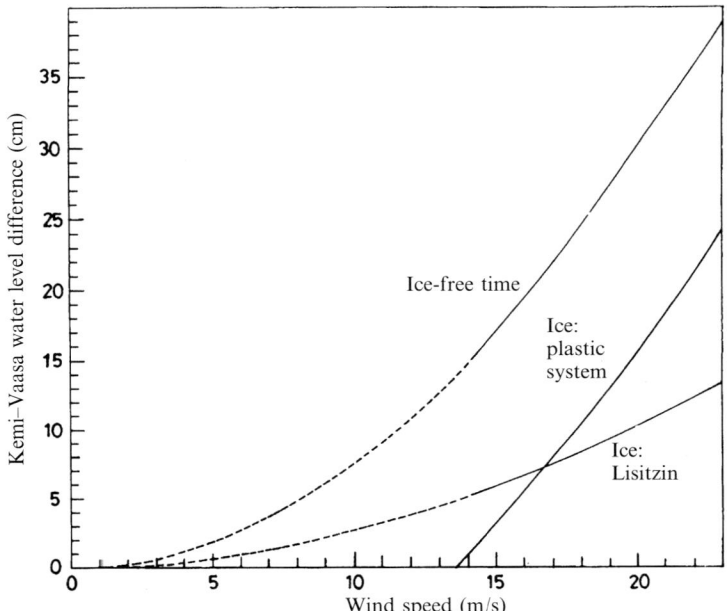

Figure 7.2. Sea surface slope vs. wind speed in the Bay of Bothnia, Baltic Sea in ice and ice-free cases. Also shown is the fit of a one-dimensional plastic ice model on the data.
From Zhang and Leppäranta (1995), modified from Lisitzin (1957).

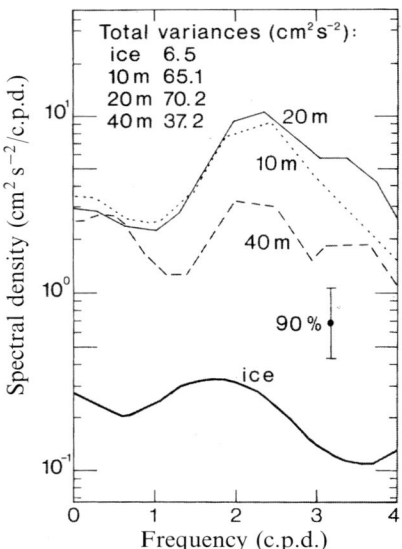

Figure 7.3. Spectra of the strain rate of ice motion and ocean currents beneath the ice in MIZEX-83 in the Greenland Sea for a 10-km area.
From Leppäranta and Hibler (1987).

Sec. 7.1] The role of internal friction 169

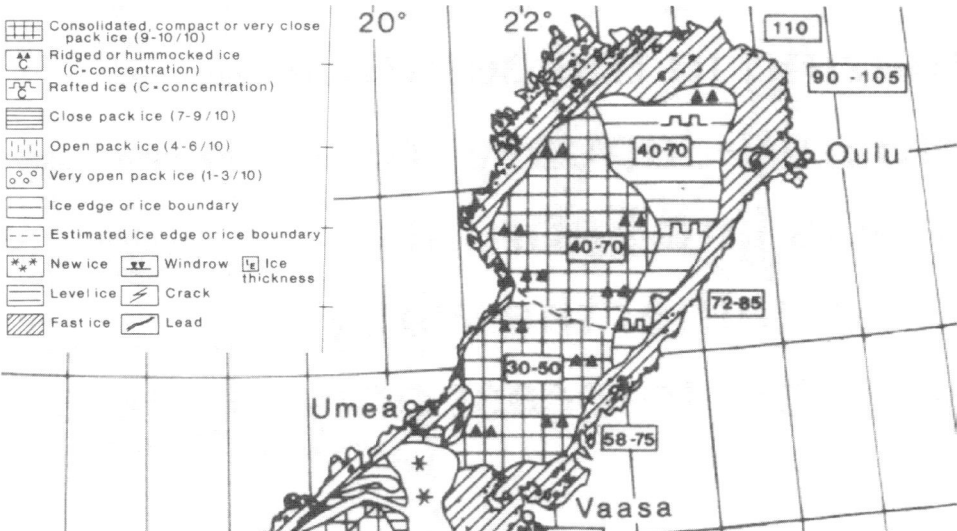

Figure 7.4. Ice conditions in the Bay of Bothnia, 4 March 1985. The ice situation was stationary for February–March 1985, but there was thickness growth of 5–10 cm. The size of the basin is about 300 km × 120 km.
Redrawn from operational ice chart by the Finnish Institute of Marine Research.

Example: non-linear internal friction of ice Several experiments have been performed in the Bay of Bothnia, the northernmost basin of the Baltic Sea. The Bay is ca 300 km × 120 km large in area. In February 1992 (Leppäranta and Zhang, 1992a), the ice was thin (10–20 cm) and strong south-westerly winds (10–20 m/s) packed the ice cover into an area 20% of the original in 6 days (Figure 5.10). Whereas, in February 1985 (Leppäranta, 1987), the ice thickness was 50 cm and similar winds did not cause any motion during 14 days, at least within the positioning accuracy of ∼ 100 m (Figure 7.4). A further case from this same basin was used to examine the free drift problem when drift ice was not in contact with fast ice boundaries (Figure 6.5). ∎

The presence of significant internal friction has been known since the advent of sea ice drift research. Nansen (1902) observed differences in the wind factor and explained them as the influence of ice compactness on the freedom of ice to move. Sverdrup (1928) collected ice drift observations on the Siberian Shelf between 1922 and 1924. He wrote that, "resistance arising from ice meeting ice in a different state of motion" was the most important factor in ice drift. He also showed that there is a strong seasonal cycle in the wind factor and deviation angle due to internal friction: between March and May they were 0.014 and 17°, between July and September they were 0.023 and 42°, respectively.

Analysing his data Sverdrup (1928) assumed the existence of a simple kinetic friction law and replaced $(\rho h)^{-1} \nabla \cdot \sigma$ by $-\kappa u$. This was not based on a physically

170 Drift in the presence of internal friction [Ch. 7

acceptable rheology but rather an empirical model for the study area. His estimate for the friction coefficient was $\kappa \approx 5 \times 10^{-4}\,\mathrm{s}^{-1}$ between March and May and $\kappa \approx 1.5 \cdot 10^{-4}\,\mathrm{s}^{-1}$ between July and September. The magnitude can be compared with the linear ice–water stress law $(\rho h)^{-1}\tau_w = -\kappa'\boldsymbol{u}$, where $\kappa' = (\rho h)^{-1}C_{w1} \approx 5\times 10^{-4}\,\mathrm{s}^{-1}$ (i.e., in Arctic seas in winter the water drag and internal friction are comparable, but in summer the latter is smaller). The truth of this is confirmed by present knowledge. In addition, Sverdrup (1928) qualitatively connected high internal friction to ridging events. Zubov (1945) also showed that the different nature of ice drift between seasons is due to the freezing of leads and the formation and evolution of ridges.

In general, the free drift assumption is good as long as the compactness of ice is less than about 0.8. However, compactness may change considerably in a short time (e.g., for a 1-cm/s difference in the normal speed between two sites 10 km apart, the total contraction would be 10% in 1 day). In an ice field with free boundaries, internal friction resists internal deformation. But when a drift ice field is in contact with a solid boundary, the influence of internal friction can be dramatic.

The formation of a proper rheological equation is needed to introduce a non-linear internal friction term with spatial derivatives into the momentum equation. The ice drift problem then becomes difficult to examine analytically, but some simple cases can be solved. In this chapter analytical models are used to examine the drift of sea ice in the presence of internal friction. A two-level (mean thickness and ice compactness, being the state variables) purely dynamic system (no thermodynamics) is employed:

$$\rho h \left[\frac{\partial \boldsymbol{u}}{\partial t} + \boldsymbol{u} \cdot \nabla \boldsymbol{u} + f\mathbf{k} \times \boldsymbol{u}\right] = \nabla \cdot \boldsymbol{\sigma} + \boldsymbol{\tau}_a + \boldsymbol{\tau}_w - \rho h g \boldsymbol{\beta} \quad (7.1\mathrm{a})$$

$$\frac{\partial h}{\partial t} + \boldsymbol{u} \cdot \nabla h = -h \nabla \cdot \boldsymbol{u}, \quad \frac{\partial A}{\partial t} + \boldsymbol{u} \cdot \nabla A = -A \nabla \cdot \boldsymbol{u} \quad (0 \leq A \leq 1) \quad (7.1\mathrm{b})$$

7.1.2 Frequency spectrum

The free drift velocity spectra were analysed in Section 6.3 using a coupled ice–ocean model. For a general forcing $F/(\rho h)$, of spectrum $p_F = p_F(\omega)$, the resulting spectrum of sea ice velocity was obtained as in (Eq. 6.37):

$$p_i(\omega) = \frac{(\lambda_i^2 + f^2 + \omega^2)}{[(\lambda_i + \lambda_w)^2 + (\omega+f)^2](\omega+f)^2} p_F(\omega)$$

In the free drift model, forcing included wind stress and sea surface slope. Let us try to examine spectral characteristics arising from the internal friction F_I by formally replacing F by $F + F_I$.

In the one-dimensional case, a general viscous model gives $\sigma \propto |\partial u/\partial x|^m$. Since $F_I = \partial \sigma/\partial x$, for a spectral velocity component $u = \hat{u}(x,y)\exp(-i\omega t)$, we get:

$$F_I \propto \left|\frac{\partial \hat{u}}{\partial x}\right|^{m-1} \frac{\partial^2 \hat{u}}{\partial x^2} \exp(-im\omega t) \quad (7.2)$$

Acceleration gives rise to a velocity component of frequency $m\omega$. In the linear case ($m = 1$) there is no frequency change, in the sublinear ($m < 1$) case there is frequency transfer toward lower frequencies, and in the superlinear case ($m > 1$) the transfer is toward higher frequencies. The sublinear behaviour could take place for very small disturbances in compact drift ice, and the superlinear behaviour could in principle take place in a floe collision rheology.

The sublinear sea ice rheology has however another distinct asymmetry: during closing the stress may be very high but during opening it is nearly zero. By coupling the ice conservation and momentum equations a new dominant feature is introduced: opening and closing frequencies are passed on to the ice velocity spectrum. Then, whatever the value of m, ice velocity is forced into the deformation frequency. Preliminary results from numerical modelling in a semi-enclosed or closed basin indeed show that, in cyclic external forcing, frequency is passed on to deformation and then further to ice velocity.

Coupling the ice conservation and momentum equations may also produce compressional waves as a result of the connection between ice strength and thickness, similarly to sound waves in compressible fluid flow. The speed of sound waves is $c^2 = \partial p/\partial \rho$ (e.g., Li and Lam, 1964). For sea ice dynamics it would read $c^2 = \rho^{-1} \partial P/\partial h$. For $P = P^* h$, $c \sim 1$–10 m/s. Such velocities have been occasionally observed in sea ice fields for progressing signals (e.g., Doronin and Kheysin, 1975; Goldstein et al., 2001). For example, wind-forced momentum may travel faster in ice than the speed at which the wind field moves. As a consequence, ice motion may commence prior to the arrival of the wind.

7.1.3 Landfast ice

Landfast ice is a solid sheet supported by islands, shoreline, and grounded ridges (Figure 7.5). Thus, by definition, $\boldsymbol{u} \equiv 0$, and we have the static equation:

$$\nabla \cdot \boldsymbol{\sigma} + \boldsymbol{F} = 0 \tag{7.3}$$

where $\boldsymbol{F} = \boldsymbol{\tau}_a + \boldsymbol{\tau}_w - \rho h g \beta$ is the forcing; this is nearly independent of ice because sea surface slope is small.

Consider a one-dimensional channel aligned with the x-axis, closed at $x = L$, and initially filled with ice of uniform thickness h_0 from $x = 0$ to L. The landfast condition $u \equiv 0$ gives:

$$\frac{\partial \sigma}{\partial x} + F = 0 \tag{7.4}$$

where $F = \tau_{ax} + \rho_w C_w \, \text{sgn}(U_w) U_w^2 - \rho h g \beta$. If $F \neq 0$ somewhere, the rheology must include an elastic or a plastic property. The solution is:

$$\sigma(x) = -\int_0^x F(x')\,dx' \tag{7.5}$$

For the landfast ice condition to hold, stress everywhere must be beneath the yield level, $-\sigma_c < \sigma(x) < \sigma_t$, where $\sigma_c = \sigma_c(h)$ and $\sigma_t = \sigma_t(h)$ are the compressive and the tensile yield strengths, respectively.

Figure 7.5. Winter roads on landfast sea ice are used in the Finnish Archipelago in the Baltic Sea. This photograph shows the road to Hailuoto Island in the north.
Photograph by Kyösti Marjoniemi, Oulu, Finland, reproduced with his permission.

Example Let us assume constant forcing $F > 0$. Then $\sigma(x) = -Fx$ and the maximum compressive stress occurs at the end of the basin, $\sigma(L) = -FL$. The ice is immobile as long as $FL < \sigma_c$. If $\sigma_c = \underline{\sigma}_c h$, where $\underline{\sigma}_c$ is the three-dimensional compressive strength, the fast ice condition is $h/L > F/\underline{\sigma}_c$. It is known that $h/L < 10^{-5}$ in drift ice basins (see Table 2.1), and since $F < 1\,\text{Pa}$ we have $\underline{\sigma}_c < 10^{-5}\,\text{Pa}$. As the local or "engineering"-scale compressive strength is $\sim 10^6\,\text{Pa}$ (Mellor, 1986), it is seen that the mesoscale or geophysical scale compressive strength is at least one order of magnitude less. ∎

In general, the stability of a solid ice sheet or the solution of the fast ice problem depends on ice thickness, size of the basin, and the limiting forcing, which is normally a representative maximum wind stress τ_{aM}. The minimum thickness of fast ice in a given basin is obtained from:

$$h_{\min} = \min\{h; \sigma_Y(h) > \tau_{aM} L\} \tag{7.6}$$

where σ_Y is the yield strength of ice. Analysis of the stability of fast ice in the Baltic Sea was performed by Palosuo (1963), and he found that the breakage of fast ice could be presented as straight lines $U_a = bh$ in a given region surrounded by islands, with b depending on the size of the basin, $b = b(L)$. The whole data set was then combined into a relationship $h/U_a = b^{-1}$ for the fast ice condition (Figure 7.6). Taking $\sigma_Y(h) \propto h^\gamma$, Eq. (7.4) tells us that $h/U_a \propto h^{1-\gamma/2} L^{1/2}$. For $\gamma = 2$, the condition would be $h/U_a \propto L^{1/2}$; however, the range in the thickness data is too

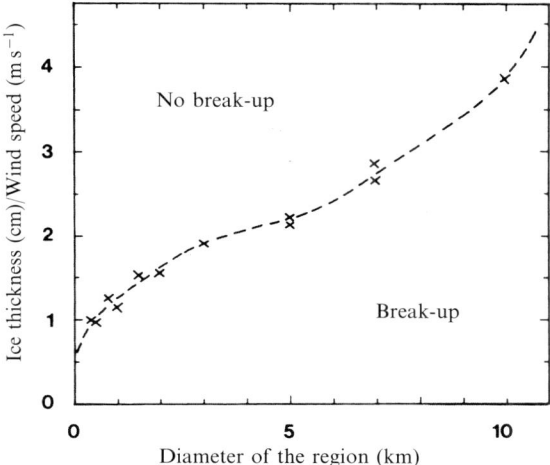

Figure 7.6. The stability of fast ice in the Baltic Sea.
From Leppäranta (1981b), after Palosuo (1963).

narrow for a good estimate of the exponent. On the other hand, it is known that a compact ice cover is normally stationary in the Baltic Sea with $h \sim 50\,\text{cm}$ and $L \sim 100\,\text{km}$ and, consequently, in Figure 7.6 the ordinate should be about 3 at $L = 100\,\text{km}$ (i.e., close to the value at $L = 10\,\text{km}$). Consequently, the graph cannot be extrapolated, and with increasing basin size the ice becomes unstable more easily.

Grounded deep ridges provide support points for the fast ice sheet. Keel depths follow exponential distribution (see Section 2.4), and the spatial density of ridges deeper than, say, H is $\mu_H = \mu\exp[-\lambda(H - h_c)]$, where μ is ridge density, h_c is cut-off depth, and the inverse of λ is the average keel depth above the cut-off size. The relationship between the density of ridges and the density of grounded ridges, μ_G, is not very clear, but one may argue that $\mu_G \sim \mu_H$ (e.g., if $\mu \sim 5\,\text{km}^{-1}$, $\lambda^{-1} \sim 5\,\text{m}$, and $H - h_c \sim 20\,\text{m}$, then $\mu_H \sim 0.09\,\text{km}^{-1}$ or, in other words, the spacing of large, 20 m deep ridges would be 11 km). Whether this spacing is enough to hold the fast ice together depends on ice thickness.

7.2 CHANNEL FLOW

In this instance the channel is aligned on the x-axis and closed at $x = L$ (Figure 7.7). The boundary conditions are $u = 0$ at $x = L$ and $\sigma = 0$ at the free ice edge(s). The channel flow is examined using a quasi-steady-state momentum equation and the two-level ice state $J = \{h, A\}$. The system of equations is written as:

$$\frac{\partial \sigma}{\partial x} + \tau_a + \rho_w C_w |U_w - u|(U_w - u) - \rho g h \beta = 0 \tag{7.7a}$$

$$\frac{\partial \{A, h\}}{\partial t} + \frac{\partial u\{A, h\}}{\partial x} = 0 \tag{7.7b}$$

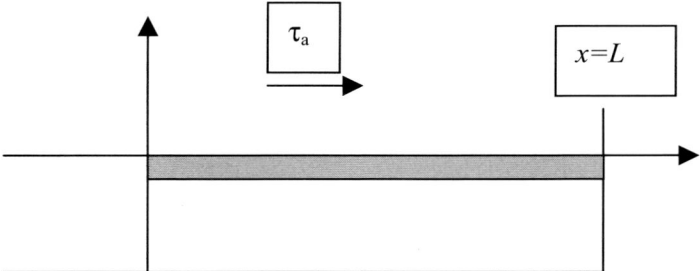

Figure 7.7. The geometry of the channel flow problem.

If the landfast ice condition is broken, ice drift commences. The free drift solution is directly obtained as $u_F = U_w + \sqrt{(\tau_a - \rho g h \beta)/(\rho_w C_w)}$.

Wind stress is assumed to be positive. It is clear that ice velocity must also be positive in the quasi-steady-state model. The steady-state solution of Eq. (7.7b) shows that uA and uh are constant across the ice field, and thus if the ice field is in contact with a rigid boundary the velocity solution is $u \equiv 0$. For the remainder of this section, we assume that the ice field is in contact with a rigid boundary and water velocity and surface slope are zero. If the ice is moving, it will drag surface water with it, but then there is a return flow at a deeper layer because of the continuity; a zero reference current for ice–water stress is therefore a reasonable assumption.

7.2.1 Creep

Whether a slow-creep regime exists is an open question. It is however used in viscous–plastic sea ice models (e.g., Hibler, 1979). Consider a slow-creep case, where the ice velocity is small enough for ice–water stress to be negligible and, for simplicity, surface slope is also ignored. Then we have:

$$\frac{d}{dx}\left(\zeta \frac{du}{dx}\right) + \tau_a = 0 \qquad (7.8)$$

where ζ is the bulk viscosity coefficient. For $\zeta = $ constant and a free edge at $x = 0$ the solution is $u = \frac{1}{2}\tau_a(L^2 - x^2)/\zeta$. For $\tau_a \sim 0.1$ Pa, $\zeta \sim 10^{12}$ kg/s, and $L = 100$ km, the ice velocity distribution would be parabolic and at its maximum at the ice edge, $u = \frac{1}{2}\tau_a L^2/\zeta \sim 0.5$ mm/s ≈ 43 m/d. For the ice velocity to remain small, viscosity should not be less than 10^{12} kg/s (as used here) if the length scale of the ice cover is of the order of 100 km. Creep with ice–water stress included will be examined in later sections.

In this creep process, velocity divergence is $-\tau_a x/\zeta$, which means that the ice accumulation rate due to convergence increases linearly toward the end of the channel. The rate is of the order of $10^{-3}h$ per day at the channel end. The steady-state solution is that all the ice becomes squeezed into the end of the channel. If viscosity were to increase with thickness as $\zeta \propto h$ or thereabouts, the change would be very slow and integration could be made stepwise in time.

The solution can be easily expanded to allow for a general non-linear viscous flow. Let us assume the power law rheology of (Eq. 4.10), $\sigma = \zeta_n |du/dx|^n (du/dx)$, which has been used widely for shear viscous glacier flow, normally with $n = -2/3$ (Paterson, 1995). The creep solution is:

$$u = \left(\frac{\tau_a}{\zeta_n}\right)^{1/(n+1)} \frac{L^b - x^b}{b} \tag{7.9}$$

where $b = (n+2)/(n+1)$. The parabolic distribution in the linear case ($n = 0$) is reproduced, and as n increases the deformation zone becomes increasingly concentrated in the channel end.

7.2.2 Plastic flow

In plastic flow, rheology is given by the yield criterion $\sigma_Y = \sigma_Y(A, h)$. In the present case our concern is compressive strength. Remember that ice continues to flow as long as internal stress overcomes the yield level. The solution of the quasi-steady-state momentum equation is formally:

$$u^2 = \frac{\tau_a}{\rho_w C_w} + \frac{1}{\rho_w C_w} \frac{d\sigma}{dx} \tag{7.10}$$

where the right-hand side must be positive, $\tau_a > -d\sigma/dx$. At the end of the channel the ice must stop and there $d\sigma/dx = -\tau_a$.

Let us start with an ice field that ranges from $x = 0$ to $x = L$, with $h = h_0 = $ constant and $A = 1$. Let us further assume $\sigma_c = P^* q(A) r(h)$, where q is the influence of ice compactness and r the influence of ice thickness on the strength, $q(0) = 0$ and $q(1) = 1$. Wind stress is taken as a constant $\tau_a > 0$. The solution to the fast ice problem shows that motion starts when $\tau_a L > \sigma_c$.

Plastic rheology allows a steady-state solution $u \equiv 0$ with a finite ice field. It is known that q is highly sensitive to compactness and that ice stress goes down one order of magnitude when compactness drops from about 1 to 0.8–0.9. This means that a very sharp ice edge, less than 1 km wide, forms at on-ice forcing (see Leppäranta and Hibler, 1985). Let us call this the ice edge zone. Beyond the edge zone, $A \equiv 1$ and thickness increases by $dr(h)/dx = \tau_a/P^*$. Ignoring ice in the edge zone, the conservation of ice volume tells us that $h_0 L = \frac{1}{2}(h_L + h_0)(L - x_0)$, where h_L is ice thickness at the channel end and x_0 is location of the ice edge.

In the steady state, wind stress must exactly balance ice strength at the channel end: $\tau_a(L - x_0) = P^* r(h_L)$. For the linear form $r(h) = h$, the location of the ice edge and ice thickness in the channel end are:

$$\left. \begin{array}{l} h_L = h_0 \left(\sqrt{\frac{1}{4} + 2X_0} - \frac{1}{2} \right) \\[6pt] x_0 = L \left(1 - X_0 \dfrac{h_L}{h_0} \right) \\[6pt] X_0 = \dfrac{\tau_a L}{P^* h_0} \end{array} \right\} \tag{7.11}$$

For $X_0 \to 1$, $h_L \to h_0$ and $x_0 \to 0$ or, in words, the fast ice condition is obtained consistently. Thus the excess of X_0 above 1 states how much deformation will take place during the adjustment process. For $\tau_a \sim 0.1\,\text{Pa}$ and $P^* \sim 10^4\,\text{Pa}$, $dh/dx \sim 10^{-5}$ or, in words, ice thickness increases by 1 m over 100 km.

Adjustment to the steady state proceeds along the following lines. The motion of ice starts to change at the boundaries. At the free boundary the ice edge zone develops by outward opening (i.e., the compactness of ice decreases to satisfy the no-stress condition). Ice build-up begins from the channel end, where the thickness slope becomes $\partial h/\partial x = \tau_a/P^*$. Then the edge zone remains static and the deformation zone spreads out from the boundary, with compactness equal to 1 and thickness gradient approaching τ_a/P^* throughout the ice field. This interpretation is consistent with the results of the numerical solution of Leppäranta and Hibler (1985), using a Lagrangian model to avoid numerical diffusion problems at the ice edge.

This steady-state solution can be generalized to spatially variable winds and other types of yield strength formulations. The steady-state compactness and thickness profile is obtained from:

$$\sigma_c(x; A, h) = \int_{x0}^{x} \tau_a(x')\,dx' \qquad (7.12)$$

which is valid as long as stress is negative. Thus, the ice compactness and thickness profile adjust to the forcing distribution. In places where stress vanishes, new free boundaries form and set drifting zones free from the pack.

Example In general, if the compressive yield strength of compact ice is $\sigma_c = \sigma_c(h)$, the steady-state ice thickness profile $h = h(x)$ provides the functional form of σ_c. Thus, the linear ice thickness profile gives $\sigma_c \propto h$, while the square-root thickness profile gives $\sigma_c \propto h^2$. The same principle holds also for the dependence of ice strength on ice compactness or on ice compactness and thickness together. ∎

Example: wind-driven frazil ice in leads or polynyas Consider a lead of width L where frazil ice is forming and there is a constant wind stress τ_a across the lead (Figure 7.8). Frazil ice drifts and accumulates at a thickness profile $h = h(x)$. In equilibrium, the ice is stationary and covers the lead. The thickness profile can be obtained from:

$$\sigma_c(h) = \tau_a x$$

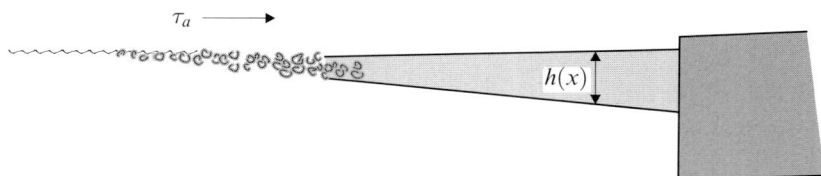

Figure 7.8. Accumulation of wind-driven frazil ice in a polynya. Typical scale is $L = 10\,\text{km}$ for the width of the polynya and $h = 10\,\text{cm}$ for the thickness of the accumulated frazil.

In particular, if $\sigma_c = P^*h$, then the thickness profile is linear, $h = (\tau_a/P^*)x$. For $P^* \sim 10\,\text{kPa}$, $\tau_a \sim 0.1\,\text{Pa}$, and $L \sim 10\,\text{km}$, the thickness at the right side of the lead is 10 cm. ∎

Ice–ocean coupling

The steady-state solution for constant forcing is straightforward. In a channel of depth H, closed at one end, the coupled system is:

$$\frac{\partial \sigma}{\partial x} + \tau_a + \rho_w C_w |U_w - u|(U_w - u) - \rho h g \beta = 0 \qquad (7.13\text{a})$$

$$-\rho_w C_w |U_w - u|(U_w - u) - \rho_w C_b |U_w| U_w - \rho H g \beta = 0 \qquad (7.13\text{b})$$

where U_w is chosen as the vertical average mean current. Wind stress on the system is balanced by ice stress and sea surface tilt. At equilibrium, $u = U_w = 0$ and Eq. (7.13b) implies that $\beta = 0$. In the absence of ice, the equilibrium condition would be $\tau_a = \rho_w H g \beta$; thus, in the steady state internal ice stress totally removes the effects of the surface pressure gradient.

Consider an ice field forced by periodic forcing at the boundary:

$$\rho h \frac{\partial u}{\partial t} = \frac{\partial \sigma}{\partial x} - \rho_w C_w |u| u \qquad (7.14)$$

with $u = u_0(-u_0)$ for $0 \leq t/T \bmod 2\pi \leq \pi$ ($\pi \leq t/T \bmod 2\pi \leq 2\pi$). If timescale T is much longer than the inertial timescale, the quasi-steady-state approach can be applied. In a stationary cycle, ice must move left and right by the same amount. Since internal ice stress is asymmetric (i.e., high stress for compression and low stress for opening), it must vanish for the stationary cycle to develop. Then, $u = u_0$ and ice compactness must be low enough for stresses to be small.

7.3 ZONAL SEA ICE DRIFT

The longitudinal boundary-zone flow offers excellent possibilities for the analysis of sea ice drift (Figure 7.9). This zone may physically represent the coastal shear zone or the marginal ice zone (MIZ), where one may anticipate transverse changes much larger than longitudinal changes. Mathematically, the y-axis is aligned along the longitudinal direction, and the situation is assumed invariant in y: $\partial/\partial y \equiv 0$ (Leppäranta and Hibler, 1985). For the ice state, the two-level description $J = \{A, h\}$ is employed.

The general solution is first derived. The y-invariant equations of ice dynamics are directly obtained from the full form: the equation of motion from Eq. (5.10), ice rheology from Eq. (4.4), and the ice conservation law from Eqs (3.14–3.15). The resulting model is:

178 Drift in the presence of internal friction [Ch. 7

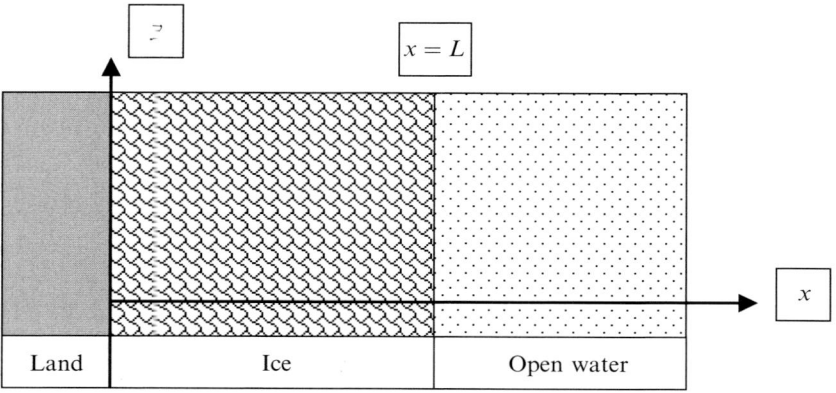

Figure 7.9. Boundary-zone configuration. The width is typically of the order of 100 km.

$$\rho h\left(\frac{\partial u}{\partial t} + u\frac{\partial u}{\partial x} - fv\right) = \frac{\partial \sigma_{xx}}{\partial x} + \tau_{ax} + \tau_{wx} - \rho h g \beta_x \quad (7.15a)$$

$$\rho h\left(\frac{\partial v}{\partial t} + u\frac{\partial v}{\partial x} + fu\right) = \frac{\partial \sigma_{xy}}{\partial x} + \tau_{ay} + \tau_{wy} - \rho h g \beta_y \quad (7.15b)$$

$$\left.\begin{array}{l} \sigma = \sigma(h, A, \dot{\varepsilon}) \\[4pt] \dot{\varepsilon}_{xx} = \dfrac{\partial u}{\partial x} \\[8pt] \dot{\varepsilon}_{xy} = \dot{\varepsilon}_{yx} = \dfrac{1}{2}\dfrac{\partial v}{\partial x} \\[8pt] \dot{\varepsilon}_{yy} = 0 \end{array}\right\} \quad (7.15c)$$

$$\frac{\partial \{A, h\}}{\partial t} + \frac{\partial u\{A, h\}}{\partial x} = 0 \quad (0 \leq A \leq 1) \quad (7.15d)$$

The atmospheric pressure gradient has been ignored in the momentum equation and the elastic part of the rheology. The boundary conditions can be taken as stationary at the land boundary and zero stress at a long distance from the land:

$$x = 0 \Rightarrow u = 0; \quad x = L \Rightarrow \sigma = 0 \quad (7.15e)$$

Equations (7.15) can be solved for free drift as shown in Chapter 6. Among these equations (taking $\sigma = 0$) is the full free drift equation, whose solution is of the form $u = u_F = u'_a + U_{wg}$, where u'_a is the wind-driven component of speed equal to a fraction α of the wind speed, and with a direction deviating by angle θ from the wind direction (see Eq. 6.3).

7.3.1 Steady-state velocity: wind-driven case

Purely wind-driven drift is considered (i.e., $U_{wg} = 0$ and $\beta = 0$). Free drift velocity can be obtained from Eqs (7.15a, b). If $u_F > 0$, then ice will drift away from the coast with free boundaries and in free drift state (see Figure 7.10) and u_F will be the true solution; the boundary condition $u = 0$ would also no longer be valid. In the case of pure wind-driven sea ice dynamics, the off-coast free drift solution results when the direction of the wind stress (or the surface wind), ϑ, satisfies:

$$-90° + \theta < \vartheta < 90° + \theta \tag{7.16}$$

(see Figure 7.10). In the northern hemisphere $\theta \sim 30°$, and therefore the wind direction must be approximately between $-60°$ and $120°$; in the southern hemisphere $\theta \sim -30°$ and the range is from $-120°$ to $60°$.

Otherwise, if condition (7.16) is not satisfied, the ice will stay in contact with the coast and internal stress will spread the friction from the coast deeper into the drift ice zone. Then, either a boundary-zone shear flow will develop or the ice will reach a stationary state (as in the channel flow case).

To solve on-shore-forced zonal flow, the ice conservation law states that uA and uh must be constant, and then the boundary condition implies $u \equiv 0$. The ice conservation law is then automatically satisfied, leaving the momentum equation and rheology for the longitudinal component of ice velocity and for the ice stress:

$$\frac{d\sigma_{xx}}{dx} + \tau_{ax} + \rho_w C_w \sin\theta_w |v|v + \rho h f v = 0 \tag{7.17a}$$

$$\frac{d\sigma_{xy}}{dx} + \tau_{ay} - \rho_w C_w \cos\theta_w |v|v = 0 \tag{7.17b}$$

$$\left.\begin{array}{l} \sigma = \sigma(h, A, \sigma) \\ \dot{\varepsilon}_{xx} = 0 \\ \dot{\varepsilon}_{xy} = \dot{\varepsilon}_{yx} = \dfrac{1}{2}\dfrac{\partial v}{\partial x} \\ \dot{\varepsilon}_{yy} = 0 \end{array}\right\} \tag{7.17c}$$

In this case, for quite general conditions:

$$\left|\frac{\sigma_{xy}}{\sigma_{xx}}\right| = \gamma = \text{constant} \tag{7.18}$$

For example, in the viscous–plastic rheology used by Hibler (1979) $\gamma = e^{-1}$, when the strain rate is given by Eq. (7.17c). Then, the ice stress derivatives can be eliminated from Eqs (7.17a, b), an algebraic equation can be obtained for v, and the solution can be easily found (see Leppäranta and Hibler, 1985). But care must be taken with the signs, since the sign for shear stress σ_{xy} depends on which side the solid boundary is and the sign for ice–water stress depends on the direction of motion. Stress σ_{xx} is negative (compressive) in all cases.

Let us first assume that the ice drifts north, $v \geq 0$ and then also $\sigma_{xy} > 0$. Equation (7.17a) can be multiplied by γ and then added to Eq. (7.17b). The result is:

$$\tau_{ay} - \rho_w C_w \cos\theta_w v^2 + \gamma(\tau_{ax} + \rho_w C_w \sin\theta_w v^2 + \rho h f v) = 0$$

This is a quadratic equation, whose solution is:

$$v = \sqrt{\frac{\tau_{ay} + \gamma\tau_{ax}}{C_N} + \left(\frac{\gamma\rho hf}{2C_N}\right)^2} - \frac{\gamma\rho hf}{2C_N} \qquad (7.19)$$

where $C_N = \rho_w C_w(\cos\theta_w - \gamma\sin\theta_w)$. In northward on-shore forcing $\tau_{ay} > 0$ and $\tau_{ax} < 0$, and thus coastal friction cuts the portion $\gamma\tau_{ax}$ out of the momentum input. Setting $v \equiv 0$ in Eq. (7.19) we get the condition $\tau_{ay} + \gamma\tau_{ax} = 0$. For a positive velocity solution, the reduced stress must be positive; that is:

$$\tau_{ay} + \gamma\tau_{ax} > 0 \qquad (7.20)$$

This means that the northward component τ_{ay} must be larger than $\gamma|\tau_{ax}|$ or the on-ice direction of the wind stress must be less than $\arctan(\gamma)$. Otherwise, a stationary ice field results in the steady state (as in the channel flow case). The following conclusion can now be presented for northward wind stress forcing:

$$0 \leq \vartheta \leq 90° + \theta \qquad \text{(free drift)}$$
$$90° + \theta \leq \vartheta \leq 180° - \arctan(\gamma) \qquad \text{(northward flow } v > 0\text{, see Eq. 7.19)}$$
$$180° - \arctan(\gamma) \leq \vartheta \leq 180° \qquad \text{(stationary ice } v = 0\text{)}$$

If $\gamma = \frac{1}{2}$, then $\arctan(\gamma) = 27°$. Thus the northward flow results from a 33° wide wind direction sector in the northern hemisphere. The deviation angle between wind and ice drift is simply equal to the on-ice angle of the wind speed; therefore, it is between 30° and 63°.

Example The characteristics of shear flow become clearer when Coriolis and boundary-layer angle effects are ignored. Then $u_F > 0$ if $\tau_{ax} > 0$. The northward flow solution can be written as:

$$v = Na\sqrt{\sin\vartheta + \gamma\cos\vartheta}|U_a|$$

where wind direction is in the range $90° \leq \vartheta < 180° - \arctan(\gamma)$. Thus $v \propto |U_a|$; however, the proportionality factor is affected by the wind direction and the strength parameter γ in addition to the drag coefficients. ∎

Second, assume that the ice drifts south, $v \leq 0$ and then also $\sigma_{xy} < 0$. The solution is found as in the northward case, and the result is:

$$v = -\sqrt{-\frac{(\tau_{ay} - \gamma\tau_{ax})}{C_S} + \left(\frac{\gamma\rho hf}{2C_S}\right)^2} + \frac{\gamma\rho hf}{2C_S} \qquad (7.21)$$

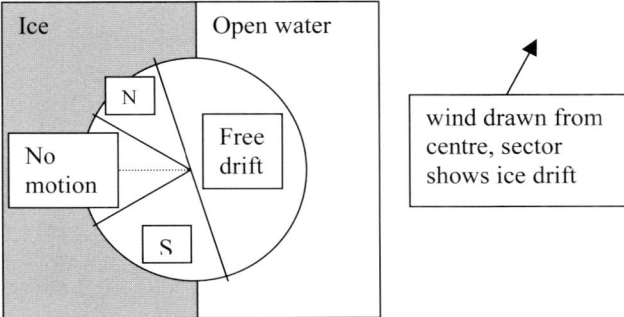

Figure 7.10. Steady-state solution of wind-driven zonal flow, northern hemisphere. The wind vector is drawn from the centre of the circle to the perimeter, and the resulting ice velocity is free drift, northward or southward boundary flow, or no motion, depending on ice strength and wind direction.
From Leppäranta and Hibler (1985).

where $C_S = \rho_w C_w (\cos\theta_w + \gamma \sin\theta_w)$. Furthermore, the following conclusion can now be presented for southward wind stress forcing:

$$180° \leq \vartheta \leq 180° + \arctan(\gamma) \quad \text{(stationary ice } v = 0\text{)}$$
$$180° + \arctan(\gamma) \leq \vartheta \leq 270° + \theta \quad \text{(southward flow } v < 0\text{, see Eq. 7.21)}$$
$$270° + \theta \leq \vartheta \leq 360° \quad \text{(free drift)}$$

The stationary sector is as in the northward case, but the flow zone is now wide in the northern hemisphere.

The general solution is illustrated in Figures 7.10 and 7.11 for the northern hemisphere case. The southern case is symmetric (i.e., the northward flow zone is wide and the southward flow zone is narrow).

It is striking that the velocity solution is independent of the exact form of rheology and even of the absolute magnitude of the stresses as long as Eq. (7.18) holds. The only rheology parameter present is γ, which describes the ratio of shear strength to compressive strength. Taking the plastic part of the viscous–plastic rheology of Hibler (1979) (Eq. 4.24), we have $\gamma = e^{-1}$ for all shear flows, with the standard level $\gamma = \frac{1}{2}$. In the floe collision model of Shen et al. (1987) (Eq. 4.27) we have:

$$\gamma = \frac{12}{\pi}\sqrt{\frac{4}{3\pi(1-\kappa)} - \frac{1}{4}}$$

with $\gamma = 0.13$ when the restitution coefficient $\kappa = 0.9$.

Note also that the zonal flow solution is valid for any wind constant in time but arbitrary in space. Thus, ice adjusts to the wind field with the wind factor reduced from the free drift factor, while direction is parallel to the coast. The time evolution of zonal flow was examined using numerical modelling by Leppäranta and Hibler (1985).

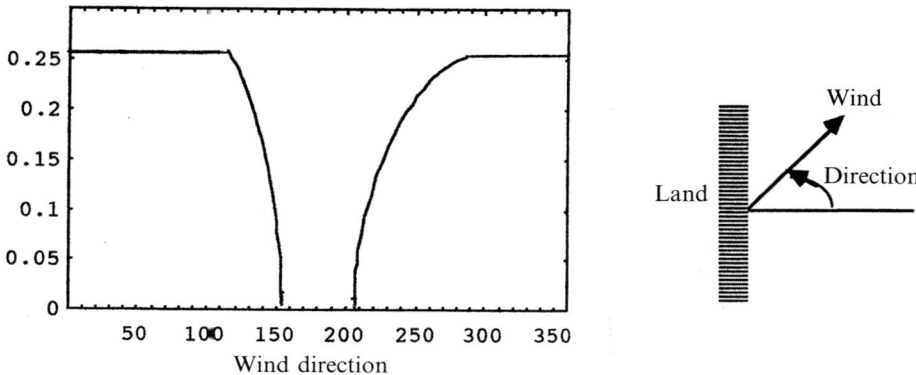

Figure 7.11. Zonal flow speed as a function of wind direction when the coastal boundary is in the east, northern hemisphere. Wind speed is 10 m/s and the strength parameter is $\gamma = \frac{1}{2}$.

7.3.2 Steady state with ocean currents

Steady-state ocean currents can be added to wind-driven zonal flow still allowing analytical solutions. It is natural to assume that ocean currents are longitudinal in the steady state, $U_{wg} = V_w \mathbf{j}$, and that sea surface slope is $\rho h g \beta_x = \rho h f V_w$. Then the steady-state equations read, taking the geostrophic current as the reference water velocity:

$$\frac{d\sigma_{xx}}{dx} + \tau_{ax} - \rho_w C_w \sin\theta_w |V_w - v|(V_w - v) - \rho h f(V_w - v) = 0 \quad (7.22a)$$

$$\frac{d\sigma_{xy}}{dx} + \tau_{ay} + \rho_w C_w \cos\theta_w |V_w - v|(V_w - v) = 0 \quad (7.22b)$$

$$\sigma = \sigma(h, A, \sigma), \quad \dot{\varepsilon}_{xx} = 0, \quad \dot{\varepsilon}_{xy} = \dot{\varepsilon}_{yx} = \frac{1}{2}\frac{\partial v}{\partial x}, \quad \dot{\varepsilon}_{yy} = 0 \quad (7.22c)$$

This is the model for relative velocity $v - V_w$, derived in exactly the same way as the wind-driven (zero ocean current) model for v. The longitudinal velocity is therefore:

$$v = V_w + v_0 \quad (7.23)$$

where v_0 is the wind-driven velocity component. Note that the solution (7.23) is valid for any value of $V_w = V_w(x)$. If $\tau_a = 0$, then ice follows the zonal flow in the ocean with $\sigma = $ constant. It is noteworthy that, even though sea ice drift is highly non-linear in the presence of internal friction, steady zonal flow allows superposition of wind-driven and ocean current-driven components.

7.3.3 Steady-state ice thickness and compactness profiles

During the time evolution of zonal sea ice flow, the ice state evolves to allow steady-state conditions to be satisfied. When longitudinal velocity has been solved, the ice stress components σ_{xx} and σ_{xy} can be solved from the momentum equation, $\sigma_{yx} = \sigma_{xy}$ holds for symmetry, and ice rheology provides the component σ_{yy}. The ice state can then be obtained from the stress distribution.

If a stationary ice field results in the steady state, integration of Eqs. (7.17) is straightforward and proceeds similarly to the channel flow problem. Otherwise, the steady-state velocity profile also influences the stress field. The y-component of the momentum balance gives:

$$\sigma_{xy}(x) = \sigma_{xy}(0) - \int_0^x (\tau_{ay} - \rho_w C_w \cos\theta_w |v|v) \, dx' \qquad (7.24)$$

The second boundary condition states that at the ice edge, $x = L$, stress must vanish. The shear stress at the coast is $\sigma_{xy}(0) = \int_0^L (\tau_{ay} - \rho_w C_w \cos\theta_w |v|v) \, dx'$.

With the present two-level ice state, ice strength depends on ice compactness and thickness. Starting at the ice edge, ice compactness first increases from 0 to 1 and thereafter a ridging zone follows. The geometric profiles of the edge zone and ridging zone depend on functional relationships between ice compactness and strength, and ice thickness and strength, formally:

$$|\sigma_{xy}(x)| = \sigma^* q(A) r(h) \qquad (7.25)$$

where σ^* is a strength constant, $0 \leq q \leq 1$, $q(0) = 0$, $q(1) = 1$, and $r(h)$ tells us to what degree the strength of compact ice depends on ice floe thickness. The inner boundary of the edge zone can be ascertained from the location where the shear stress in Eq. (7.24) is in absolute value equal to $\sigma^* r(h_0)$.

Example In the plastic rheology of Hibler (1979), shear stress is shown in the steady state as:

$$\sigma_{xy} = \text{sgn}\left(\frac{\partial v}{\partial x}\right) \frac{P^* h}{e} \exp[-C(1-A)]$$

Assume a constant wind stress toward the sector where the steady state solution is the northward zonal flow. Ignoring the Coriolis and oceanic drag turning angle, we have:

$$\tau_{ay} - \rho_w C_w v^2 = -\gamma \tau_{ax}$$

and, consequently, from Eq. (7.24) we have $\sigma_{xy}(x) = \sigma_{xy}(0) + \gamma \tau_{ax} x$. The shear stress at the fast ice boundary is $-\gamma \tau_{ax} L$. Let us assume that initially the ice cover consists of homogeneous floes of thickness h. The ridging condition is $-\tau_{ax} L > P^* h$. If it is satisfied, then ice builds up at the coast and the ice thickness at the boundary will be $h(0) = -\tau_{ax} L / P^*$. Ridging takes place in the zone $-\tau_{ax}(L-x) > P^* h$, while beyond that the ice field opens up. Then, integration of the ice state out from the coast proceeds exactly as in the one-dimensional case. Consequently, the channel model and shear flow give similar ice thickness and compactness profiles, but in the shear zone there may be steady-state, non-zero longitudinal velocity. Adding Coriolis and boundary-layer turning angle effects would modify the profiles, but qualitatively the result would be similar. The full solution with time evolution is shown in Figure 7.12.

184   Drift in the presence of internal friction   [Ch. 7

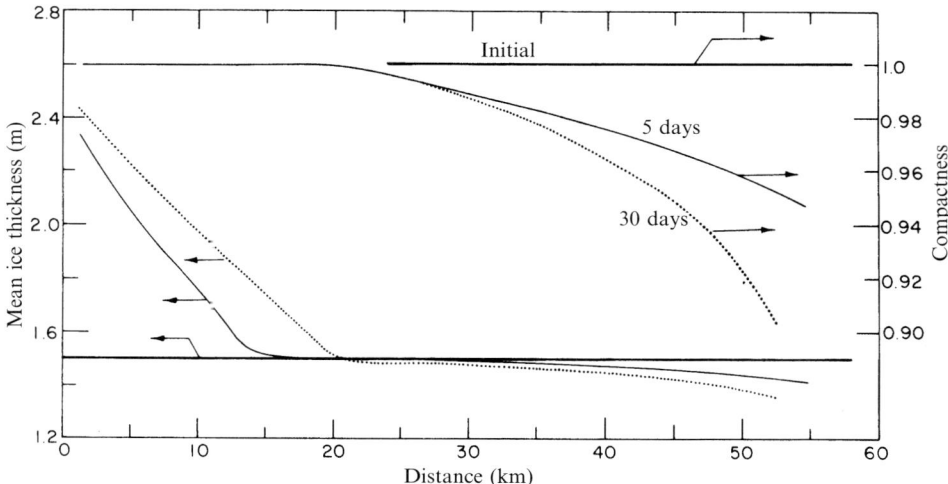

Figure 7.12. The profile of ice thickness and ice compactness from the fast ice boundary to the ice edge. Numerical solution of time evolution, the solution at 30 days is close to the steady state.
From Leppäranta and Hibler (1985).

The ice edge zone draws a very sharp ice compactness profile because of the high sensitivity of ice strength to ice compactness. ∎

7.3.4 Viscous models

The general viscous model was given by Eq. (4.9): $\sigma = \sigma(\dot{\varepsilon}, p; \zeta, \eta) = (-p + \zeta \operatorname{tr} \dot{\varepsilon})\mathbf{I} + 2\eta \dot{\varepsilon}'$. In steady-state zonal flow, $\operatorname{tr} \dot{\varepsilon} = 0$ and:

$$\left. \begin{aligned} \frac{\partial \sigma_{xx}}{\partial x} &= -\frac{\partial p}{\partial x} \\ \frac{\partial \sigma_{xy}}{\partial x} &= \frac{\partial}{\partial x}\left(\eta \frac{\partial v}{\partial x}\right) \end{aligned} \right\} \qquad (7.26)$$

Here pressure depends on ice-state variables and viscosity additionally depends on strain-rate invariants.

Consider the steady-state zonal flow problem (Eq. 7.17). Ignoring the Coriolis and the oceanic turning angle:

$$\frac{dp}{dx} + \tau_{ax} = 0$$

$$\frac{d}{dx}\left(\eta \frac{dv}{dx}\right) + \tau_{ay} - \rho_w C_w |v| v = 0 \qquad (7.27)$$

A pressure gradient is therefore needed to balance on-ice forcing. If it does not exist, the true steady state is never achieved, but the ice continues to creep toward the shore and the wind stress is balanced by the bulk viscous stress. The longitudinal equation allows a general solution if ice–water stress can be ignored similarly to the channel flow case (Eq. 7.9).

Example: the Laikhtman model A historical landmark in sea ice dynamics is the work of Laikhtman (1958), which presented the first frictional sea ice model. He used a linear viscous model for the zonal flow at a land boundary. He took constant viscosity, ignored the Coriolis and turning angle, and assumed the wind and water stresses to be linear. The momentum equation in the *y*-direction is then:

$$\eta \, d^2v/dx^2 - \rho_w C_{w1} v + \rho_a C_{a1} V_a = 0$$

With the boundary conditions $v = 0$ at $x = 0$ and $v \to v_\infty = [\rho_a C_{a1}/(\rho_w C_{w1})] V_a$ (free drift) as $x \to \infty$, the solution is:

$$\left. \begin{array}{c} v = v_\infty \left[1 - \exp\left(-\dfrac{x}{L}\right) \right] \\ L = \sqrt{\dfrac{\eta}{\rho_w C_{w1}}} \end{array} \right\}$$

where L is a length scale. For $C_{w1} \sim C_w v_\infty$, $v_\infty \sim 0.5 \, \text{m/s}$ and $\eta \sim 10^{10} \, \text{kg/s}$, the width is $L \sim 140 \, \text{km}$. ∎

Under constant forcing the linear viscous case thus gives a gently sloping coastal boundary zone, while in plastic flow the steady-state boundary is a sharp slip line, although, during the adjustment process, plastic flow has a spatial structure. When viscous rheology becomes increasingly sublinear, the plastic solution is approached.

7.3.5 Marginal ice zone

The marginal ice zone (MIZ) is located along the open ocean edge of sea ice cover, its width usually referred to as 100 km (Figure 7.13). In the MIZ, ice dynamics has a number of qualitative features:

- Off-ice forcing results in ice edge dispersion.
- Formation of ice bands, parallel to the MIZ.
- Oceanic eddies form and ice tracks them.
- On-ice forcing gives a compact ice edge.
- Little ridging takes place.
- Intensive air–ice–sea interaction takes place.

On-ice and off-ice forcing cases are therefore highly dissimilar. Ice compactness and thickness increase inward from the ice edge. Because of the looseness of its ice, the outermost MIZ can move for short periods at high velocity compared with the bulk-zone flow. Sometimes these high-velocity cases have been termed "ice edge jets".

The zonal flow problem also applies to MIZ ice dynamics. If the free drift solution is off-ice, the MIZ will diffuse into a free drifting, open ice field, where

Figure 7.13. An ice edge in the Greenland Sea.

ice velocity consists of the wind effect superposed on the current field. For on-ice forcing, the steady-state solution is a stationary ice field or a longitudinal flow, which runs parallel to the ice edge and moves at a reduced free drift velocity. In such conditions, disturbances in the ice floes at the ice edge could give rise to transient ice edge jets.

Integrating the shear stress (Eq. 7.24) from the ice edge, we have:

$$\sigma_{xy}(x) = \int_x^L (\tau_{ay} - \rho_w C_w \cos\theta_w |v|v) \, dx' \tag{7.28}$$

Then $A = 1$ at $x = L_1$; the inner boundary of the MIZ can be taken as $\sim L_1$ since little ridging occurs there. Let us take the plastic rheology of Hibler (1979). Further, let us assume that the ice cover initially consists of homogeneous floes of thickness h_i. Starting at the ice edge, ice compactness is first obtained from the equation:

$$\text{sgn}\left(\frac{\partial v}{\partial x}\right) \frac{P^* h}{e} \exp[-C(1-A)] = \int_x^L (\tau_{ay} - \rho_w C_w \cos\theta_w |v|v) \, dx' \tag{7.29a}$$

This is integrated to $A = 1$ where $x = L_1$. Thereafter $A \equiv 1$ and h is obtained from:

$$h = h_i + \text{sgn}\left(\frac{\partial v}{\partial x}\right) \frac{e}{P^*} \int_{L_1}^x (\tau_{ay} - \rho_w C_w \cos\theta_w |v|v) \, dx' \tag{7.29b}$$

Taking a northward flow and ignoring the Coriolis and turning angle, we have $L_1 = L + P^* h_i / \tau_{ax}$ (note: $\tau_{ax} < 0$). A sharp ice edge results because of the exponential factor in the strength–compactness relation. If $P^* = 10\,\text{kPa}$, $e = 2$, $h_i = 1\,\text{m}$, $C = 20$, and $\tau_{ay} - \rho_w C_w \cos\theta_w |v| v = 0.1\,\text{Pa}$, compactness is equal to 0.8 at a distance of 0.7 km from the ice edge. A sharp ice edge under steady on-ice wind is well known from observations (i.e., here the compactness increased from 0 to 0.8 over the first 0.7 km). But to go from 0.8 to 1.0 an additional distance of 49.3 km is needed. A linearly increasing ice thickness profile results at point L_1 and extends inward to the ice pack, the thickness slope being equal to $-\tau_{ax}/P^*$.

Since an ocean current field parallel to the ice edge can be superposed on the wind-driven MIZ solution, any jet in the ocean beneath the ice at the ice edge is added to the ice velocity. Ice drift strongly interacts with atmospheric and oceanic boundary layers in the MIZ. In particular, with compact ice edges there is a very rapid change in surface characteristics. The surface roughness of sea ice is different from the surface roughness of the open ocean, the drag coefficient usually being larger over sea ice. Surface temperature also changes across the ice edge. As far as ocean forcing is concerned, there is a seasonal difference in the forcing of the ocean across the ice edge, giving rise either to divergence or convergence. In winter, freezing of the sea and ice growth cause brine formation and reduce the stability of stratification in the water, while in summer ice melt yields the opposite effect. As a consequence, an ice edge front may form where dynamic instabilities form, resulting in ice edge eddies being formed.

Example: upwelling at the ice edge   Assume a constant north wind that is parallel to the ice edge. Ekman transport in the ocean surface layer is $\mathbf{i}\tau_{0y}/f$ (e.g., Cushman-Roisin, 1994), where $\boldsymbol{\tau}_0 = \mathbf{j}\tau_{0y}$ is surface stress. Now $\tau_{0y} = \rho_a C_{aw} V_a^2$ over open ocean and $\tau_{0y} \approx \rho_a C_{ai} V_a^2 + d\sigma_{xy}/dx$ over ice, where C_{aw} and C_{ai} are the air drag coefficients over open water and sea ice, respectively. Since usually $C_{aw} < C_{ai}$, wind stress transmitted to the ocean is larger over ice if $|d\sigma_{xy}/dx|$ is small. But when $|d\sigma_{xy}/dx|$ is large enough, wind stress will be larger over open water, resulting in divergence and consequent upwelling at the ice edge. ■

Consequently, we can compile Table 7.1 for upwelling (+) and downwelling (−) at the ice edge. It works for either hemisphere, whether the ice is to the left or right of the wind, and whether the internal friction of ice is large or small. The large ice friction case also applies to the fast ice edge.

7.3.6 Circular ice drift

A y-invariant ice flow can close into a circle (termed a "circular ice drift"), and then it is natural to use spherical co-ordinates. Let us consider a polar cap and choose spherical co-ordinates $r = r_e$, Z, λ (Figure 7.14). The steady-state momentum

188 **Drift in the presence of internal friction** [Ch. 7]

Table 7.1. Upwelling (+) and downwelling (−) at the ice edge.

MIZ location		Ice friction	
Arctic seas	Antarctica	Large	Small
Left of wind	Right of wind	+	−
Right of wind	Left of wind	−	+

equation is:

$$\frac{1}{r_e \sin Z} \frac{\partial \sigma_{\lambda\lambda}}{\partial \lambda} + \frac{1}{r_e} \frac{\partial \sigma_{Z\lambda}}{\partial Z} + \frac{2}{r_e} \sigma_{Z\lambda} \cot Z + \tau_{a\lambda} + \tau_{w\lambda} + \rho h 2\Omega \cos Z u_Z$$
$$-\rho g \beta_\lambda = 0 \quad (7.30a)$$

$$\frac{1}{r_e \sin Z} \frac{\partial \sigma_{Z\lambda}}{\partial \lambda} + \frac{1}{r_e} \frac{\partial \sigma_{ZZ}}{\partial Z} + \frac{1}{r_e}(\sigma_{ZZ} - \sigma_{\lambda\lambda}) \cot Z + \tau_{aZ} + \tau_{wZ} - \rho h 2\Omega \cos Z u_\lambda$$
$$-\rho g \beta_Z = 0 \quad (7.30b)$$

The y-co-ordinates that transform into circles correspond to zenith angle circles $Z = $ constant (or latitude circles), and for the situation to be invariant along these circles we have $\partial/\partial \lambda \equiv 0$. When the ice state is fully adjusted and the ice conservation law has also achieved a steady state, the zenith angle velocity component is 0, $u_Z \equiv 0$, as the east component becomes 0 in the Cartesian system. The ice conservation law is then automatically satisfied, and the equilibrium momentum equation becomes:

$$\frac{1}{r_e} \frac{d\sigma_{Z\lambda}}{dZ} + \frac{2}{r_e} \sigma_{Z\lambda} \cot Z + \tau_{a\lambda} + \tau_{w\lambda} = 0 \quad (7.31a)$$

$$\frac{1}{r_e} \frac{d\sigma_{ZZ}}{dZ} + \frac{1}{r_e}(\sigma_{ZZ} - \sigma_{\lambda\lambda}) \cot Z + \tau_{aZ} + \tau_{wZ} + \rho h 2\Omega \cos Z(u_\lambda - U_{w\lambda}) = 0 \quad (7.31b)$$

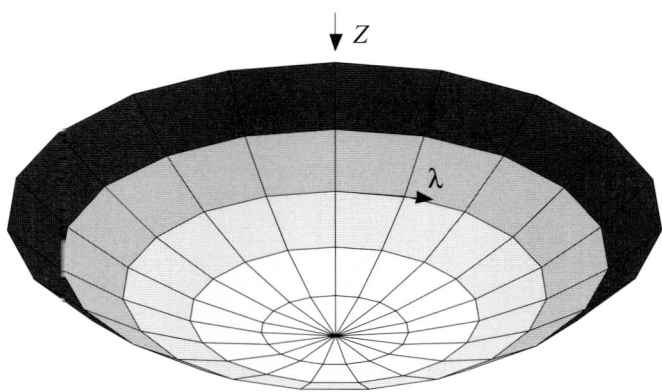

Figure 7.14. Configuration of zonal flow on a southern polar cap.

Sec. 7.3] **Zonal sea ice drift** 189

Now the strain-rate components are:

$$\dot{\varepsilon}_{ZZ} = 0, \quad \dot{\varepsilon}_{\lambda\lambda} = 0$$
$$\dot{\varepsilon}_{Z\lambda} = \frac{1}{2}\left(\frac{1}{r_e}\frac{du_\lambda}{dZ} - \frac{u_\lambda}{r_e}\cot Z\right) \tag{7.32}$$

This problem is then solved using the plastic rheology of Hibler (1979) and, as a result, $\sigma_{ZZ} = \sigma_{\lambda\lambda} = -\frac{1}{2}P$ and $\sigma_{Z\lambda} = \frac{1}{2}\text{sgn}(\dot{\varepsilon}_{Z\lambda})P/e$. Equations (7.31) then become:

$$\text{sgn}(\dot{\varepsilon}_{Z\lambda})\frac{1}{2er_e}\frac{dP}{dZ} + \frac{2}{r_e}\text{sgn}(\dot{\varepsilon}_{Z\lambda})\frac{P}{2e}\cot Z + \tau_{a\lambda} + \tau_{w\lambda} = 0 \tag{7.33a}$$

$$-\frac{1}{2r_e}\frac{dP}{dZ} + \tau_{aZ} + \tau_{wZ} - \rho h 2\Omega\cos Z(u_\lambda - U_{w\lambda}) = 0 \tag{7.33b}$$

Let us now ignore the Coriolis and oceanic turning angle. For a westerly wind $\tau_{a\lambda} > 0$, $\tau_{aZ} = 0$, we have $P = $ constant from Eq. (7.33b). Equation (7.33a) now gives:

$$\text{sgn}(\dot{\varepsilon}_{Z\lambda})\frac{P}{r_e e}\cot Z + \tau_{a\lambda} - \rho_w C_w u_\lambda^2 = 0 \tag{7.34}$$

The cotangent term geometrically restricts the motion of ice. Approaching the poles, $|\cot Z| \to \infty$ and, therefore, there must be a rigidly rotating polar cap. Since $P/(r_e e) \sim 10^{-3}$ Pa, usually we need $|\cot Z| < 100$ or the width of the rigid cap is $\sim 2r_e|\text{arccot}Z| \sim 100$ km. As the strength of ice increases with thickness, the width of the rigid cap must also increase with thickness.

In the general case, the pressure gradient is eliminated and the resulting equation is:

$$\text{sgn}(\dot{\varepsilon}_{Z\lambda})\left[\frac{P}{r_e e}\cot Z + \tau_{a\lambda} + \tau_{w\lambda}\right]H + \frac{\tau_{wZ}}{e} + \frac{\tau_{aZ}}{e} - \rho h 2\Omega\cos Z(u_\lambda - U_{w\lambda})/e = 0$$
$$\tag{7.35}$$

This equation gives $u_\lambda = u_\lambda(Z)$. Then, ice compactness and thickness can be integrated using the Z-momentum equation. The solution for ice velocity is schematically shown in Figure 7.15.

Example: Antarctic drift The Antarctic ice drift problem (Figure 7.16) can be idealized as a continent symmetric around the pole, with $-90° \leq Z \leq Z_0$, an east wind zone for $Z_0 \leq Z \leq Z_1$, and a west wind zone for $Z_1 \leq Z \leq Z_2$. Assume zero geostrophic water current. Because Coriolis acceleration is to the left of motion in the southern hemisphere, west wind drift is diffusive while east wind drift is compressive. Therefore, the west wind drift zone is in a near-free drift state, but significant friction is present in the east wind drift zone. In the latter zone $\tau_{a\lambda} < 0$, $\tau_{aZ} = 0$, $\text{sgn}(\dot{\varepsilon}_{Z\lambda}) = -1$. Eliminating the pressure gradient from the momentum equation gives:

$$\frac{P}{r_e e}\cot Z - \tau_{a\lambda} - \rho_w C_w(\cos\theta_w + e^{-1}\sin\theta_w)u_\lambda^2 - \rho h 2\Omega\cos Z e^{-1}u_\lambda = 0$$

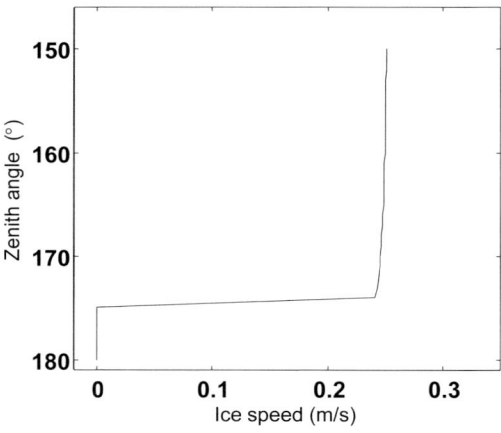

Figure 7.15. The steady-state solution of circular ice drift.

Figure 7.16. Sea ice cover in the Antarctic is more divergent than in the Arctic, characteristic features being the presence of polynyas and the co-existence of icebergs inside the pack ice.
Reproduced with permission from Professor Hardy B. Granberg, Sherbrooke, Quebec.

Ignoring the Coriolis and the oceanic turning angle, we have:

$$\frac{P}{r_e e}\cot Z - \tau_{a\lambda} - \rho_w C_w u_\lambda^2 = 0$$

A non-zero solution exists if:

$$|\tau_{a\lambda}|r_e e/P > |\cot Z|$$

When velocity has been resolved, the compactness and thickness profiles are integrated as in the rectangular case. When Coriolis acceleration is added, the right-hand side of the above equation increases and a stronger wind is needed to initiate ice motion. ∎

The circular ice drift model also serves as an analytic tool for examining sea ice flows on other Earth configurations (e.g., the distribution of land and water surfaces) and other (theoretical) planetary bodies[1] as well.

7.4 MODELLING OF ICE TANK EXPERIMENTS

The calibration and validation of continuum models have been based on full-scale observations, mainly encompassing the extent, compactness, and kinematics of drift ice. As to whether small-scale tank experiments could be useful has not been much examined. Such experiments might be used to investigate questions like the redistribution of drift ice thickness, drift ice rheology, and the dependence of the yield function on the state of the drift ice field. But, how well can ice tank tests be reproduced using mathematical sea ice models? An effort was made by Ovsienko et al. (1999b) to model ice tank experiments using a numerical model. The result was quite promising in that an elastic–plastic drift ice model could reproduce the outcome of the tank test.

7.4.1 Drift ice dynamics in a tank

An ice tank is a small, elongated rectangular basin, in which the x-axis is aligned with its major axis. The mathematical modelling of tank ice dynamics is based on a two-level system $J = \{A, h\}$ using the equations of ice dynamics given in Eqs. (7.1). In tanks the Coriolis term can be ignored, and forcing and motion are longitudinal along the x-axis:

$$\rho h\left(\frac{\partial u}{\partial t} + u\frac{\partial u}{\partial x}\right) = \frac{\partial \sigma_{xx}}{\partial x} + \frac{\partial \sigma_{yx}}{\partial y} + \tau_{ax} + \rho_w C_w |U_w - u|(U_w - u) - \rho h g\beta \quad (7.36a)$$

$$\frac{\partial \{A, h\}}{\partial t} + \frac{\partial u\{A, h\}}{\partial x} = 0 \quad (0 \leq A \leq 1) \quad (7.36b)$$

---

[1] Planetary bodies with similar "sea ice" dynamics to those on Earth are not known. However, the ice on Europa could be considered a floating glacier.

192   Drift in the presence of internal friction   [Ch. 7

The boundary conditions depend on the experiment configuration. The initial conditions are usually $u(x,0) = 0, J(x,0) = J_0(x) = \{A_0(x), h_0(x)\}$. The difference from the channel flow model is the inclusion of shear friction, which may be important for narrow tanks (Hopkins and Tuhkuri, 1999). The system can be forced by wind, water flow, or from the boundary. We will now analyse the boundary forcing case, mainly because ice tank experimental data exist for this case.

In the boundary forced case, an ice field is compressed by being pushed at velocity u_b by a moving boundary toward a fixed boundary (Figure 7.17). Then, wind stress, water surface slope, and water flow velocity are all zero, and the momentum equation is:

$$\rho h \left(\frac{\partial u}{\partial t} + u \frac{\partial u}{\partial x} \right) = \frac{\partial \sigma_{xx}}{\partial x} + \frac{\partial \sigma_{yx}}{\partial y} - \rho_w C_w |u|u \qquad (7.37)$$

Let us assume that the ice field is initially homogeneous, $J_0 = $ constant, and its length is L_0. The origin is at the fixed boundary, and the x-coordinate increases toward the moving boundary located at $L = L(t)$, $dL/dt = u_b$. Momentum advection and ice–water stress scale as U^2 and their ratio is $h/(C_w L) \sim 1$ for $h \sim 10$ cm, $C_w \sim 10^{-2}$, and $L \sim 10$ m. In fast-velocity experiments they are both important, while in slow-velocity experiments they become very small. Local acceleration is small if the velocity of the moving boundary is held constant.

Consequently, in the simplest experiment—slow, fixed boundary speed and enough tank width for low shear friction—compressive stress is the dominant factor, and Eq. (7.37) gives the solution $\sigma_{xx} \approx $ constant. Then, according to the boundary condition of a moving boundary:

$$\sigma_{xx} = -\frac{F}{b} \qquad (7.38)$$

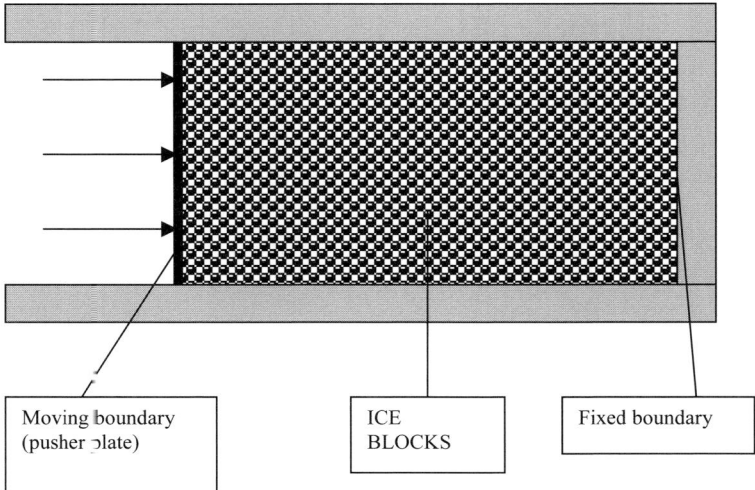

Figure 7.17. Ice tank experiment with boundary forcing.

where b is the width of the tank and $F = F(t)$ is the pushing force required at the moving boundary.

For a general rheology, let us take the power law as $\sigma_{xx} = \zeta\,du/dx$ where $\zeta - 1 = \zeta_n(A, h)|du/dx|^n$ is the bulk viscosity. The cases $n = 1$, $n = 0$, and $n = -1$ correspond, respectively, to floe collision rheology (Shen et al., 1986), linear viscous rheology, and plastic rheology. In these cases it is natural to assume that ice velocity is monotonously decreasing with x, $u(0) = 0$, $u(L) = u_b < 0$, and thus $du/dx < 0$. We have:

$$\zeta_n \left|\frac{du}{dx}\right|^{n+1} = -\frac{F}{b} \tag{7.39}$$

The case $n = -1$ corresponds to plastic ice rheology and ζ_{-1} is compressive yield strength σ_c. Then, under compressive deformation, internal ice stress corresponds to the minimum yield strength $\sigma_c = \sigma_c(J)$ in the tank, $\sigma_{xx} = -\min\{\sigma_c(x)\,|\,0 \leq x \leq L\}$. Since the ice state is initially homogeneous, stress equals yield stress everywhere. Ice fails in a uniform manner, and the resulting uniform deformation necessitates a linear velocity profile. For $n \neq -1$, a linear velocity profile $u = -u_b(x/L)$ also results if the ice state is constant in space. Therefore, the solution is continuous for $n \to -1$.

As regards the linear velocity profile, the ice conservation law is:

$$\frac{\partial J}{\partial t} + u_b \frac{x}{L}\frac{\partial J}{\partial x} = -J\frac{u_b}{L} \quad (0 \leq A \leq 1) \tag{7.40}$$

If the ice state is constant in space, then by $\partial J/\partial t = -Ju_b/L$ it must remain so (i.e., $J = J(t)$). Consequently, boundary forcing of an ice field with its state constant in space results in a rubble field that thickens evenly (no particular pressure ridge features can form). This result is quite general with all viscous power laws, including that of plastic flow. If the initial condition were changed to some $J_0 \neq$ constant, the weakest points would fail first and eventually the situation $J(t_1) = $ constant would be faced. A uniformly growing rubble field would result after time t_1.

Ice volume conservation requires that $hL = h_0 L_0$ at any time. Deformation of the ice field proceeds first by compacting to the maximum packing A^*, and thereafter rubble formation takes place and the actual ice thickness increases. We have for the compaction and rubble growth phases:

$$A(t) = \frac{A_0 L_0}{L_0 + u_b t} \leq A^*, \quad h = h_i A, \quad 0 \leq t \leq t^* \tag{7.41a}$$

$$h(t) = \frac{h_i L^*}{L^* + u_b t}, \quad A = A^*, \quad t > t^* \tag{7.41b}$$

where $h_i = h_0/A_0$ is the initial ice floe thickness and $L^* = L_0 A_0/A^*$ is the length of the ice field at the time t^* when the free paths in the ice floe field have all closed up.

If the time evolution of the forcing is known, it is possible to obtain information on rheology. Since the ice state depends on time, so does the strength. Therefore, these tank experiments do not provide a unique solution for viscosity ζ_n and power n.

Instead, if n is given, viscosity together with its functional form can be determined. With the plasticity assumption, $\sigma_c(A, h) = F/b$ the yield function can be examined for the dependence of σ_c on A in the compaction phase and on h in the rubble growth phase.

7.4.2 Case study

Let us take the tank experiments of Tuhkuri and Lensu (1997) for a case study. The tank was 40 m × 40 m with a depth of 2.8 m. It was first frozen, then test areas 6 m × 20 m were cut into the ice sheet, and the ice in each test area was broken into floes with a fixed thickness (h_i), diameter, and shape. In one experiment, the ice floe field was forced from one boundary by a pusher plate set to a fixed velocity u_b (Figure 7.17). The initial compactness was $A_0 = 0.7$, with 10% of the ice being brash. The only measured data concerned the force history at the pusher plate. The water flow velocity U_w was zero, and since the test regions were inside a large tank it is assumed that no significant surface slope was built up. These experiments have been simulated by a discrete particle model with good agreement (Hopkins and Tuhkuri, 1999). The continuum modelling approach is also very useful since it is the standard means of addressing the sea ice dynamics problem.

The outcome of 17 tests was presented in Tuhkuri and Lensu (1997). Nine tests were carried out with a variable ice thickness (28–54 mm) and velocity of the pusher plate (15–52 mm/s), while the shape and size of ice floes were fixed (circular and of diameter 400 mm). In eight tests the velocity of the pusher plate was nearly the same (10–13 mm/s), and ice floes had variable thickness (36–57 mm), shape (square or irregular), and diameter (300–600 mm). The force at the pusher plate was 50–200 N at $A = 1$, which had increased up to 300–1,500 N at $h \approx 8h_i$ at the time the tests drew to a close.

Thus the relevant scales were $u \sim 30$ mm/s and $h \sim 50$ mm. Inertial force is much less than $\frac{1}{2}bL\rho h \Delta u/\Delta t \sim 40$ N; this would result if half of the test area would change the velocity in 1 s by half the maximum u_b used in the tests. By using a similar magnitude analysis, advective acceleration and ice–water stress are found to be less than 10 N. Consequently, the slow model with weak ice–water interaction is realistic. Also friction effects at the channel edges are relatively small for these cases, but they would become significant at half of the present channel width (Hopkins and Tuhkuri, 1999).

The outcome of each experiment shows two phases. First comes the compaction phase, which lasts until time $t^* = (A^* - A_0)L_0/u_b$; for $A^* = 1$ and $u_b = 50$ mm/s this is 120 s. The second phase is the piling up of ice floes into a rubble field right up to the end of the experiment, when the rubble thickness is nearly ten times the floe thickness. The experiment outcome agrees with the mathematical model in that, for boundary-forced deformation, no pressure ridges form and the rubble thickens evenly. Figure 7.18 shows one experiment, the essential features of which are adequately repeated in the other experiments as well.

The basic statistics of experiments with uniform circular ice floes is shown in Table 7.2. In the compaction phase the force at the pushing plate and also the

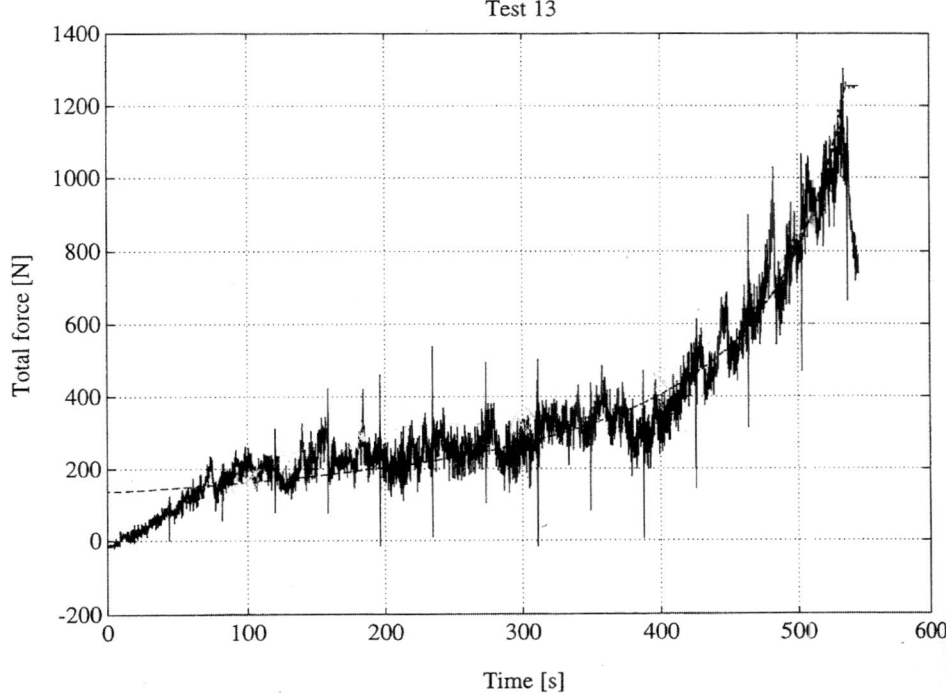

Figure 7.18. The force at the pusher plate in ice tank test #13 with pusher plate velocity $u_b = 33$ mm/s; the dashed line shows the fit of the analytical interpretation model.
Reproduced from Tuhkuri and Lensu (1997), with permission from Helsinki University of Technology.

Table 7.2. The measured force (in Newtons) at the pusher plate at $A = 1(F_1)$ and at $h = 4h_i(F_4)$. The diameter of the circular ice floes was 400 mm and the pusher plate velocity was 15–52 mm/s.

	Ice thickness					
	28 mm		38 mm		54 mm	
Pusher plate speed	F_1	F_4	F_1	F_4	F_1	F_4
15 mm/s	80	300	150	450	220	1,000
33 mm/s	50	170	140	330	280	830
52 mm/s	50	180	70	240	100	600
Mean	60	220	120	340	200	810

compactness increase almost linearly. The stress level is half that at $A = 1$ when compactness is within 0.8–0.9. This is different from common sea ice rheologies, where strength increases quickly with compactness. In the widely used strength formulation of Hibler (1979), $\sigma_c \propto \exp[-C(1 - A)]$, $C = 20$, which means one order of magnitude strength increase for an increase of 0.15 in the compactness. In addition, the floe collision rheology of Shen et al. (1986) has a singularity at $A = A^*$ and therefore a very high sensitivity to compactness as the maximum packing density is approached.

In the rubble growth phase, the force at the pusher plate depends on the thickness of the rubble and the speed of the pusher plate. In the cases $h_i = 28$ mm and 54 mm, strength increases in proportion to the thickness of the accumulated rubble field, but for $h_i = 38$ mm the increase in strength is a little slower. But a remarkable feature is that a decrease in strength is seen when the speed of the pusher plate is increased. This was explained by Tuhkuri and Lensu (1997) as a decrease in the floe–floe friction coefficient.

Let us now assume a general power law for ice strength at $\zeta_n = \zeta_{n\kappa} h^\kappa$ where $\zeta_{n\kappa}$ is a constant. The strain rate is $\dot{\varepsilon} = u_b/L$, and thus Eq. (7.39) gives:

$$\zeta_{n\kappa} \frac{(L_0 h_i)^\kappa |u_b|^{n+1}}{(L_0 + u_b t)^{\kappa+n+1}} = -\frac{F}{b} \qquad (7.42)$$

The form of the experimental curves is in very good agreement with this law when we take $\kappa + n + 1 = 1$ (see Figure 7.18).

In the plastic law $n = -1$ and therefore $\kappa = 1$. Averaging over the different pusher plate speeds for a representative plastic flow gives a yield function of compact ice in agreement with Hibler (1979), $\sigma_c = -P^* h$. The estimate for P^* becomes 0.35 kPa for $h_i = 28$ mm, 0.45 kPa for $h_i = 38$ mm, and 0.62 kPa for $h_i = 54$ mm; the overall average is $P^* = 0.47 \pm 0.29$ kPa. Including the pusher plate speed, the data suggest that $n + 1 \approx -\frac{1}{2}$ and therefore $\kappa = \frac{3}{2}$. This tells of a strain-rate weakening property of the ice field. The strain here is pure compression, and the ice hardening effect due to thickness increase (power $= \frac{3}{2}$) overcomes the weakening due to more rapid compression (power $= -\frac{1}{2}$). Therefore, the result is stable material behaviour.

The potential energy of the rubble is given as $\frac{1}{2} g[\rho(\rho_w - \rho)/\rho_w] h^2 \sim 10$ N/m (Rothrock, 1975a). This is very small in comparison with the work done by the pusher plate. Consequently, the resistance of the ice field to rubble formation is almost purely due to friction. According to Hopkins and Tuhkuri (1999), the failing of this ice field is due to floe underturning and rafting, and more pushing force is needed when the ice–ice friction coefficient increases or the aspect ratio of floes increases. With a fixed floe diameter, the aspect ratio is proportional to ice thickness. Therefore, part of the $\frac{3}{2}$ power of κ is due to the increase in aspect ratio.

Experiments with different kinds of floes were made at low speed (10–13 mm/s) and then compared with the lowest speed in the fixed floe cases. The results show good correspondence. Further inspection of the experiment results did not indicate any systematic changes in ice strength with the size and shape of the floes. However, the range of floe diameters and therefore aspect ratios was not large.

Scaling of tank experiments to nature is partly open to question. Ice strength in these experiments was in the range 10–30 Pa, while in nature it is 1,000 times as much, the difference being largely due to the thickness of the ice. The evolution of ice thickness distribution is unnatural in the confines of a tank since the system is forced from a boundary rather than by a distributed forcing as in nature. For constant forcing over the whole ice field, stress would peak at one boundary where pressure ridge formation would begin. Nevertheless, the tank results show the feasibility of ice tank experiments for the progress of knowledge in sea ice dynamics and for parameterization of sea ice dynamics models. The design of further experiments should be based on setting hypotheses and questions from sea ice dynamics science.

In these tank tests real floating ice blocks were used, although the ice is not the same as in nature. In ice tank technology, solid ice is softened by using dissolved substances in the water, which freezes for the tank ice (see Tuhkuri and Lensu, 1997). One line of research could be physical analogue experiments (Philippe Blondel, personal communication). For example, layers of sand, mixed with honey and engine oil, are used as scaled-down versions of tectonic plates (e.g., to model the collision of India with the Eurasian continent and creation of the Himalayas).

The advantage of analytical modelling is its ability to provide a full picture of the physical problem. Because of necessary simplifications, the outcome is not quantitatively exact, but the main qualitative characteristics and their physical background become understandable. In addition, analytical solutions serve as excellent validation cases for the numerical model technology. Having learned the analytical methods, the next step is to go into the world of numerical models.

8

Numerical modelling

8.1 NUMERICAL SOLUTION

This chapter contains a numerical modelling of sea ice drift. The idea is to introduce the numerical technology used in sea ice modelling and present standard-type cases of model applications for ice forecasting and climate investigations. The reader with an understanding of the fundamental laws of sea ice dynamics, together with the free drift solution and analytical one- and 1.5-dimensional models, will be able to interpret the numerical model outcome in this chapter. From the perspective of our present knowledge, continuum numerical models provide the best quantitative solution to ice drift, up to the highest spatial resolution the floe size allows.

Numerical modelling of sea ice dynamics began in the 1960s in the Soviet Union and in the USA (Campbell, 1965; Doronin, 1970). The first models were linear and matched the computational technology of the time. During the 1970s the modern theory of sea ice dynamics was developed, and at the end of the decade continuum sea ice models were close to what they are now, at least in terms of the description of physical processes. Thereafter, coupled ice–ocean models have been developed, and spatial resolution has been continuously improving as well. The main applications of these models have been short-term ice forecasting for shipping and oil drilling, and recently for weather forecasting and long-term simulations of climate change.

8.1.1 System of equations

In a way the classical Nansen–Ekman drift law is already a model of some sort. But numerical models must be *full models* (i.e., closed system models), which can solve the sea ice circulation problem for the whole basin under consideration. The modeller chooses the ice state and rheology, and the conservation laws for the ice

William J. Campbell (1930–1992) was a pioneer in numerical modelling of sea ice dynamics. His PhD thesis at the University of Washington (Campbell, 1965) showed the first Arctic Ocean sea ice drift model. He then worked with the US Geological Survey, focusing his research in the 1970s on remote sensing of sea ice.
Courtesy of *Puget Sound* magazine.

state and momentum close the system. A full drift ice model thus consists of four basic elements:

$$\text{Sea ice state } J = \{J_1, J_2, J_3, \ldots\} \quad (8.1a)$$

$$\text{Sea ice rheology } \sigma = \sigma(J, \varepsilon, \dot{\varepsilon}) \quad (8.1b)$$

$$\text{Conservation of momentum } \rho h \frac{Du}{Dt} = \nabla \cdot \sigma + F_{\text{ext}} \quad (8.1c)$$

$$\text{Conservation of ice } \frac{DJ}{Dt} = \psi + \phi \quad (8.1d)$$

Elements (8.1a) and (8.1b) constitute the heart of a sea ice model and are reflected in the model attributes: one speaks of a three-level ($\dim(J) = 3$) viscous–plastic sea ice model, etc. A modeller has to make three key choices:

(1) The choice of ice state (i.e., how many different thickness or other morphological properties of the ice pack are needed).
(2) With a given ice-state definition, the ice conservation law must be split into as

many equations as there are ice-state variables and a choice needs to be made on how the ice state is redistributed in mechanical deformation.
(3) The choice of rheology depends on what features need to be solved by the model. A realistic rheology is necessary for closure of the model (e.g., the free drift approach could give reasonable velocities in a central basin, but would produce seriously biased ice conditions at the coast).

When ice state, its redistribution function, and ice rheology have been chosen, the conservation laws for ice and momentum are given their specific form and can be solved for evolution of ice conditions.

The full sea ice dynamics problem includes three unknowns: ice state, ice velocity, and ice stress. The number of independent variables is $N = j + 2 + 3$, where $j = \dim(J)$ is the number of levels in the ice state. Any proper ice state has at least two levels, $j \geq 2$ and thus $N \geq 7$.

The first numerical methodology choice would be between *discrete particle models* and *continuum models*; but in mesoscale and large-scale sea ice dynamics all workable models are still in the continuum world. The present chapter therefore focuses on the continuum world too. An effort to construct a compromise model between continuum and discrete particle models was made by Rheem et al. (1997), for the purpose of trying to solve ice drift and ice forcing on structures with the same model.

The inertial timescale of the ice is of the order of 1 hour, much less than the advective timescale, and therefore a quasi-steady-state approach for the momentum equation is feasible. It is applied in particular to the seasonal sea ice zone where ice thickness is less than about 1 metre.

The model parameters can be grouped into four categories:

(i) Drag parameters of the atmospheric and oceanic boundary layers.
(ii) Sea ice rheology parameters.
(iii) Ice-state redistribution parameters.
(iv) Numerical design parameters.

The primary *geophysical parameters* of sea ice dynamics models are the drag coefficients and the compressive strength of ice. Drag coefficients together with Ekman angles tune the free drift velocity of thin ice or strong winds, while compressive strength tunes the length scale of the ice flow in the presence of internal friction. These are also the main tuning parameters of sea ice models. Secondary geophysical parameters come from the chosen sea ice rheology—What more is required than compressive strength?—and from the ice-state redistribution scheme. Ice-state redistribution parameters are important, but their fixing really suffers from lack of good data. One parameter needed in low-level ice states is the demarcation thickness h_0; ice must be defined as being equal to or thicker than h_0 (see Eq. 2.24) (otherwise the term "mean thickness" does not represent the strength of ice well).

A numerical model includes *numerical design parameters*, which include artificial practical formulations due to limitations of numerical approximation and methodological parameters to fix the resolution and to keep the numerical solution stable.

This includes first of all the choice of grid and its size. This also stems from the fact that the system to be solved is highly non-linear and the stability of the solution may require smoothing techniques.

Sea ice dynamics–thermodynamics models also have heat budgets; but, here we focus on dynamics, because the thermodynamic growth rates of ice-state variables are already known. Although sea ice has salt inclusions, the salinity level is much less than in ocean water. Therefore, sea ice models have no internal salt budgets; it is assumed that in freezing all salts are rejected and in the melting of ice the meltwater is fresh.

There is a numerical modelling line, which is based on the free drift theory but also includes the ice conservation law. These models have singularities, the influence of which is made smaller by using properly smoothed forcing fields. A model based on free drift with semi-empirical corrections on the coast was presented by Nikiforov et al. (1967), and the incompressible inviscid solution was presented by Rothrock (1975c). However because of the high level of the internal friction of sea ice and because it is not possible to properly close the ice motion without internal friction, these non-friction models are not further discussed in this chapter.

8.1.2 Numerical technology

In the continuum approach, there are different steps along which to proceed. The first step is the choice between the *finite element* and *finite difference* approaches. The latter has by far been the dominant technique in sea ice dynamics, and only that will be used below. Finite element models could be preferable in small time- and space-scales, when the geometry of boundaries has a pronounced influence on dynamics.

Eulerian/Lagrangian frames

The next step is between *Lagrangian* and *Eulerian* frames (i.e., respectively, whether the numerical grid is fixed in the medium or in space). The latter has been most frequently used in sea ice dynamics models. But, the ice drift problem has the peculiarity of moving boundaries that are difficult to take care of in a Eulerian system. In common numerical techniques it is assumed that the medium (and ice state) is evenly spread within each grid cell, and this leads to strong numerical diffusion at open boundaries, when the usual first- or second-order approximations are used for spatial derivatives. In particular, a diffuse ice edge results, which may cause serious mistakes when the location of the boundary is important.

Consequently, more advanced methods have been used for the ice conservation law. The solution to the ice edge is significantly improved if the location of the ice edge is kept as a variable in the model. It is then advected by the velocity field and boundary conditions are precisely taken care of by this real ice edge.

Ovsienko (1976) employed a Eulerian–Lagrangian technology. He solved the momentum equation in a Eulerian grid[1] and the ice conservation law in a

---

[1] A Eulerian grid is fixed in space, while a Lagrangian grid is fixed in the medium.

Lagrangian grid using the *particle-in-cell* method (Harlow, 1964). The ice field is represented by a large number of particles with given thickness and size. For each Eulerian grid cell they are summed for compactness and mean thickness, and then the momentum equation can be solved. Interpolating an individual velocity for each particle, these can be advected in space for a new configuration and state distribution of the ice field. Then the new ice state is obtained for the Eulerian grid by summing from the advected configuration. The particles are virtual; with their large number they may represent a sea ice continuum, but they do not interact mechanically as is the case with discrete particle models. A similar approach was taken by Flato (1993) for a short-term model in the Beaufort Sea and Shen et al. (1993) for an ice dynamics model in the Niagara River.

Grids

The suitability of the continuum approach for drift ice was discussed in Section 2.2, and the length scale hierarchy was set as $d \ll D \ll \Lambda$, where d is the floe size, D is the continuum particle size, and Λ is the gradient scale. The grid size of numerical models fits in as:

$$d \ll D \sim \Delta x \ll \Lambda \tag{8.2}$$

Below D, continuum physics is no longer valid, so choosing $\Delta x \ll D$ would introduce artefacts. This problem has not been well examined in drift ice modelling; but it is now of major concern since the grid size in ocean models has become much smaller than D (in other words, we need to find a way of constructing high-resolution, coupled ice–ocean models). On the other hand, $\Delta x > D$ is allowed; but because D is relatively large, there is normally enough computational power to work with $\Delta x \sim D$. Although there is not much freedom to choose a proper grid size, in practice the grid size has been between 10 km and 100 km. In ocean dynamics, on the other hand, the choice of Δx is mainly determined by the computational power available and the resolution required—and the problem is to develop a solution in these conditions (e.g., Kowalik and Murty, 1993).

In each grid cell, ice-state variables are taken as averages, as by definition are the probabilities of the classes of ice thickness distribution. If spatial distribution is strongly non-uniform in a grid cell, additional effects could arise that are not included in the thickness distribution theory: cracks and leads may have a preferred orientation; how a given percentage of open water is distributed in a grid cell could make a difference; different ice types may occur in patches; etc.

In principle a grid can have any geometry; but, in practice, most sea ice models by far use a regular rectangular grid, the use of which makes the construction of numerical algorithms straightforward. In the very first models all quantities were calculated in the same grid points, but since then staggered grids have been employed (Figure 8.1). Triangular grids would be more flexible and could follow the solid and open boundaries of sea ice fields better.

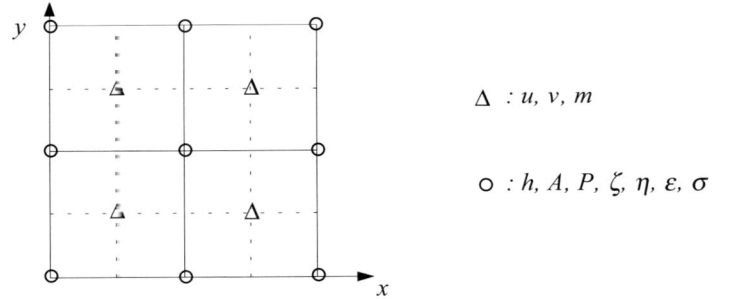

Figure 8.1. Spatial grid (Arakawa B type) normally employed in the numerical modelling of sea ice dynamics.

Time steps are determined by the stability criterion set by the required resolution and method of numerical time integration. In practice, time steps have been from half an hour to 1 day.

From the physics point of view, since the timescale of ice inertia is $T_1 = \rho h/(\rho_w C_w U)$ (see Section 5.3.2), approximately $10^3 h$ s/m, a safe time step would be $10^2 h$ s/m to resolve acceleration and deceleration events. However, in most cases this high temporal resolution is not needed. The highest frequency of interest is the Coriolis frequency, where ice drift follows inertial oscillations together with the mixed layer of the ocean. In high polar regions the Coriolis period is about 12 hours; this takes the necessary time step down to 1–2 hours. At such time steps the inertia of thick multi-year ice may be significant, and the inertial term consequently needs to be included in the model. Whereas in the seasonal sea ice zone, the quasi-steady-state approach works well.

Initial and boundary conditions

Initial ice-state conditions are needed in purely dynamic cases. However, because the inertial timescale of sea ice is small, of the order of 1 hour, in non-steady-state models, initial ice velocity can be taken as zero. In quasi-steady-state models the steady-state solution of the momentum equation for the given initial ice state and forcing serves as the initial field.

For boundary conditions there are two approaches. First, the more physical approach is to define the ice region by $\Omega = \Omega(t)$ and its boundary curve by $\Gamma = \Gamma(t)$ (e.g., Ovsienko, 1976). Along Γ, on an open boundary the normal stress is zero, $\sigma \cdot \mathbf{n} = 0$, and ice drift moves the boundary along a solid boundary, the ice cannot go to land. $\mathbf{u} \cdot \mathbf{n} \leq 0$, and if $\mathbf{u} \cdot \mathbf{n} < 0$ then $\sigma \cdot \mathbf{n} = 0$.

The second approach, which is more practical, is to define open water as ice with zero thickness and avoid the existence of the open boundary. At a solid boundary the no-slip condition $\mathbf{u} = 0$, normal for viscous fluids, is employed.

Numerical integration

The numerical solution progresses by solving the momentum equation and the ice conservation law in turn. First, with a fixed ice state, the momentum equation is solved. Because of the high ice strength and inclusion of inertia, direct, explicit (Euler) time marching would require a very short time step. Therefore, more advanced methods are used. In the quasi-steady-state approach this problem is avoided.

But, in any case, because the momentum equation is non-linear, a series of iterations is normally needed. With a given ice velocity, the equation of motion is linearized and a new, corrected velocity field is obtained by a matrix algorithm, such as the over-relaxation method (Ames, 1977). Then the corrected velocity is inserted for a new linearized equation and the matrix algorithm is used again. The whole cycle is repeated until the desired accuracy is obtained.

When ice velocity has been solved, the ice conservation law is integrated for temporal change in ice state. Different kinds of techniques have been employed in this integration. Doronin (1970) took upstream spatial derivatives (i.e., one-sided derivatives based on the direction from which ice was entering the grid cell). Hibler (1979) integrated the ice conservation law explicitly with a second-order differencing scheme. Ovsienko (1976) used the particle-in-cell method, which efficiently tracks the location of the ice edge and the internal boundaries of the ice field.

With explicit time stepping of the ice conservation law, the principal stability condition for the whole system comes from the Courant–Friedrichs–Lewy criterion:

$$\frac{\Delta x}{\Delta t} \geq \sqrt{2}|\boldsymbol{u}| \tag{8.3}$$

This puts the time scale below 1 day for a 100-km grid or below 3 hours for a 10-km grid.

8.1.3 Calibration/Validation

The calibration and validation of sea ice models has been made based on ice concentration, ice thickness, and ice velocity. The best validation data exist for concentration, since it can be easily detected by satellite remote sensing. Velocity data are also available from drifter station time series and from sequential satellite mapping (Figure 8.2). The thickness of ice is the main problem; suitable information contains only statistical mean fields. An essential improvement would be to run sea ice models against high-resolution changes of ice thickness fields, since they integrate the mechanical growth of ice thickness and could tell us how the ice state is redistributed in mechanical deformation.

Satellite remote sensing provides excellent velocity fields, which could serve the redistribution problem directly. By using observed velocities directly, ice-state integration would progress without systematic errors in velocity. But, validation would still require thickness data.

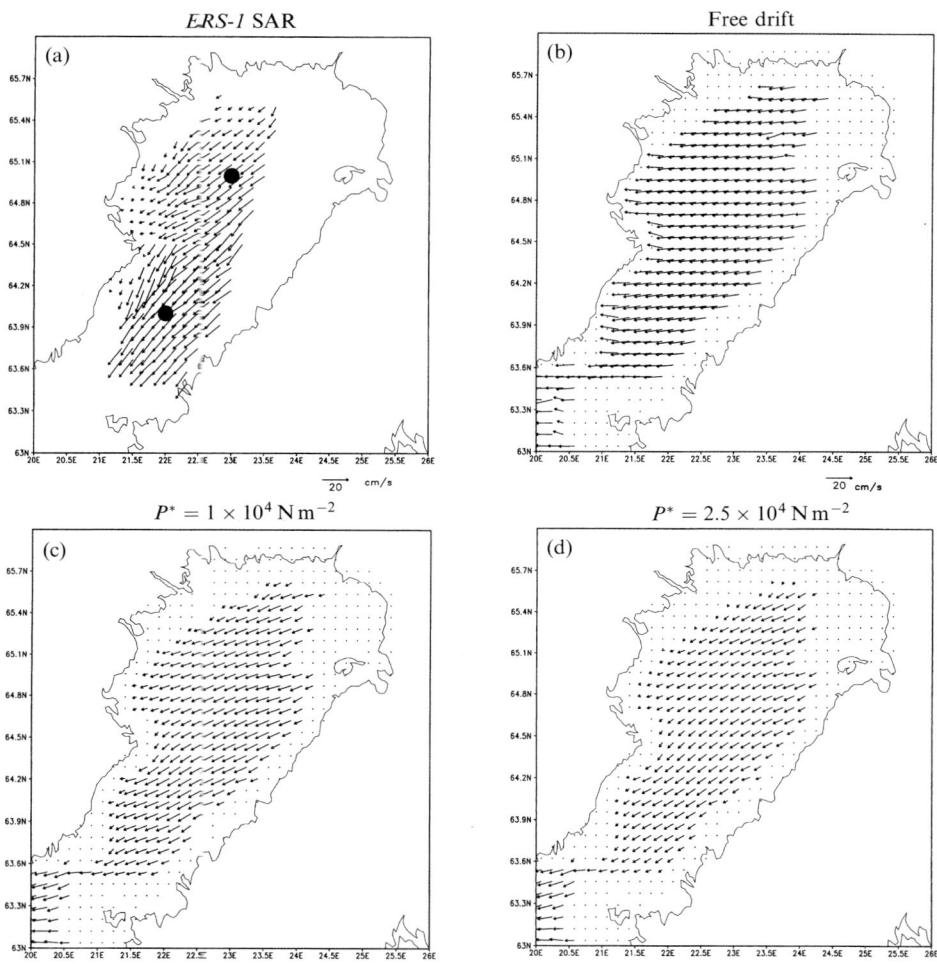

Figure 8.2. Calibration runs for ice strength against ice velocity fields obtained from *ERS-1* SAR data in the Bay of Bothnia, Baltic Sea. The ice is thin (10–50 cm) and has low strength. The black dots here have no relevance.
From Leppäranta et al. (1998).

8.2 EXAMPLES OF SEA ICE DYNAMICS MODELS

8.2.1 The Campbell and Doronin models

The first models in the 1960s were linear viscous models. Campbell (1965) presented a steady-state circulation model of ice and water for the whole Arctic Basin, while Doronin (1970) presented a prognostic short-term sea ice dynamics model for the Kara Sea. Campbell (1965) was able to reproduce the Transpolar Drift Stream and

Sec. 8.2] Examples of sea ice dynamics models 207

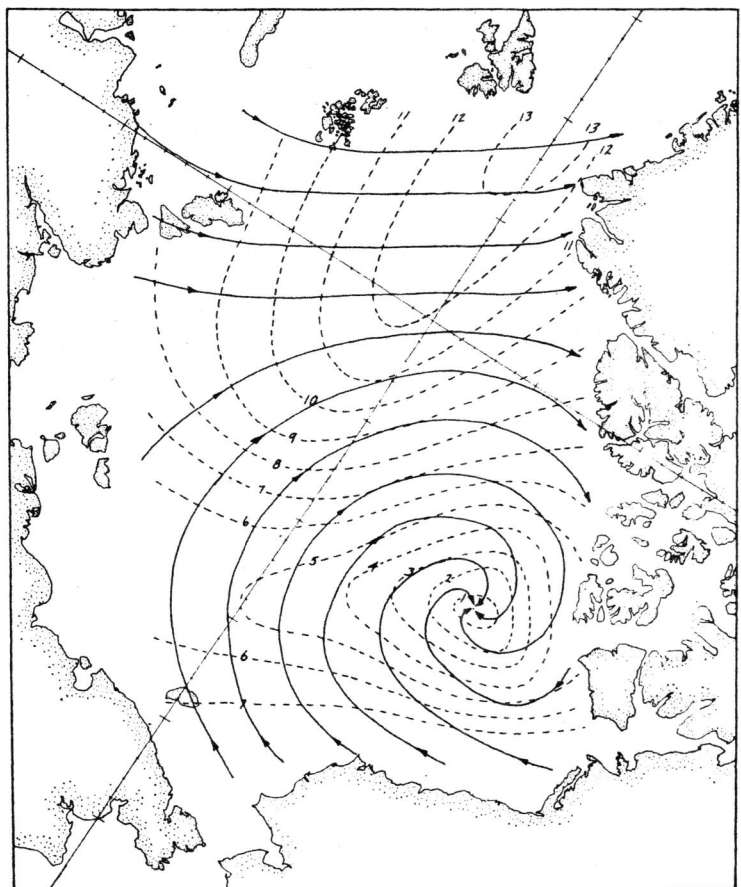

Figure 8.3. Steady-state ice circulation in the Arctic Basin according to the linear viscous model of Campbell (1965). Solid lines show ice drift streamlines, and dashed lines show ice drift speed isolines (in cm/s).
Reproduced with permission from the American Geophysical Union.

the Beaufort Sea Gyre (Figure 8.3). The result was significantly different from earlier free drift approaches. There is too much convergence in the Beaufort Sea Gyre, as noted by Campbell himself (a problem that arises from using a linear viscous rheology). Air–ice and ice–water stresses were determined from Prandtl-type boundary-layer models. The viscosity of sea ice was formulated as $\eta = \rho h K_i$, where K_i is the eddy viscosity coefficient of ice floes; the best fit resulted in $K_i = 3 \times 10^8 \, \text{m}^2/\text{s}$, corresponding to dynamic viscosity of about $3 \times 10^{11} \, \text{kg/s}$ for 1 m thick ice. A total of 260 grid points covered the central Arctic Ocean, corresponding to a grid size of about 150 km. The conjugate gradient method was used to iterate for the solution of two ice velocity components and the stream function of water flow.

The Doronin (1970) model was a two-level, linear viscous quasi-steady-state model:

$$J = \{h, A\} \tag{8.4a}$$

$$\boldsymbol{\sigma} = 2\eta\dot{\boldsymbol{\varepsilon}}' \quad (\eta \propto A) \tag{8.4b}$$

$$f\mathbf{k} \times \boldsymbol{u} = \nabla \cdot \boldsymbol{\sigma} + \boldsymbol{\tau}_a + \boldsymbol{\tau}_w - \rho h g \boldsymbol{\beta} \tag{8.4c}$$

$$\frac{\mathrm{D}\{h, A\}}{\mathrm{D}t} = \{0, -A\nabla \cdot \boldsymbol{u}\} + \{\phi_h, \phi_A\} \tag{8.4d}$$

Atmospheric and oceanic drag forces were taken linear, like the solutions from Ekman velocity profiles. Boundary-layer parameters could be fixed so that the solutions agreed with usual free drift conditions when internal friction became small. The best fit linear viscosity of sea ice was obtained as 3×10^8 kg/s for $A = 1$, relatively low but the simulations were made for summer ice conditions. The numerical solution of ice velocity and compactness was obtained simultaneously by the over-relaxation technique. Then the thickness of ice was determined by thermal changes plus advection (advection for a given grid point was taken by the upstream spatial derivative). The grid size was 100 km and the time step was 1 day.

The model was applied for summer conditions in the Kara Sea, the ice state changing due to melting, advection, and dynamic opening and closing. This model structure introduced the first full sea ice model as a closed system.

8.2.2 Hibler model

Most present drift ice dynamics models are based on Hibler's (1979) model. It is therefore a very good benchmark for other models. The Hibler model is a two-level viscous–plastic model, with ice state as given in the Doronin (1970) model. The full momentum equation is solved. The original model application was to study seasonal ice conditions in the Arctic Ocean:

$$J = \{h, A\} \tag{8.5a}$$

$$\left.\begin{array}{l} P = P^* h \exp[-C(1-A)] \\[4pt] \boldsymbol{\sigma} = \zeta \operatorname{tr} \dot{\boldsymbol{\varepsilon}} \mathbf{I} + 2\eta\dot{\boldsymbol{\varepsilon}}' - \tfrac{1}{2} P \mathbf{I} \\[6pt] \zeta = \dfrac{P}{2\max(\Delta, \Delta_0)}, \quad \eta = e^{-2}\zeta, \quad \Delta_0 = \sqrt{\dot{\varepsilon}_I^2 + (\dot{\varepsilon}_{II}/e)^2} \end{array}\right\} \tag{8.5b}$$

$$\rho h \left[\frac{\mathrm{D}\boldsymbol{u}}{\mathrm{D}t} + f\mathbf{k} \times \boldsymbol{u}\right] = \nabla \cdot \boldsymbol{\sigma} + \boldsymbol{\tau}_a + \boldsymbol{\tau}_w - \rho h g \boldsymbol{\beta} \tag{8.5c}$$

$$\frac{\mathrm{D}\{h, A\}}{\mathrm{D}t} = -\{h, A\}\nabla \cdot \boldsymbol{u} + \{\phi_h, \phi_A\} + D_1 \nabla^2 \{h, A\} + D_2 \nabla^4 \{h, A\}, \quad 0 \leq A \leq 1 \tag{8.5d}$$

When the elliptic yield curve is assumed, the plastic flow rule can be analytically solved, written in viscous law form. Then the whole viscous–plastic rheology can be

Table 8.1. Parameterization of the viscous–plastic sea ice model of Hibler (1979). The parameter groups are: I—atmospheric and oceanic drag parameters; II—rheology parameters; III—ice-state redistribution parameters; and IV—numerical design parameters.

	Parameter	Value
I	Air–ice drag coefficient, C_a	1.2×10^{-3} (geostrophic wind)
	Air–ice Ekman angle, θ_a	$25°$ (geostrophic wind)
	Air–water drag coefficient, C_w	5.5×10^{-3} (geostrophic current)
	Air–water Ekman angle, θ_w	$25°$ (geostrophic current)
II	Compressive strength constant, P^*	5 kPa; replaced later by 25 kPa
	Aspect ratio of yield ellipse, e	2
	Strength constant for opening, C	20
	Maximum viscous creep rate, Δ_0	$2 \times 10^9 \, \text{s}^{-1}$
III	Demarcation thickness, h_0	50 cm
IV	Spatial grid size, $\Delta x = \Delta y$	125 km
	Time step, Δt	1 day
	Harmonic diffusion coefficient, D_1	500 m^2/s
	Biharmonic diffusion coefficient, D_2	7.81×10^{12} m^4/s

expressed in the compact form (8.5b) (see Section 4.3.4) and the momentum equation is solved in its full form, including both local and advective acceleration. In the ice conservation law two additional terms are introduced: harmonic and biharmonic diffusion with their corresponding diffusion coefficients D_1 and D_2, respectively. These diffusion terms are needed for the numerical stability of the solution, the stability problem being created by the highly non-linear form of the momentum equation.

The original model parameters are shown in Table 8.1. The boundary-layer parameters were based on Arctic Ice Dynamics Joint Experiment (AIDJEX) results (Brown, 1980; McPhee, 1982) and are representative of the central Arctic Ocean. For geostrophic wind the thin-ice wind factor is given as 1.7% and the deviation angle as zero, corresponding to 2.8% and $25°$ for the surface wind.

There are four rheology parameters. The compressive strength constant is the main tuning parameter. It defines the internal length L scale of sea ice dynamics by $FL \sim P^*h$, where F is the forcing. In the original model (Hibler, 1979) P^* was tuned low, at 5 kPa, because 8-day average winds were used for the forcing; later the level was found to be higher, 20–30 kPa being the normal level. The aspect ratio e and the strength reduction constant for opening C have almost always been fixed to the original values. Physically, e is the ratio of compressive stress to shear stress in pure shear and C^{-1} is the e-folding value of the strength of changes in compactness. It is obvious that $1 < e \ll \infty$ and $C \gg 1$ (i.e., shear strength is significant but less than compressive strength and both strengths are very sensitive to the compactness of ice).

This viscous–plastic rheology attempts to approximate a plastic medium, and the viscous part is here to make strain rates solvable at subyield stresses. There is no physical support for linear viscous behaviour at very small strain rates; therefore, the maximum viscous strain rate needs to be much lower than typical strain rates. The level $\Delta_0 = 2 \times 10^{-9}\,\mathrm{s}^{-1}$ is low enough (see Section 3.2.3) and corresponds to extension/contraction of 1 m over a distance of 20 km in 1 day. Another viewpoint would be to regard rheology as plastic and take the viscous submodel as a numerical design parameter.

With a two-level ice state, redistribution is straightforward; only the demarcation thickness is needed as a model parameter. It is taken as 50 cm, much less than typical ice thicknesses in the model basin, as it should.

The grid size was taken as 125 km, sufficiently large for continuum modelling and not too large to resolve the main features of ice circulation in the central Arctic Basin. The smallest geometry to be resolved is the outflow of ice through the Fram Strait. In fact, a realistic grid size here should be within a factor of about $\frac{1}{2}$ up to 2 from that used. The time step was 1 day, small enough for stability, and since the time step of the forcing data was 8 days, a shorter time step would not provide much more new information. The ratios of the diffusion terms to the mechanical deformation term are $D_1/(U\Delta x) \approx 0.04$ and $D_2/[U(\Delta x)^3] \approx 0.04$ for $U \sim 10\,\mathrm{cm/s}$. Thus for low ice velocities the diffusion terms become comparable with the deformation terms. More exactly, Eulerian time differencing gives:

$$\Delta_t[D_1 \nabla^2\{h, A\}] \approx \frac{D_1 \Delta t}{(\Delta x)^2} \Delta^2\{h, A\} \qquad (8.6)$$

where Δ_t stands for temporal change and Δ^2 is the second-order spatial difference, $[\Delta^2 Q]_{ij}$ being equal to the mean of the differences $Q_{i\pm 1, j\pm 1} - Q_{ij}$ for a quantity Q. For $\Delta x = 125$ km and $\Delta t = 1$ d, Eq. (8.6) gives $2.76 \cdot 10^{-3} \Delta^2\{h, A\}$. Thus, at each time step, thickness and compactness are smoothed by a factor of 2.76×10^{-3} toward the mean of the values in the surrounding grid points, which means that the timescale for homogenization of a stationary ice field is ≈ 360 d. The biharmonic smoothing operator is analogous, with $D_2 \Delta t/(\Delta x)^4 \approx 2.76 \cdot 10^{-3}$ (in words, the smoothing is equally effective). To prevent instabilities a minimum value is specified for non-linear viscosities, $\zeta > \zeta_{\min} = 4 \times 10^8$ kg/s; and $\eta = \zeta/e^2$ holds. These limits are much lower than the normal level of viscosities.

One time step proceeds as follows. The momentum equation is solved repeatedly from linearized equations by over-relaxation until the desired accuracy is obtained. Then the ice conservation equation is integrated for a new ice state. The original model was validated against observed average thickness and velocity fields: the Transpolar Drift Stream and the Beaufort Sea Gyre come out well, but more importantly the mean ice thickness field is good. An independent comparison for ice velocity which gave a good result was made by Zwally and Walsh (1987). Sensitivity studies of ice model dynamics have been made by Holland et al. (1993), among others.

The stability of the numerical solution has been examined by Gray and Killworth (1995). They illustrate cases when the non-zero tensile strength contained in the elliptic yield curve, although small, may lead to instabilities. In these cases, ice–water stress was ignored; but, had it been included, the instabilities would not form. Consequently, the instability problem may exist only in the most extreme situations. Instabilities could be totally avoided by limiting the whole yield curve to the third quadrant in the principal stress space (Gray and Killworth, 1995). Later, Gray (1999) analysed the case of uniaxial divergent flow and showed that compactness distribution develops into a deepening finger-shaped form, but does not actually become unstable.

Further developments of the model include the use of a multi-level thickness distribution for ice state (Hibler, 1980a). The rheology is inconsistent in having stress for an ice field at rest, but this has been removed: the original rheology gives $\boldsymbol{\sigma} = -\frac{1}{2}P\mathbf{I}$ for $\dot{\boldsymbol{\varepsilon}} = 0$, evidently incorrect, and the revised form has $\boldsymbol{\sigma} \to 0$ as $\dot{\boldsymbol{\varepsilon}} \to 0$ (Hibler, 2001). Flato and Hibler (1990) designed a cavitating fluid version in which shear stresses are ignored, $e = \infty$, leading to less computationally expensive solutions. An anisotropic extension was developed by Hibler and Schulson (2000) to examine the dynamics of oriented lead and crack systems.

The model has been used for climate studies and short-term regional ice forecasting. It has been shown to be feasible with about the same parameterization over a wide range of time- and spacescales. Hunke and Dukewicz (1997) developed an elastic–viscous–plastic scheme for the numerical solution of Hibler's (1979) model. It is easily applicable to parallel computing, in particular.

8.2.3 The AIDJEX model

The AIDJEX model is the product of the extensive AIDJEX programme performed in the 1970s (Coon, 1980). The new features introduced by the AIDJEX group were an elastic–plastic rheology (Coon et al., 1974) and the concept of thickness distribution (Thorndike et al., 1975). The system of equations reads:

$$J = \Pi \tag{8.7a}$$

$$\left. \begin{array}{l} F(\sigma_I, \sigma_{II}, P^*) \leq 0, \quad P^* = c_p(1+\Gamma) \int_0^\infty h^2 \psi'_- \, dh \\ F < 0 : \boldsymbol{\sigma} = (M_1 - M_2)\, \mathrm{tr}\, \boldsymbol{\varepsilon}\mathbf{I} + 2M_2 \boldsymbol{\varepsilon}',\ \dot{\boldsymbol{\varepsilon}} - \dot{\boldsymbol{\Omega}} \cdot \boldsymbol{\varepsilon} + \boldsymbol{\varepsilon} \cdot \dot{\boldsymbol{\Omega}} = \dot{\boldsymbol{\varepsilon}} - \dot{\boldsymbol{\varepsilon}}_p \\ F = 0 : \dot{\boldsymbol{\varepsilon}}_p = \lambda \dfrac{\partial F}{\partial \boldsymbol{\sigma}} \end{array} \right\} \tag{8.7b}$$

$$\rho h \left[\frac{D\boldsymbol{u}}{Dt} + f\mathbf{k} \times \boldsymbol{u} \right] = \nabla \cdot \boldsymbol{\sigma} + \boldsymbol{\tau}_a + \boldsymbol{\tau}_w - \rho h g \boldsymbol{\beta} \tag{8.7c}$$

$$\frac{D\Pi}{Dt} + \phi_h \frac{\partial \Pi}{\partial h} = \Psi - \Pi \nabla \cdot \boldsymbol{u} \tag{8.7d}$$

Ice state is defined as ice thickness distribution. The yield function F has the shape of a teardrop (Figure 4.4). Its size is defined by strength P^*, which depends on the ridging function ψ'_-, buoyancy parameter $c_p = \frac{1}{2}\rho g(\rho_w - \rho)/\rho_w$, and the parameter Γ equal to the ratio of frictional losses to the production of potential energy in ridging (see Section 4.3.3). Inside the yield curve, rheology gives elastic behaviour, while on the yield curve plastic deformation $\dot{\varepsilon}_p$ takes place according to the associated flow rule. The magnitude of mesoscale elastic constants is 10–100 MN/m, $M_1 \approx 2M_2$ (Pritchard, 1980b). The elastic part necessitates tracking any change from the reference configuration, which makes the numerical solution quite complicated. The momentum equation is solved in its full form, and the ice conservation law is integrated for the whole thickness distribution. Note that ice state as ice thickness distribution is taken in finite-dimensional form (i.e., as a histogram presentation).

The boundary-layer parameters were based on the AIDJEX results and are basically as in Hibler (1979). Ice rheology has two elastic constants. Strength is integrated directly from the thickness redistribution by determining the work needed to form new ridges. The coefficient Γ is equal to zero if all energy sinking at ridging is due to potential energy increase. However, with frictional losses added, it becomes $\Gamma > 0$. The thickness redistribution function has the following parameters: the band used for ridging and how large ridges will form. The usual solution is to take the lower 15% of the thickness distribution and transform them to k-multiple ice thicknesses; originally k was 5 (Coon, 1974) but later it was increased to 15 (Pritchard, 1981). The grid size was 50–100 km in the original application in the Beaufort Sea.

8.2.4 The Baltic Sea model

Sea ice model development in the Baltic Sea has progressed in Finland parallel to work in the Arctic. The first model was a linear viscous four-level model, used initially for short-term sea ice forecasting (Leppäranta, 1981a). In the late 1980s, modelling collaboration was commenced with Chinese scientists, working in the Bo Hai Sea. A three-level viscous plastic model was designed for both seas (Wu and Leppäranta, 1990; Leppäranta and Zhang, 1992b). All modelling until then had forecasting as its purpose. Haapala and Leppäranta (1996) developed a dynamic–thermodynamic regional sea ice climate model, where the dynamic part was in the one used by Leppäranta and Zhang (1992b).

The present Baltic Sea ice dynamics model differs from the Hibler (1979) model in that a three-level ice description is included, and the momentum equation is quasi-steady-state with sea surface tilt ignored. The system of equations reads:

$$J = \{h_u, h_d, A\} \tag{8.8a}$$

$$\left. \begin{array}{l} \boldsymbol{\sigma} = \left(\zeta - \dfrac{P}{2}\right)\mathbf{I} + 2\eta\dot{\boldsymbol{\varepsilon}}', \quad P = P^*\boldsymbol{h}\exp[-C(1-A)] \\[6pt] \zeta = \dfrac{\dot{\varepsilon}_I P}{2\max(\Delta,\Delta_0)}, \quad \eta = e^{-2}\zeta, \quad \Delta_0 = \sqrt{\dot{\varepsilon}_I^2 + (\dot{\varepsilon}_{II}/e)^2} \end{array} \right\} \tag{8.8b}$$

$$\rho h f \mathbf{k} \times \boldsymbol{u} = \nabla \cdot \boldsymbol{\sigma} + \boldsymbol{\tau}_a + \boldsymbol{\tau}_w \tag{8.8c}$$

$$\frac{\mathrm{D}\{h_u, h_d, A\}}{\mathrm{D}t} = \{0, \psi_d, -A\nabla\cdot\boldsymbol{u}\} + \{\phi_u, \phi_d, \phi_A\} + D_1\nabla^2\{h_u, h_d, A\}, \quad 0 \leq A \leq 1 \tag{8.8d}$$

Table 8.2. Parameterization of the Baltic Sea viscous–plastic sea ice model (Leppäranta and Zhang, 1992b; Haapala and Leppäranta, 1996). The parameter groups are I—atmospheric and oceanic drag parameters; II—rheology parameters; III—ice state redistribution parameters; and IV—numerical design parameters.

	Parameter	Value
I	Air–ice drag coefficient, C_a	1.8×10^{-3} (surface wind)
	Air–ice Ekman angle, θ_a	0 (surface wind)
	Air–water drag coefficient, C_w	3.5×10^{-3} (vertically averaged current velocity)
	Air–water Ekman angle, θ_w	17° (vertically averaged current velocity)
II	Compressive strength consant, P^*	25 kPa
	Aspect ratio of yield ellipse, e	2
	Strength constant for opening, C	20
	Maximum viscous creep rate, Δ_0	$2 \times 10^9\,\mathrm{s}^{-1}$
III	Demarcation thickness, h_0	10 cm
IV	Spatial grid size, $\Delta x = \Delta y$	18 km
	Time step, Δt	6 h
	Harmonic diffusion coefficient, D_1	500 m²/s
	Biharmonic diffusion coefficient, D_2	0

where h_u and h_d are the thicknesses of undeformed and deformed ice, respectively, and ψ_d is the ridging function:

$$\psi_d = \begin{cases} -(h_u + h_d)\nabla \cdot \boldsymbol{u}, & \text{if } \nabla \cdot \boldsymbol{u} < 0 \text{ and } A = 1 \\ 0, & \text{otherwise} \end{cases} \quad (8.9)$$

and $\{\phi_u, \phi_d, \phi_A\}$ is the thermodynamic change of ice state. The early four-level version split the thickness of deformed ice into mean ridge size and ridge density, but since observations showed that the mean ridge size varied very little it was possible to go down to three levels. It has been possible to validate the model outcome of the volume of ridged ice against observations with good results.

The model parameters were determined from Baltic Sea field data and model experiments (Table 8.2), while the boundary-layer parameters were based on field experiment results (Joffre, 1984; Leppäranta and Omstedt, 1990). Both drag coefficients are in essence larger in the Arctic, by factors of 1.3 for the air drag coefficient and 1.6 for the water drag coefficient, due to the greater roughness of Arctic ice. For the surface wind the asymptotic thin-ice wind factor is 2.6% and the deviation angle is 17°; this wind factor is a little less than in the Arctic, but because the ice is much thinner in the Baltic the real (observed) wind factor is higher.

Compressive strength was tuned to 25 kPa, and the range in different numerical experiments against field data was 10–50 kPa (Zhang and Leppäranta, 1995; Leppäranta et al., 1998). The aspect ratio e and the strength reduction constant for opening were equal to the original values (Hibler, 1979).

The demarcation thickness was 10 cm, less than typical ice thicknesses. There was also a physical background in that ice thinner than about 10 cm underwent rafting under compression and offered much less resistance to deformation, as validated from basin-wide ice kinematics data (Leppäranta et al., 1998).

This model was recently examined in detail for dynamics in different basins, in particular for scaling and the influence of coastal geometry and islands (Leppäranta and Wang, 2002; Wang et al., 2003). It worked well down to a bay size of 15 km with thin ice (10 cm) moving under strong winds (25 m/s).

8.3 SHORT-TERM MODELLING APPLICATIONS

In short-term modelling the timescale ranges from 1 hour to 10 days. The approach is often purely dynamic (i.e., thermodynamics is neglected). The initial field of ice state must therefore be given. The objective of short-term modelling includes basic research into drift ice dynamics, coupled ice–ocean–atmosphere modelling, simulations to examine the influence of ice dynamics on planned marine operations (Figure 8.4), and ice forecasting.

8.3.1 Research work

The knowledge of sea ice dynamics can be improved by numerical experiments. This is especially true of rheology and thickness redistribution problems. Direct evidence is in short supply for these questions. However, with accurate indirect information, model simulations may provide information about background physics: as an example, sea level variations in the Gulf of Bothnia in ice conditions can provide data on the losses of kinetic energy in mechanical deformation (Figure 8.5).

Figure 8.4. Shipping in ice-covered seas has penetrated deeper and deeper into the ice pack. The first ship to reach the North Pole was the Soviet nuclear ice-breaker *Arktika* in summer 1977.
Reproduced with permission from the Russian State Museum of the Arctic and Antarctic, St Petersburg.

Sec. 8.3]   **Short-term modelling applications**   215

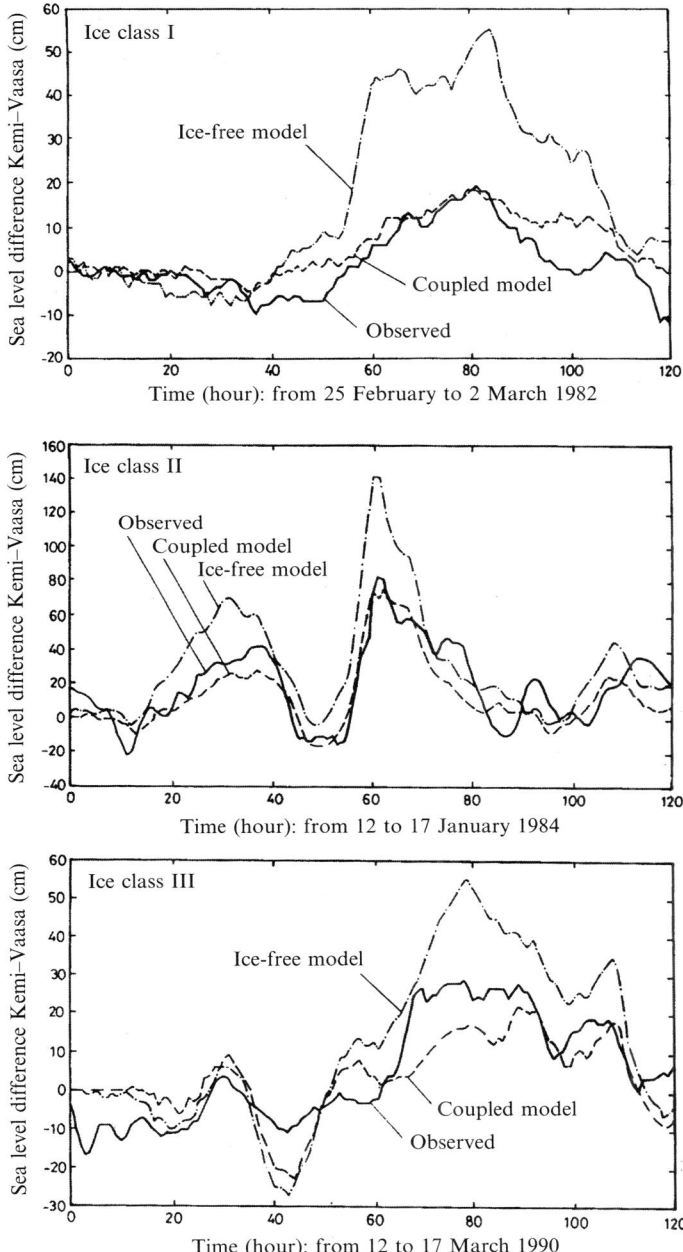

Figure 8.5. Simulated sea level elevation in the Gulf of Bothnia, Baltic Sea. Ice-free model and viscous–plastic coupled ice–ocean model have been employed. Ice classes: I—severe, II—normal, III—mild.
From Zhang and Leppäranta (1995).

The question of ice–tide interaction has been investigated using short-term ice–ocean dynamics models. In the presence of internal friction, ice introduces an additional damping effect on tides. Kowalik (1981) examined the situation in the Arctic Ocean using a non-linear sea ice model, and results showed good overall agreement between the tides in the real and ice-free Arctic Ocean. The influence of ice was mostly seen in coastal regions. Between the ice and water velocities there is a time lag, which is a complicated function of internal ice friction and ice–water stress parameters. Tidal amplitude appears damped in the ice, and, due to the non-linearity of sea ice rheology, a residual ice drift results from the tidal cycle. Kowalik and Proshutinsky (1995) examined the topographic influence on tides in the Barents Sea, and the model predicted trapping of ice dynamics at Bear Island by the residual tidal flow (basically in agreement with observations).

A large amount of short-term sea ice modelling work has been done in the marginal ice zone (MIZ). In the neighbourhood of the ice edge, an intensive air–ice–ocean interplay takes place and the location of the ice edge introduces a discontinuity in the characteristics of the air–sea interface, together with velocity and temperature.

The situation allows a 1.5-dimensional approach[2] that assumes alongshore, or longitudinal, variations to be much less than transverse variations and, consequently, alongshore derivatives can be ignored (see Section 7.3). The two-level viscous–plastic model was solved analytically for steady-state conditions and numerically for non-steady-state cases by Leppäranta and Hibler (1985). Comparison between the continuum viscous–plastic model and a discrete particle model was made by Gutfraind and Savage (1997), and the best agreement was obtained when the Mohr–Coulomb yield criterion was used in the viscous–plastic model. This is to be expected since the physical basis is similar in the Mohr–Coulomb and discrete particle models.

Similar 1.5-dimensional coupled ice–ocean models have been used to further examine the MIZ and coastal dynamics. A general coupled model for coastal seas was presented by Overland and Pease (1988) with a barotropic ocean. One major topic is upwelling at the ice edge. The problem can be approached by analytical modelling of the steady state (see Section 7.3), where the surface stress difference across the ice edge may have either sign, resulting in upwelling or downwelling analogous to coastal ocean dynamics. A more detailed oceanic case allowing an analytical solution was studied by Van Hejst (1984). Other ice edge process models include banding (Häkkinen, 1986) and ice edge eddies.

Regional coupled ice–ocean modelling has been one major question, aiming at research and forecasting purposes. The Baltic Sea and the Bo Hai Sea are discussed below. In other regions, models for the Sea of Okhotsk have been presented by Rheem et al. (1997), for the Labrador Sea by Ikeda (1985), Keliher and Venkatesh (1987), and Fissel and Tang (1991), and for Hudson Bay by Wang et al. (1994).

---

[2] In these 1.5-dimensional models there are two space co-ordinates, but the dependent variables are allowed to vary only along one of the co-ordinates.

Example (Lehmuskoski and Mäkinen, 1978) Sea ice is a good platform for gravity measurements over the ocean. This has been utilized by the Finnish Geodetic Institute in the Baltic Sea. However, because of Coriolis acceleration, the eastward component of sea ice velocity affects the apparent local gravity. Therefore, short-term ice dynamics simulations have been made to correct gravity measurements. ∎

8.3.2 Sea ice forecasting

Sea ice conditions can change over short timescales due to dynamics. Leads up to 20 km wide may open and close in a single day, and heavy pressure may build up in the compression of compact ice. These processes have a strong influence on shipping, oil drilling, and other marine operations. Changes in ice conditions, such as the location of the ice edge, are important for weather forecasting over a few days. Consequently, short-term ice forecasts are crucial for ice-covered seas, in particular in the seasonal sea ice zone when human activities are at their height.

Sea ice drift has long been predicted using wind factor and deviation angle rules (e.g., Vasiliev, 1985). However, it is only with numerical models that the full sea ice problem can be solved for realistic basin-wide dynamics. The first prognostic sea ice model was a two-level linear viscous ice model for the Kara Sea (Doronin, 1970), applicable to sea ice forecasting. The Hibler and AIDJEX models (Coon, 1980; Hibler, 1979) have also been applied to the ice forecasting problem.

The key areas of short-term modelling research are now ice thickness distribution and evolution, and the use of satellite synthetic aperture radars (SARs) for ice kinematics. The scaling problem and in particular the downscaling of stress from geophysical to local (engineering) scale is examined by combining scientific and engineering knowledge and developing ice load calculation and forecasting methods. The physics of drift ice is quite well represented in short-term ice-forecasting models, in the sense that other questions are more critical for their further development. In particular, the data assimilation problem has not been much examined for sea ice models.

Baltic Sea

In the Baltic Sea the main harbours have been kept open all year since 1970. Winter shipping has been assisted by 20–25 ice-breakers, and even then transportation systems have suffered from delays. This was the catalyst for an extensive research programme in the 1970s in Finland and Sweden, organized by a joint winter navigation research board. One of the research aims was to develop an ice-forecasting system based on mathematical modelling (Udin and Ullerstig, 1976; Leppäranta, 1981a). The first real ice-forecasting season was winter 1977 (Figure 3.6), based on a four-level extension of the Doronin (1970) model (Leppäranta, 1981a). Initially, the forecast period was 30 hours, limited by the period of available wind forecasts. The grid size was 27 km and the time step was 6 hours, the same as the time step in the wind forecast. The ice forecast, which was sent to the ice-breakers by fax,

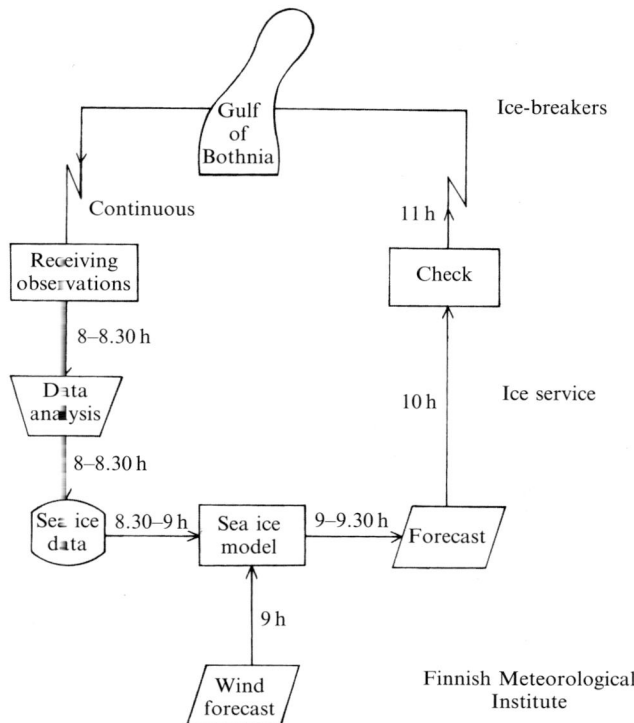

Figure 8.6. The structure of the Finnish ice-forecasting system in winter 1977 in the Finnish Institute of Marine Research.
From Lepparanta (1981a).

contained ice speed, direction of ice drift, change in ice concentration, and qualitative pressure field.

The outcome in the first winter was marginally positive (Lepparanta, 1981a). Comparison with real ice conditions showed satisfactory results, and, in answer to a questionnaire, the ice-breaker personnel concluded that the most important quantity was ice drift velocity and its direction with respect to the fast ice boundary. This is understandable since the opening and closing of leads at the fast ice boundary is the process that is key to shipping (Figure 8.7).

The system was continued for subsequent winters along with model improvement, and, as a result, lengthening of the forecast period. A viscous–plastic three-level model was first adopted for operational ice forecasting in 1992 (Lepparanta and Zhang, 1992b). This model was also integrated into the ice-forecasting system in Sweden (Omstedt et al., 1994). With the progress of satellite remote-sensing techniques, initial ice compactness became better mapped. However, the thickness of ice has continued to be a difficult quantity, with updates coming from occasional ship reports. From the outset any new initial ice thicknesses were constructed from a hindcast combined with new observations.

Figure 8.7. Convoy of ships assisted by an ice-breaker in the Baltic Sea. Ice dynamics determines the length of a convoy.

Bo Hai Sea

Another region subject to modelling has been the Bo Hai Sea, China (Wu et al., 1998). A three-level dynamic–thermodynamic sea ice model has been employed for 3–5-day sea ice-forecasting since 1990 by the National Research Centre for Marine Environment Forecasts, Beijing. The model is linked to numerical weather forecasting. Its dynamics is similar to the Baltic Sea ice model (Wu and Leppäranta, 1990). The principal motivation behind the Bo Hai Sea ice model is to serve the oil-drilling operation of the Bo Hai Oil Corporation in the basin.

Ice motion in the Bo Hai Sea is forced by strong, cold winds from the north and strong tides of up to 1 m/s (Wu et al., 1998). The ice is fairly thin, typically 25 cm. The model is initialized by a composite analysis of *NOAA* satellite images, aerial reconnaissance, and ground data as well as real-time data from an oil-drilling platform in Liaodong Bay. The grid size is about 10 km. Ice forecasts are sent to users by fax and email (Figure 8.8). Between 1991 and 1996 some 30–80 ice forecasts were prepared each winter, and verification showed that error in location of the ice edge was within 10 km in 80–90% of all cases for 3-day forecasts and in 50–80% of all cases for 5-day forecasts.

Arctic seas

In Canada, the Atmospheric Environment Service started ice forecasting in 1979, intially in the Beaufort Sea (Neralla et al., 1988). The grid size was then 42 km, with a time step of 3 hours and total length of forecast period 48 hours. At that time a similar ice-forecasting model also developed for Lake Erie in the USA (Chieh et al.,

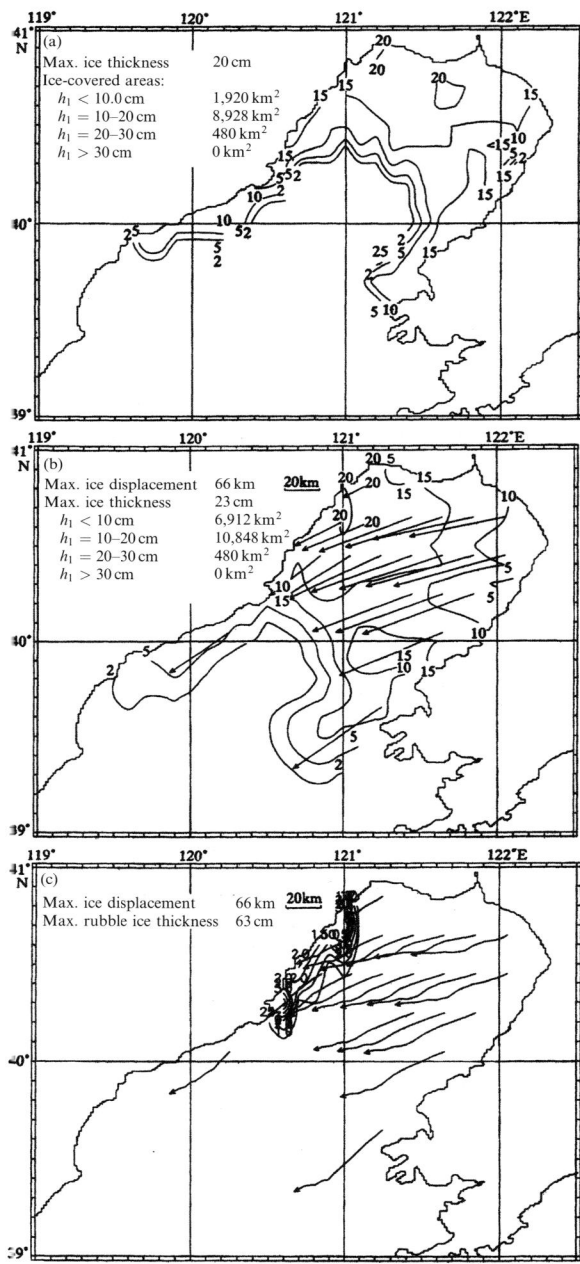

Figure 8.8. Five-day ice forecast for the Liaodong Bay, Bo Hai Sea, from 13 January 1993: (a) initial ice state; (b) ice thickness and velocity forecast; and (c) ice rubble thickness and track forecast.
From Wu et al. (1998).

1983). The Arctic and Antarctic Research Institute (AARI), St Petersburg, has investigated the ice-forecasting problem since its formation in the 1920s, and presently a numerical model-based forecast is prepared for the Arctic seas on a regular basis (see http://www.aari.nw.ru/).

Oil spills

Oil spills are difficult problems under normal circumstances but in ice conditions they are particularly so (Figure 8.9). Effective oil spill-cleaning methods do not exist, and it is difficult to keep track of where the oil is going. Oil may penetrate into the ice sheet and drift with the ice or drift on the surface of openings and beneath sea ice. A simple modelling approach is an oil advection model with ice and surface current, with random diffusion superposed, using a Monte Carlo method. An advanced, physical model treats oil as a viscous medium with density and viscosity dependent on the type of oil (e.g., Venkatesh et al., 1990). Venkatesh et al. state that, for ice compactness greater than 30%, oil practically drifts with the ice, and that in slush or brash the thickness of the oil film can be much larger than in open cold water. When compactness reaches the 80% level the oil is trapped between ice floes and above 95% the oil is forced beneath the ice.

Figure 8.9. Oil spill in the Gulf of Finland.

Photograph by Jouko Pirttijärvi, reproduced with permission from the Finnish Environment Centre, Helsinki.

Modelling the boundaries of an oil spill calls for an advanced advection scheme. Ovsienko et al. (1999a) used the particle-in-cell technique in their model, which has been applied to several areas of Arctic seas.

8.4 LONG-TERM MODELLING APPLICATIONS

In long-term modelling the timescale ranges from 1 month to 100 years. The approach is dynamic–thermodynamic, and sometimes even the dynamics is ignored. Initial conditions are arbitrary in very long timescales, but relevant for ice state in monthly problems. The objective of long-term modelling includes basic research into drift ice geophysics, ice climatology investigations, and coupled ice–ocean–atmosphere climate modelling.

The role of ice dynamics in the climatology problem is to provide transport of ice to regions where it would not be formed. This transport modifies the ice boundary and, therefore, the air–sea fluxes of momentum, heat, and matter. It also transports latent heat and freshwater. Equally important is differential ice drift: leads open and close, resulting in major changes to air–sea heat fluxes, and the mechanical accumulation of ice blocks, like ridging, adds large amounts to the total volume of ice.

Long-term sea ice modelling has increased as a result of climate research, giving computational possibilities to have more realistic coupled ice–ocean and atmosphere–ice–ocean models. Initially, only thermodynamic models were available for forecasting the times of freezing and ice break-up and for the evolution of ice thickness. But it became rapidly clear that a realistic ice dynamics is needed for ice transport and for the opening and closing of leads. Large amounts of heat are transmitted through leads from the ocean to the atmosphere. By freezing and melting the ice has a major influence on the hydrographic structure of the ocean. Therefore, the motion of ice has an important role since ice often melts in a different region from the one in which it forms.

8.4.1 Arctic regions

The first realistic sea ice climate model for the Arctic Ocean was that by Hibler (1979). The mean annual velocity from the simulation together with the mean thickness field corresponded well when validated against submarine data[3] (Figure 8.10; see also Figures 2.12 and 3.17). The model predicted that there is always at least a small amount of open water or thin ice present in the central Arctic Ocean and, further, that with dynamics the annual cycle of mean thickness has about the same amplitude as it does without dynamics, but the level drops by about 1 metre. Maximum ice thicknesses were 5–8 m due to mechanical deformation off the northern coast of Greenland.

Almost ten years later, coupled ice–ocean climate models with full dynamic–

---

[3] Hibler (1979) used submarine data received from L.A. LeSchack by private communication; some of these data are reported in LeSchack et al. (1971).

Sec. 8.4] Long-term modelling applications 223

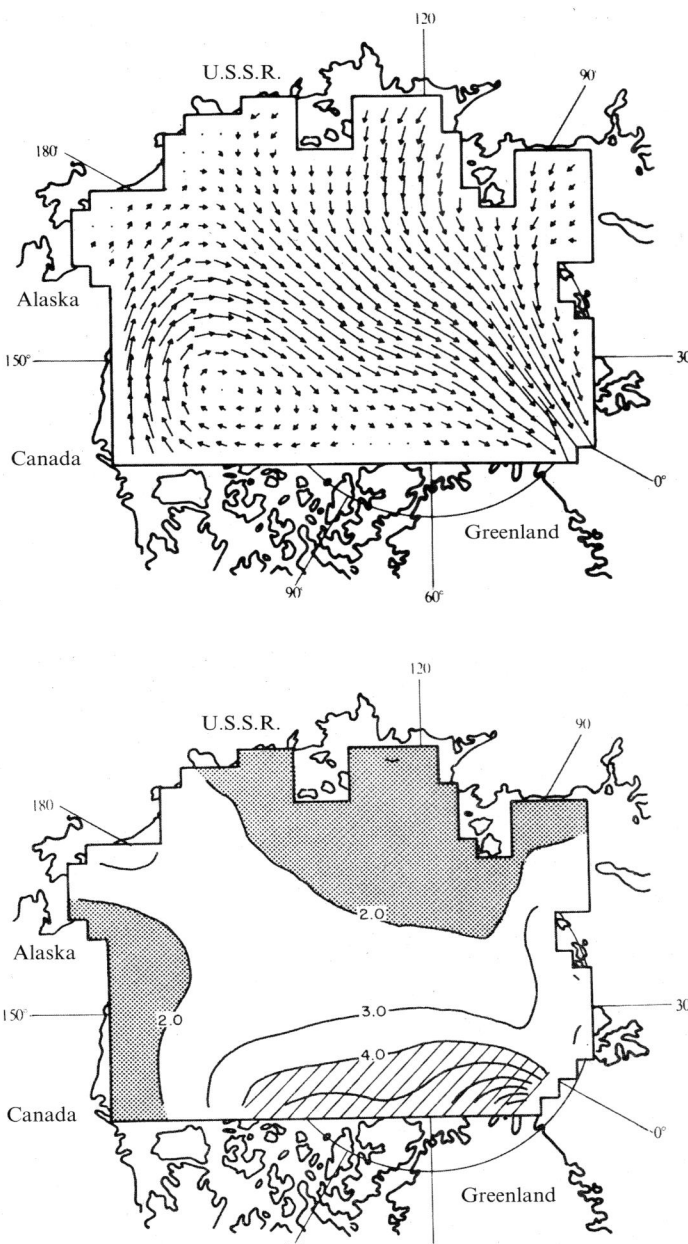

Figure 8.10. The average annual ice velocity (*top*) and April thickness contours (*bottom*) produced by the model of Hibler (1979).
Reproduced from Hibler (1979), with permission from the American Meteorological Society..

thermodynamic ice physics were introduced and the seasonal ice cycle was improved in a diagnostic model (Hibler and Bryan, 1987). In more recent models, resolution, numerical technology, and coupling have all been improved, but the ice physics itself is not much different from what it was in the ice-only models of the 1980s. Better model bases have also improved the outcome for ice state and motion (e.g., Zhang et al., 1998; Arberter et al., 1999; Hunke and Zhang, 1999).

8.4.2 Antarctic regions

Sea ice models in Antarctic seas have been basically free drift models or viscous–plastic Hibler-type models. Hibler and Ackley (1983) applied the Arctic model to Antarctic regions, with the essential changes of much greater strength, $P^* = 27.5$ kPa, and higher demarcation thickness, $h_0 = 1$ m. The high strength was explained by temporal spacing of the wind data, while the higher demarcation thickness, which may seem strange, was explained on the basis of the large amounts of frazil ice in the Antarctic and, therefore, a large demarcation thickness would be required by thermodynamics. But this means that the internal friction of ice thinner than 1 m does not influence model ice dynamics. The grid size used by the model was 222 km. Stössel et al. (1990) used the model for the entire Southern Ocean.

Timmermann et al. (2002) employed the Hibler model for long-term simulations in the Weddell Sea. The boundary-layer parameters were the following: the ice–air drag coefficient was 1.32×10^{-3} for the surface wind, the ice–ocean drag coefficient was 3×10^{-3}, and the turning angle was $-10°$ for the upper layer velocity in the ocean model. The thickness of the upper layer was 8–40 m, smaller in shallow areas. The thin-ice wind factor then becomes 2.4%; this is a bit low for the Antarctic, as is the deviation angle, but it can be to some degree explained by the fact that the upper layer water current was the reference for water stress, not the geostrophic flow. The compressive strength constant was taken as $P^* = 20$ kPa. The model grid size was about 165 km, and the model sea ice outcome was calibrated with drift buoy data for ice velocity (Kottmeier and Hartig, 1986) and upward-looking sonar data for ice thickness (Strass and Fahrbach, 1998) (Figure 8.11). There is a strong convergence region in the south-west part of the basin, and advection of the ice shows up in larger ice thickness northward along the Antarctic Peninsula. The width of the compressive region east of the Peninsula is around 500 km.

8.4.3 Baltic Sea

A dynamic–thermodynamic model was developed for the Baltic Sea for investigations into the ice climatology of the region (Haapala and Leppäranta, 1996). Its dynamics differs from that of the Hibler model: a three-level ice description is included, the momentum equation is quasi-steady-state, and the surface pressure gradient is ignored. It is a Hibler-type, three-level (compactness, undeformed ice, deformed ice), dynamic–thermodynamic ice model coupled to an ocean model with shallow-water circulation and four-layer, vertical heat transfer. The model was

Sec. 8.4] Long-term modelling applications 225

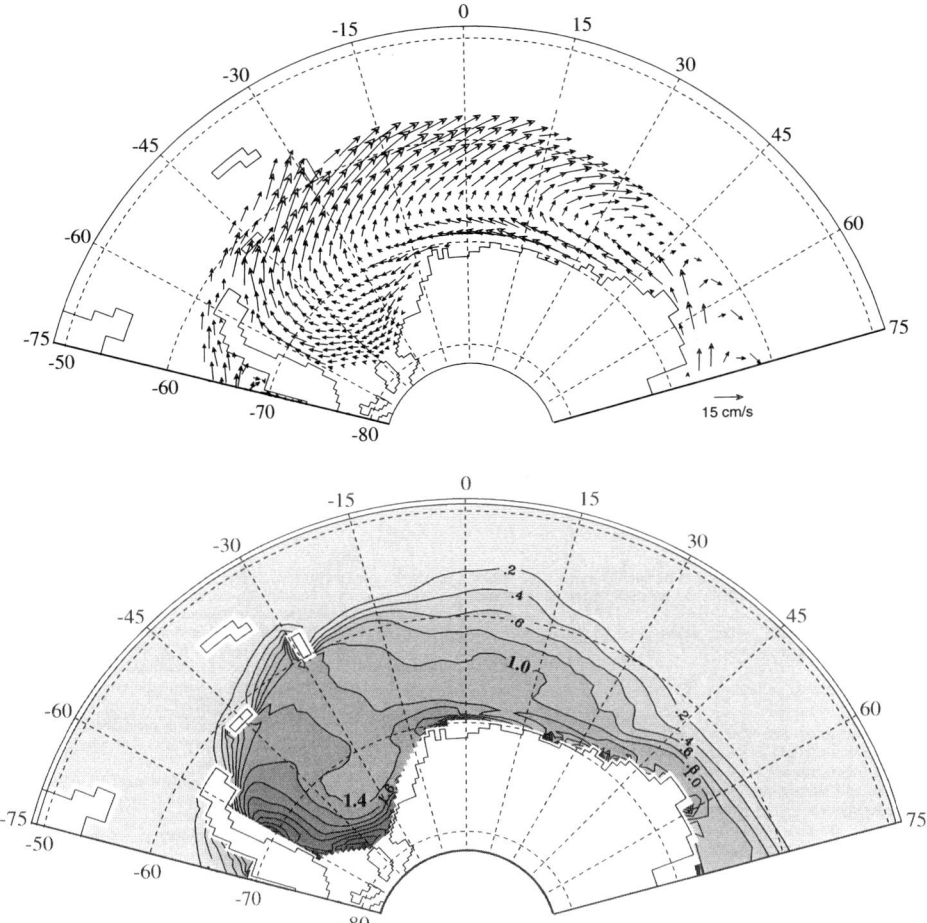

Figure 8.11. Climatological (*top*) sea ice velocity and (*bottom*) sea ice thickness in the Weddell Sea.
From Timmermann et al. (2002), with permission from the American Geophysical Union.

forced by prescribed atmospheric evolution and had a grid size of 18 km. It was calibrated using three different winter categories (mild, normal, and severe) and was used for simulations of future ice seasons under different atmospheric climate scenarios (with an end date of 2100).

Figure 8.12 gives an example of the outcome, in which model calibration is shown for the normal ice season of 1983/1984. It shows total ice thickness, the thickness of deformed ice, and their comparison with an operational ice chart at the time of the maximum ice extent. Overall agreement is quite good; the main discrepancy is at latitudes 61–62°N where there was an open-water region

226   **Numerical modelling**   [Ch. 8

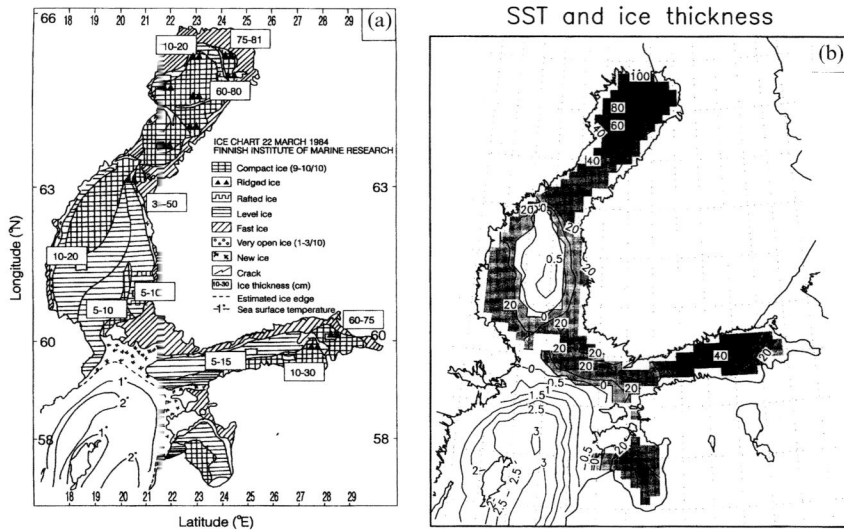

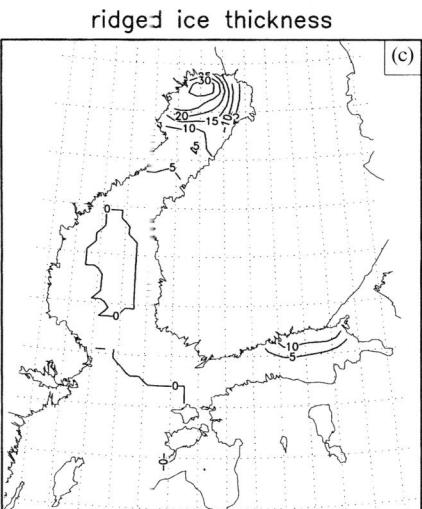

Figure 8.12. Ice conditions and sea surface temperature (SST) on 22 March 1984 in the Baltic (normal winter): (a) observed situation according to an operational chart; (b) modelled SST (°C) in open sea (white area) and total ice thickness (cm); and (c) modelled thickness of deformed ice.
From Haapala and Leppäranta (1996).

in the model result (due to excessive vertical convection in the model). A thermodynamic-only model would give smoothly increasing thicknesses from the ice edge to 60–90 cm in the eastern and northern corners of the Baltic.

In the open-water region mentioned above the real ice thickness was only

Sec. 8.4] Long-term modelling applications 227

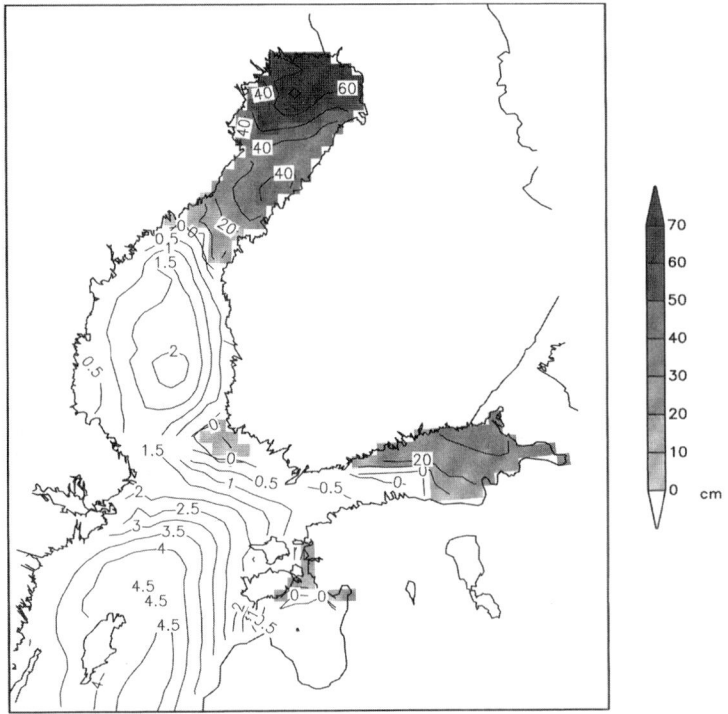

Figure 8.13. Average ice situation at the annual maximum in the Baltic Sea, 2050 ± 15 years, based on a regional sea ice model forced by an expected climate-warming scenario. Grey tones show ice thicknesses, and the white area is the open sea with numbers showing the surface temperature (°C).
From Haapala and Leppäranta (1997a).

5–10 cm, and so model bias was not too bad. Deformed ice comes out well in the model. Significant values are shown north of 63°N and east of 26°E; these correspond to the ridged areas shown qualitatively by the black triangles in the operational ice chart. Numerical values cannot be compared with quantitative data from the same season, but they correspond to the volume of ridged ice in the basin observed in another winter (Lewis et al., 1993). Comparison with the amount and distribution of deformed ice is sensitive to the sea ice dynamics representation in a model. Figure 8.13 shows the model outcome for the year 2050 forced by a weather generator with suggested climate warming as the input. It is seen that predicted ice coverage is less than half the true normal winter level between 1960 and 1990 (Figure 8.12).

This completes our review of the current status of the numerical modelling of sea ice drift. Although the solution is rather good, there still are some open questions regarding its physics, numerical technology, and coupling with atmosphere and ocean. Interesting research topics for the modelling problem as well as for the whole sea ice drift problem are outlined in Chapters 9 and 10.

9

Use and need for knowledge on ice drift

9.1 SCIENCE

The ice drift problem contains interesting *basic research* questions. It is a two-dimensional problem in the branch of mechanics called geophysical fluid dynamics. The medium is a compressible fluid, whose dynamic characteristics are its largely variable strength (low to very high), low inertia, and the irreversibility of mechanical deformation processes. In ice-only dynamics, adjustment of the ice mass to external forcing takes place, ruled by the strength of ice and the forcing characteristics. As ice is relatively thin, the direct Coriolis effect on it is rather weak, in contrast to the atmosphere and ocean, and, therefore, Coriolis acceleration gives second-order modifications.

The principal open questions in sea ice dynamics are the sea ice rheology, redistribution of ice in deformation, and scaling. Rheology seems to be an ever-lasting topic, such as turbulence in fluid dynamics. As far as thickness distribution is concerned, many theories and models exist, but the lack of good data is a major barrier to progress. The thickness-mapping problem is common to most sea ice research. The scaling of sea ice dynamics is becoming better understood and is closely connected to the rheology and redistribution of ice. The theory and models of sea ice dynamics have been utilized for ice dynamics in large lakes and rivers, the essential feature being that floating ice breaks into floes and then drifts (Wake and Rumer, 1983; Shen et al., 1993).

The dynamics of sea ice is intimately coupled with the upper ocean. The dynamic ice–ocean interaction is dictated by skin friction and form drag due to hummocks and ridges at the ice–water interface. At extrema the ice may drift along with surface currents or may act as a stationary lid providing frictional resistance to surface currents. In the adjustment of ice–ocean dynamics, atmospheric kinetic energy is transferred through drifting ice to the ocean, where it is modified by ocean dynamics, and a part is returned to the ice.

So, the drift of sea ice receives many of its characteristic features from atmospheric and oceanic boundary layers in the interplay of the three media. The principal problem in this interplay is the parameterizing of the ice–ocean drag force; it is much less understood than the atmospheric drag force on the ice. In ice–ocean modelling the question of spatial resolution has yet to be answered: How can we construct high-resolution models when the continuum physics basis breaks down for sea ice?

Another interaction problem that has only recently been considered is that between sea ice and icebergs. The large-scale drift of compact sea ice forces the drift of icebergs (Lichey and Hellmer, 2001), while icebergs act as moving islands for sea ice drift. Modelling of coupled sea ice–iceberg dynamics would be highly desirable both for short-term and long-term questions.

But the underlying scientific motivation for the examination of sea ice dynamics is indirect: to understand the entire *atmosphere–ocean interface* and consequently the physics of the atmosphere–ocean system at high latitudes. The exchange of momentum, heat and freshwater between the atmosphere and the ocean takes place in sea ice fields, where the interface experiences changes in location and in structure due to ice motion.

Sea ice has an important role in the evolution of hydrographical conditions in the upper ocean. In the presence of ice the ocean surface temperature is locked into the freezing point, and the amplitude of the annual surface temperature cycle is small. The salinity of new ice is about half that of seawater and the salinity of multi-year ice is around 1‰ (e.g., Weeks, 1998a). Therefore, as sea ice grows, brine is absorbed by the surface layer, which decreases the stability of the stratification, while as sea ice melts the surface layer becomes more fresh and stable.

The location of the sea ice edge introduces a discontinuity in the air–ice–ocean interface and, consequently, particular dynamic and thermodynamic phenomena follow. In polar oceans the ice edge and oceanic polar front locations constitute a coupled problem, just like the one in the austral winter in Antarctica. Whereas in small basins the location of the ice edge is basically controlled by seasonal atmospheric conditions, in the Baltic Sea the winter ice extent is well correlated with the North Atlantic Oscillation (Tinz, 1996).

The transport of ice takes sea ice to regions where it would not be formed by thermodynamic processes alone. Good examples are the southward flows of ice in the East Greenland Current and in the Labrador Current. Within the ice pack, ice transport influences the local age of ice and thus the distribution of first-year and multi-year ice. Along the whole Siberian Shelf, ice drifts north and the local ice is therefore predominantly first-year ice, much thinner than it would be without the ice motion.

The Fram Strait is the main channel of ice outflow from the central Arctic Ocean (e.g., Rudels, 1998). Ice transport is $0.1 \times 10^6 \, m^3/s$, about equal to river run-off into the basin. Ice outflow is about 5% of the outflow of polar water, but with its latent heat the ice has more influence on the heat exchange between the central Arctic Ocean and the Greenland Sea. The flow integrates to about $3 \times 10^3 \, km^3$ of ice in a year; melting this in an area of $3,000 \, km \times 1,000 \, km$ would correspond to precipitation of 1,000 mm per year.

Due to differential sea ice drift, leads open and close, and hummocks and ridges form: this is crucially important and makes the sea ice drift problem a serious challenge, since differential motion is difficult to solve with good accuracy. Drift ice also plays a role in the life history of polynyas, as thin ice and frazil ice are easily transported due to their low strength.

Leads and polynyas constitute an extremely sensitive element in the atmosphere–ocean physics in polar seas. For the transfer of momentum, the strength of ice and therefore its mobility and stiffness are highly sensitive to leads. Heat flux in winter through open water may be up to 100 times as effective as through the ice, and in summer open-water surfaces absorb solar radiation up to 10 times as effectively as the ice cover. Consequently, heat flux is extremely sensitive to sea ice dynamics. Similarly, leads and polynyas also play an important role in the exchange of moisture between the atmosphere and the ocean.

Ridges and hummocks are thick accumulations of sea ice blocks, and with scales of 100 km their mean thickness may be two or three times as much as that of thermally grown ice. These accumulations have high strength and are no longer deformable; due to their large volume they provide thermal inertia for the melting of ice. The principal problem with our understanding of ridging and hummocking is to discover how ice is broken and rearranged into a new configuration (in particular, how the thickness distribution changes during deformation). Research on this problem is severely limited by the inadequacy of existing technical methodologies to map the thickness of sea ice.

The global sea ice cover is an important factor in the *climate system*. Atmosphere–ocean heat exchange is to a great extent directly affected by sea ice area and compactness at high latitudes. The summer heat budget is also present in the cryospheric albedo problem, with changes in snow and ice surface area having a positive feedback. Growth of the snow and ice area decreases the absorption of solar radiation by the Earth, which leads to further cooling and further growth of snow and ice area, etc.; the opposite is true when snow and ice area decreases. Sea ice extent can be changed by thermodynamic or dynamic forcing, while other components of the cryosphere are influenced by thermodynamics only. The drift of ice involves the transport of latent heat and freshwater; freshwater flux into the oceanic surface layer has major consequences. Corresponding to considerable precipitation, the melting of sea ice increases the stability of stratification and polar deep convection slows down. This is potentially a serious problem in the northern part of the North Atlantic Ocean and has led scientists to closely monitor the Atlantic Conveyor.

Sea ice is an important topic in the *environmental research* of ice-covered seas (Thomas and Dieckmann, 2003). During freezing and ice growth, impurities are taken by the ice sheet from the seawater, sea bottom and atmospheric fallout (Pfirman et al., 1995; Lange and Pfirman, 1998). These impurities are transported with the ice and released into the water column during the short melting season. The drifting of sea ice may transport ice and these impurities over long distances, as takes place in the Arctic Ocean from the Siberian coast to the northern part of the North Atlantic Ocean (e.g., Volkov et al., 2002).

In the ecology of polar seas, the location of the ice edge and its ice-melting processes is a fundamental boundary condition for summer productivity (e.g., Gradinger, 1995). It appears that hydrographical and light conditions for primary production are excellent in the ocean surface layer close to the ice edge and, consequently, the dynamics of sea ice in the marginal ice zone (MIZ) is crucial for marine biology in the region (Figure 9.1). Sea ice also supports its own internal biota in brine pockets, less critically connected to ice dynamics.

A recent research area for sea ice dynamics is palaeoclimatology and palaeoceanography (Bischof, 2000). The data archive of drift ice and icebergs exists in marine sediments and the influence of drift ice on ocean circulation has been shown to have been an active agent in global climate history.

9.2 PRACTICE

In the practical world there are three major concerns connected with sea ice dynamics. First, sea ice models have been applied to tactical navigation to provide short-term forecasts of ice conditions (e.g., Leppäranta, 1981a). Second, ice forcing on ships and fixed structures is affected by the dynamic behaviour of the ice (e.g., Sanderson, 1988). Third, the question of pollutant transport by drifting sea ice has become an important issue (Pfirman et al., 1995; Lange and Pfirman, 1998). In particular, assessment of the risk of oil spills and the need for oil clean-up operations require proper oil transport and dispersion models for ice-covered seas (e.g., Ovsienko et al., 1999a).

Sea ice information services are operational routine systems to support shipping and other marine operations, such as the drilling of oil wells in ice-covered seas. Figure 9.2 shows some modern ice information products, based on satellite information, overlain on a marine chart. Ice information services exist in all countries with seasonal sea ice zones (e.g., the Arctic and Antarctic Research Institute, St Petersburg for the eastern Arctic and the Canadian Ice Service, Ottawa for the western Arctic). Due to the drift of sea ice, ice conditions can change quickly; therefore, ice charts need to be updated on a daily basis. Updating is done on the basis of remote sensing and occasional surface observations. However, with a proper sea ice model it should be possible to assimilate a model outcome using real data. Closely connected to real-time ice charting is short-term (1–10 days) ice forecasting. With today's ice chart as the initial situation, ice forecasts may be produced using a mathematical model up to the time limit in the available weather forecast (i.e., up to about a week). Such a forecasting model could also be applied to the assimilation problem.

Ice forecasting has become a routine activity for ice information services, and the demand for these forecasting systems is increasing. The development of the Northern Sea Route along the Eurasian coast of the Arctic Ocean will increase the shipping there, and the drilling of oil wells is expanding in the seasonal sea ice zone in the Sea of Okhotsk, the Barents Sea, and the Kara Sea.

The field of *ice engineering* attempts to evaluate forces on fixed structures and ships in an ice-covered sea. These are connected to the mobility and motion of sea

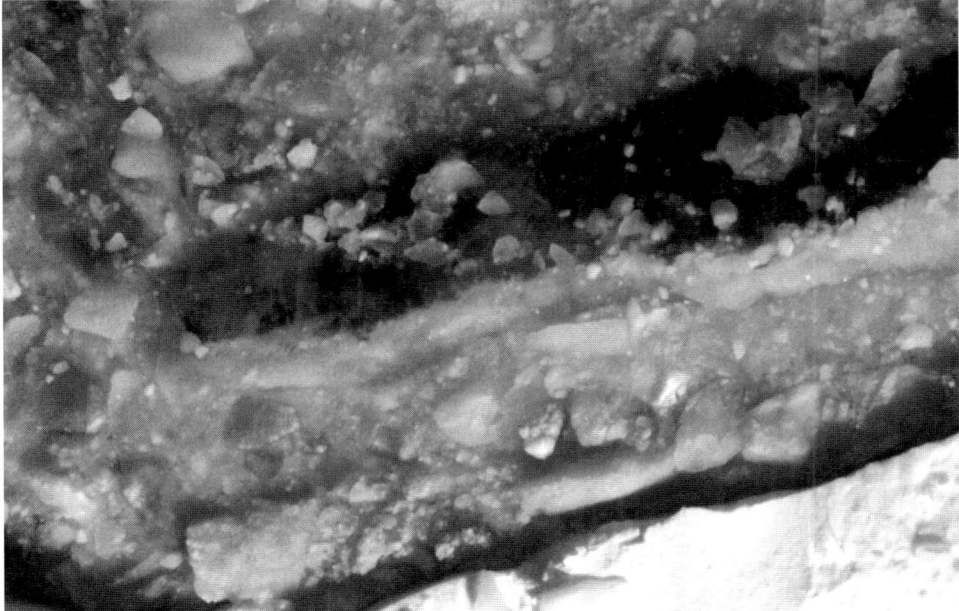

Figure 9.1. High algae productivity exists in the ocean close to the sea ice edge. (*Top*) Diatoms in Antarctica imparting a brownish–greenish colouring to the ice pack. (*Bottom*) Close-up of sea ice algae.
Reproduced with permission from Dr Johanna Ikävalko.

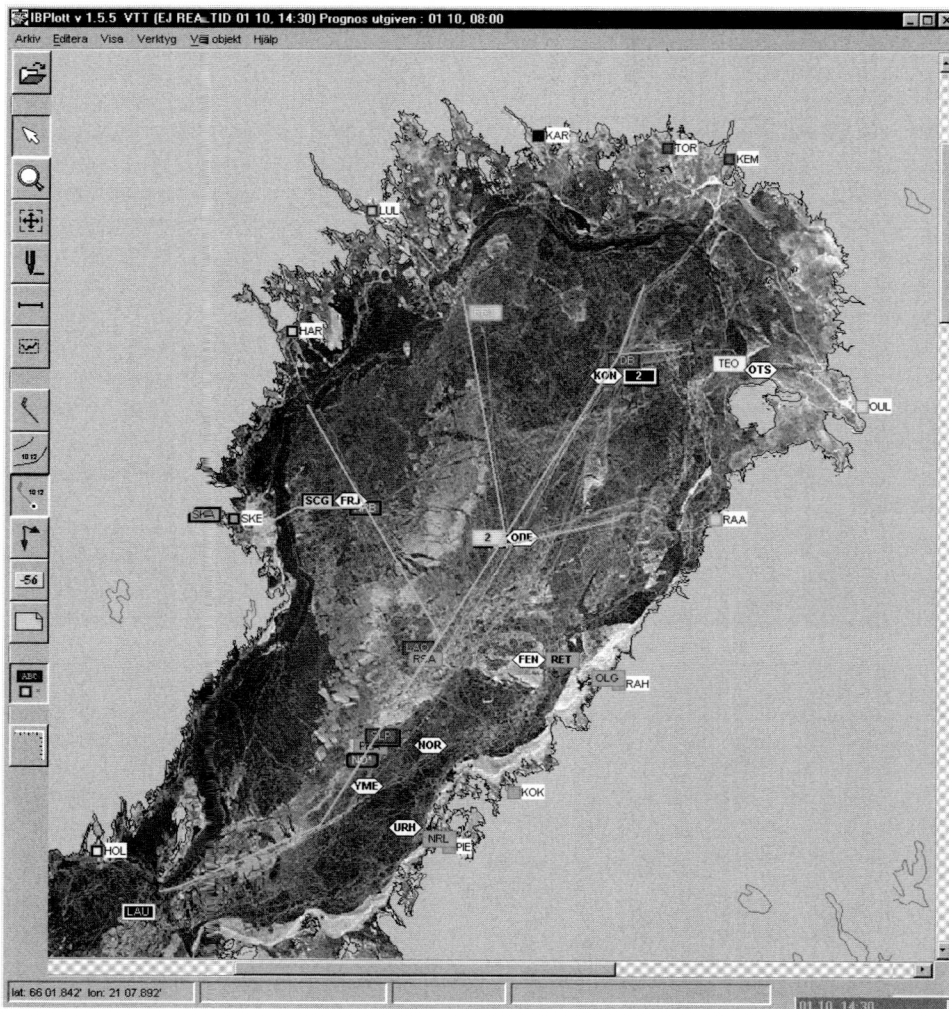

Figure 9.2. Modern sea ice information products include navigation data (depth contours, ship routes, lighthouses, ships) overlain on ice information. The picture shows the ice conditions (*Radarsat* image) and traffic situation in the Bay of Bothnia on 10 January 2003. Merchant ships and ice-breakers are displayed as symbols on top of a radar image, and the recommended routes determined by the ice-breakers are indicated. This IBPlott application is part of the distributed traffic information system IBNet used by Swedish and Finnish ice-breakers.
Prepared by Mr Robin Berglund and reproduced with permission from the VTT Technical Research Centre of Finland.

ice, and thus the problem is closely linked to sea ice dynamics (Figure 9.3). Recently, great strides have been taken in understanding the scaling between engineering and geophysical—local scale and mesoscale—sea ice mechanics. Large loads are connected with hummocks and ridges, particularly in the seasonal sea ice zone,

Figure 9.3. Illustration of the geophysics–engineering scaling question: the local ice forcing on the lighthouse is an ice engineering problem, while the mesoscale drift of ice past the lighthouse is a geophysical problem. The photographs are of the Baltic Sea.
The topmost photograph was taken by Dr Tuomo Kärnä. It is reproduced with permission from the VTT Technical Research Centre of Finland.

and progress there is still suffering from the inadequate knowledge of geophysical, full-scale processes.

Sea ice formation and evolution, including its dynamics, have seen great progress in regional oceanographic knowledge of ice-covered seas in recent years, as reflected in the book about the Kara Sea by Volkov et al. (2002). For a long time, an ice-covered sea was something to avoid in observational programmes because of major practical difficulties and in oceanographic research because of the large difference between the physics of ocean waters and sea ice.

9.3 FINAL COMMENTS

The first chapter of this book gave a brief historical overview of sea ice dynamics research and an introduction to the problem of sea ice drift with its applications. Chapters 2–4 presented the drift ice medium, material properties, kinematic properties, and rheology. Chapter 5 investigated external forcing from the atmosphere and ocean, and Newton's second law was applied to obtain the equation of motion for sea ice. Chapters 6–8 pondered solutions to the sea ice drift problem for three different circumstances: free drift, analytical modelling, and numerical modelling.

Chapter 2 provided a description of drift ice material, ice floes, and ice thickness, leading to the concept of *ice state*, a set of relevant quantities for the mechanical behaviour of drift ice. Chapter 3 dealt with ice velocity observations and proposed a theoretical framework for ice kinematics analysis. Then, the *ice conservation law* was derived for ice state. Chapter 4 treated the *drift ice rheology*, a difficult but necessary question to understand the motion of drift ice. It was recognized that the medium possesses a significant internal stress field, which influences its dynamics as an internal force field. Chapter 5 derived the *momentum equation* of ice dynamics and then went on to discuss the principal external forcing from the atmosphere and ocean, as well as magnitude and dimension analyses of the momentum equation. Altogether, ice state, ice conservation law, ice rheology and equation of motion constitute the physical basis for the closed system of equations for the sea ice drift problem.

Chapter 6 examined free drift, or drift in the absence of internal friction, leading to simple drift rules: ice velocity is equal to the wind-driven drift superposed on the ocean current beneath the layer of frictional influence of ice. Chapter 7 studied ice drift in the presence of internal friction, based on analytical one- and 1.5-dimensional models and application of plastic rheology. Chapter 8 gave the modern-numerical modelling solution to sea ice drift, starting with an introduction to numerical modelling techniques. Chapters 6–8 considered the idea of stepwise progression to derive solutions for increasingly complicated systems. The reader following this line will be able to understand the whole solution, how it reflects the behaviour of real sea ice, and how one should interpret the outcome of numerical sea ice dynamics models. Finally, Chapter 9 briefly discussed some consequences of the ice drift phenomenon. Chapter 10 gives a collection of study problems. Chapter 11 lists the references.

The drift of sea ice is an essential element in the dynamics of polar oceans, those essential and fragile components of the world's climate system. The drift transports sea ice over long distances, even to regions where ice is not formed by thermodynamic processes. At the same time the ice transports latent heat and freshwater; the influence of icemelt on the salinity of the mixed layer is equivalent to considerable precipitation. The ice cover forms a particular air–sea interface, which is modified drastically by differential ice motion. Consequently, sea ice dynamics is a key factor in air–sea interaction processes in the polar oceans. This is of renewed interest owing to the increasingly growing concern about man-made global warming.

The drift of ice is also a major ecological and environmental factor. Ice margin regions are known for their high biological productivity because of favourable light and hydrographical conditions. Therefore, their location—to some degree determined by the drift of ice—is of deep concern in marine ecology. Pollutants accumulate in the ice sheet, originating from the water body, sea bottom, and atmospheric fallout, and they are transported within the ice over long distances. A particular pollutant question is oil spills.

Sea ice has always been a barrier to winter shipping. For the purpose of sea ice monitoring and forecasting, ice information services collect and distribute ice charts and forecasts. The drift of ice shifts the ice edge, opens and closes leads, and forms pressure zones, all of which are key to ice navigation. The timescale of remarkable changes is 1 day. Presently, expansion of the Northern Sea Route requires further development in sea ice-mapping and forecasting services.

All in all, the reading of Chapters 1–8 provides a complete overview of the problems associated with measuring, geophysics, and the modelling of sea ice drift. The principal idea has been to include the whole story of sea ice dynamics in a single book, from material state through dynamics laws to mathematical models. There was a crying need for a synthesis of these research endeavours, because sea ice dynamics applications have been increasing and, apart from review papers, no book dedicated to this topic exists in English.

10

Study problems

This chapter contains a collection of study problems for students and other readers who want to test their knowledge, to look at possible applications, and to teachers who want to find ready-made exam questions. The level of these exercises ranges from third-year undergraduate to graduate students.

CHAPTER 1

1. In the event of all sea ice melting, how would the sea surface level be affected?

2. Assume that the sea ice melt rate is cuT_w, where c is a heat exchange coefficient, u is ice velocity and T_w is water temperature. An ice floe with thickness h_0 starts to drift with constant speed from the origin along the x-axis. How far would it get, if water temperature increases linearly with space, $T_w = ax$? What would be the optimum speed? Work through the numerical example: $h_0 = 1\,\text{m}$, $c = 10^{-6}\,°\text{C}^{-1}$, $a = 1°\text{C}/100\,\text{km}$, $u = 10\,\text{cm/s}$.

3. A strong wind blows across a basin totally covered by 1 m thick sea ice and drives the ice to drift at 20 cm/s. How wide will the lead open on the lee side after 6 hours? On the opposite (windward) side, how thick will the deformed ice zone be at the same time, if its width is 1 km and the ice is evenly packed?

4. An ice floe is drifting free in the Antarctic Circumpolar Current at 60°S. The ocean current is 10 cm/s eastward along the latitude circle and wind is 10 m/s from the west. Determine the total displacement of the floe after 6 months.

CHAPTER 2

1. With an ice thickness of 50 cm, how large would an ice floe need to be for you to stand on it with dry feet?

2. For uniform, circular ice floes, show that the most open and dense packings, where floes are locked together, are $\pi/4 \approx 0.79$ and $\pi/(2\sqrt{3}) \approx 0.91$.

3. Show that the median, mean, and root-mean-square values are $\lambda^{-1}\log 2$, λ^{-1}, and $\lambda^{-1}\sqrt{2}$, respectively, for a random variable with exponential distribution—the probability density is $p(x; \lambda) = \lambda\exp(-\lambda x)$, $x \geq 0$. What is the 95% fractile $x_{0.95}$, defined by $\text{Prob}(x \leq x_{0.95}) = 0.95$?

4. Assume that the surface temperature is negative and averages Θ during the nine winter months and for the three summer months the radiation balance is $\alpha Q_s + Q_{nL} > 0$, where α is the albedo, Q_s is mean incoming solar radiation, and Q_{nL} is mean net long-wave radiation. Assume further that ice grows according to Stefan's law in winter and melts by radiation in summer:
   (a) Determine the conditions necessary for survival of the ice over summer and the equilibrium thickness of multi-year ice.
   (b) What is the sensitivity of multi-year ice thickness to Θ, α, and Q_{nL}?
   (c) Work through the numerical example: $\Theta = -20°C$, $\alpha = 0.3$, $Q_s = 300\,\text{W/m}^2$, and $Q_{nL} = -50\,\text{W/m}^2$.

5. The ratio of sail height to keel depth is $1:5$ and $1:3$ for first-year and multi-year sea ice ridges. How is the difference explained geometrically?

CHAPTER 3

1. (a) An ice floe is drifting along latitude ϕ with speed u. How long does it take to complete the full latitude cycle? Work through the numerical example: $\phi = 70°S$, $u = 10\,\text{cm/s}$.
   (b) An ice floe starts from the North Pole at velocity v along the zero meridian on 1 January. When does it meet the polar front with warm waters at $75°N$ travelling at $v = 5\,\text{cm/s}$?

2. (a) Derive the mathematical expressions of eigenvalues and eigenvectors for a two-dimensional symmetric tensor.
   (b) Show that the sum of the squared components of a two-dimensional symmetric tensor is invariant.

3. Analyse the strain-rate tensor of zonal flow: full expression, eigenvalues and eigenvectors, and invariants. How does a 10 km square, oriented in xy co-

ordinate axes, change in 1 day, if $\partial u/\partial x = 10^{-6}\,\mathrm{s}^{-1}$ and $\partial v/\partial x = 5 \times 10^{-6}\,\mathrm{s}^{-1}$, where x is the transverse co-ordinate?

4. Consider a stochastic ice drift model $du/dt = -\lambda(u - U_w) + \varepsilon$, where λ is constant, u and U_w are ice and ocean current velocities, and ε is white noise with variance σ^2. Examine the spectrum of ice velocity for periodic ocean currents.

5. Derive the conservation law for ice compactness in spherical co-ordinates on the Earth's surface.

CHAPTER 4

1. Lake Ladoga is a lake in North Europe of size 150 km × 100 km. In principle, it is possible to put all the people in the world on the ice of the lake by allocating each one a 1-m² spot. But, under what conditions would this be physically possible?

2. How long can an icicle with a uniform cross-section grow? Take the tensile strength of ice equal to 1 MPa.

3. For a Newtonian viscous fluid, $\boldsymbol{\sigma} = -p\mathbf{I} + \eta\dot{\boldsymbol{\varepsilon}}'$, where p is hydrostatic pressure and $\eta =$ constant is viscosity, show that stress divergence consists of a pressure gradient and internal friction as $\nabla \cdot \boldsymbol{\sigma} = -\nabla p + \eta \nabla^2 \mathbf{u}$.

4. Derive the relationship between stress and strain rate for a square yield curve:

$$F(\sigma_1, \sigma_2) = \begin{cases} (0, \sigma_2), & 0 \geq \sigma_2 \geq -\sigma_y \\ (\sigma_1, -\sigma_y), & 0 \geq \sigma_1 \geq -\sigma_y \\ (-\sigma_y, \sigma_2), & 0 \geq \sigma_2 \geq -\sigma_y \\ (\sigma_1, 0), & 0 \geq \sigma_1 \geq -\sigma_y \end{cases}$$

where σ_y is the uniaxial compressive yield strength (the normal flow rule is assumed).

CHAPTER 5

1. Gravity measurements are normally difficult to take in the open ocean. Lehmuskoski and Mäkinen (1978) have made gravity measurements on drifting ice in the Baltic Sea. The drifting platform introduced a change in the apparent gravity via the Coriolis effect. Estimate the magnitude of this effect (called the Eötvös effect in gravimetry) when the ice velocity is 10 cm/s.

Study problems

2. Derive the sea surface tilt term with the horizontal surface as the zero reference.

3. Show that the following relationship holds between roughness length and the drag coefficient: $z_0/z = \exp(\kappa/\sqrt{C_z})$.

4. Estimate the magnitudes of the forces in the ice equation of motion by magnitude analysis, as well as possible, when:
   (a) wind speed = 20 m/s, ice speed = 40 cm/s;
   (b) wind speed = 10 m/s, ice speed = 10 cm/s; and
   (c) wind speed = 2 m/s, ice speed = 1 cm/s.

   The geostrophic current is 5 cm/s and ice thickness is 1 m or 3 m.

5. Examine the ice drift problem and assume $C_w = 0$. What are the dominant forces and key dimensionless numbers?

CHAPTER 6

1. In a horizontal channel, ice is moving at steady-state speed u driven by wind stress. When the wind ceases, how far will the floe travel before stopping.

2. Assume an open channel flow, water depth H, and channel inclination β. How is the total discharge affected when some of the water is frozen into a sea ice lid of thickness h which drifts freely.

3. For a perturbation approach, let $U = U + E$, where U is the mean level of the drift and E is the first-order perturbation. Derive the free drift solution including first-order terms. How does it compare with the linear solution?

4. The Newton–Raphson iteration scheme for a pair of non-linear equations $F(X, Y) = 0$ is written:
$$(X, Y)^{(n+1)} = (X, Y)^{(n)} - F^{(n)}[(DF)^{(n)}]^{-1}$$
where DF is the gradient 2×2 matrix of F. Use this to solve the wind factor and deviation angle from the wind-driven free drift equation:
   (a) Write the iteration scheme in explicit form.
   (b) Work through the numerical example: wind = 3, 5, 10, 15 m/s and ice thickness = 0.1, 0.5, 1, 2, 3, 6 m.

5. Solve the one-dimensional, non-steady-state free drift equation:
$$\rho h \frac{du}{dt} = \rho_a C_a \, \text{sgn}(U_a) U_a^2 + \rho_w C_w \, \text{sgn}(U_w - u)(U_w - u)^2$$
with U_w = constant and $u(t = 0) = u_0$.

CHAPTER 7

1. Construct the solution of a one-dimensional channel drift forced by a constant current velocity.

2. (Laikhtman problem) Solve a coastal ice drift subject to viscous rheology:

    $$\eta \frac{d^2 v}{dx^2} - c_1 v + c_a V_a = 0$$

    where $\eta \approx 10^8$–10^{10} kg/s, v is alongshore ice velocity, x is the distance from the shore, c_1 and c_a are linear drag coefficients, and V_a is wind speed. Take $V_a = 10$ m/s and find suitable values for c_1 and c_a.

3. Solve the coastal zone plastic ice flow problem when the ice–ocean interaction is ignored.

4. Assume a 30°, on-ice surface wind in the northern hemisphere, with the marginal ice zone (MIZ) to the left of the wind. Solve the steady-state plastic flow for thickness and compactness distribution, when the initial ice thickness is 1 m, concentration is 80%, and width of the MIZ is 200 km:

    (a) wind speed is 10 m/s;
    (b) wind speed increases from 0 to 20 m/s across the MIZ.

5. Solve the steady-state plastic circular flow in polar co-ordinates. Compare the solution for the East Wind Drift in Antarctica with the solution in spherical co-ordinates given in the text.

CHAPTER 8

1. Write a consistent, discrete formula for the internal friction of drift ice assuming plastic flow with an elliptic yield curve. Use Taylor's polynomial to obtain approximations for the derivatives. *Note*: "consistent" means that the original differential form results when the grid size approaches zero.

2. Write numerical model equations for a one-dimensional plastic channel flow that include the ice thickness conservation law and the momentum equation. Integrate the model for 1 day when the channel length is 200 km and the grid size is 10 km. The on-ice wind speed is 10 m/s, initial ice thickness is 1 m, and ice concentration is 100%. The quasi-steady-state approach may be taken.

3. Take a rectangular basin 200 km × 100 km, and ignore Coriolis acceleration, Ekman angle in the ocean, and geostrophic ocean current. Solve numerically the steady-state plastic flow for ice of thickness 1 m and concentration 100%,

wind speed 10 m/s along the major axis of the basin, and a yield curve that is (a) square and (b) elliptic.

4. Solve numerically the ice drift past a circular island of radius r using a linear viscous and a plastic rheology. Choose the grid size equal to $r/3$.

11

References

Adams, R.A. (1995) *Calculus: A Complete Course* (3rd edn, 1090 pp.). Addison-Wesley, Don Mills, Ontario.
Agnew, T.A., Le, H., and Hirose, T. (1997) Estimation of large scale sea ice motion from SSM/I 85.5 GHz imagery. *Annals of Glaciology* **25**, 305–311.
Ahlfors, L.V. (1966) *Complex Analysis* (331 pp.). McGraw-Hill, New York.
Albright, M. (1980) Geostrophic wind calculations for AIDJEX. In: R.S. Pritchard (ed.), *Sea Ice Processes and Models* (pp. 402–409). University of Washington Press, Seattle.
Ames, W.F. (1977) *Numerical Methods for Partial Differential Equations* (2nd edn, 365 pp.). Academic Press, New York.
Andreas, E.L. (1998) The atmospheric boundary layer over polar marine surfaces. In: M. Leppäranta (ed.), *Physics of Ice-Covered Seas* (Vol. II, pp. 715–773). Helsinki University Press, Helsinki.
Andreas, E.L. and Claffey, K.J. (1995) Air–ice drag coefficients in the western Weddell Sea: 1. Values deduced from profile measurements. *Journal of Geophysical Research* **100**(C3), 4821–4831.
Andreas, E.L., Tucker, W.B., III, and Ackley, S.F. (1984) Atmospheric boundary layer modification, drag coefficient, and surface net heat flux in the Antarctic marginal ice zone. *Journal of Geophysical Research* **89**(C1), 649–661.
Appel, I.L. (1989) Variability of thickness of ice and snow in Arctic seas and its parametrization. *Meteorologiya i Gidrologiya* **3**, 80–87.
Arberter, T.E., Curry, J.A., and Maslanik, J.A. (1999) Effects of rheology and ice thickness distribution in a dynamic–thermodynamic sea ice model. *Journal of Physical Oceanography* **29**, 2656–2670.
Armstrong, T., Roberts, B., and Swithinbank, C. (1966) *Illustrated Glossary of Snow and Ice*. Scott Polar Research Institute, Cambridge, UK.
Arya, S.P.S. (1975) A drag partition theory for determining the large-scale roughness parameter and wind stress on the Arctic pack ice. *Journal of Geophysical Research* **80**(24), 3447–3454.
Babko, O., Rothrock, D.A., and Maykut, G.A. (2002) Role of rafting in the mechanical distribution of sea ice thickness. *Journal of Geophysical Research* **107**(C8), 27.

Banke, E.G., Smith, S.D., and Anderson, R.J. (1980) Drag coefficients at AIDJEX from sonic anemometer measurements. In: R.S. Pritchard (ed.), *Sea Ice Processes and Models* (pp. 430–442). University of Washington Press, Seattle.

Belliveau, D.J., Bugden, G.L., Eid, B.M., and Calnan, C.J. (1990) Sea ice velocity measurements by upward-looking Doppler current profilers. *Journal of Atmospheric and Oceanic Technology* **7**, 596–602.

Bischof, J. (2000) *Ice Drift, Ocean Circulation and Climate Change* (215 pp.). Springer-Praxis, Chichester, U.K.

Blondel, Ph. and Murton, B.J. (1997) *Handbook of Seafloor Sonar Imagery* (314 pp.). Wiley-Praxis, Chichester, UK.

Bourke, R.H. and Garret, R.P. (1987) Sea ice thickness distribution in the Arctic Ocean. *Cold Regions Science and Technology* **13**, 259–280.

Brennecke, W. (1921). Die ozeanographischen Arbeiten der Deutschen Antarktischen Expedition 1911–1912. *Archiv der Deutschen Seewarte*, **XXXIX** (Nr. 1) [in German].

Brown, R.A. (1980) Boundary-layer modeling for AIDJEX. In: R.S. Pritchard (ed.), *Sea Ice Processes and Models* (pp. 387–401). University of Washington Press, Seattle.

Bukharitsin, P.I. (1986) Calculation and prediction of rafted ice thickness in navigable regions of the northwestern Caspian Sea. *Meteorologiya i Gidrologiya* **4**, 87–93.

Bushuyev, A.V., Volkov, N.A., Gudkovich, Z.M., and Loshchilov, V.S. (1967) Rezul'taty ekspeditsionnykh issledovanii dreifa i dinamiki ledianogo pokrova Arkticheskogo Basseina Vesnoi, 1961g. [Results of expedition investigations of the drift and dynamics of the ice cover of the Arctic Basin during the spring of 1961]. *Trudy Arkticheskii i Antarkticheskii Nauchno-issledovatel'skii Institut* **257**, 26–44 [Engl. summary *AIDJEX Bulletin* **3**, 1–21].

Campbell, W.J. (1965) The wind-driven circulation of ice and water in a polar ocean. *Journal of Geophysical Research* **70**, 3279–3301.

Campbell, W.J. and Rasmussen, L.A. (1972) A numerical model for sea ice dynamics incorporating three alternative ice constitutive laws. In: T. Karlsson (ed.), *Sea Ice: Proceedings of an International Conference* (pp. 176–187). National Research Council, Reykjavik.

Carsey, F. (ed.) (1992) *Microwave Remote Sensing of Sea Ice* (Geophysical Monograph No. 68, 462 pp.). American Geophysical Union, Washington, DC.

Cavalieri, D., Gloersen, P., and Zwally, J. (1999–updated regularly) Near real-time DMSP SSM/I daily polar gridded sea ice concentrations. Edited by J. Maslanik and J. Stroeve. NSIDC, Boulder, CO. (http://nsidc.org)

Chamberlain A.C. (1983) Roughness lengths of sea, sand and snow. *Boundary Layer Meteorology* **25**, 405–409.

Charnock, H. (1955) Wind stress on water surface. *Quarterly Journal of Royal Meteorological Society* **81**, 639.

Chieh, S.-H., Wake, A., and Rumer, R.R. (1983) Ice forecasting model for Lake Erie. *Journal of Waterway, Port, Coastal and Ocean Engineering*, **109**(4), 392–415.

Colony, R. and Thorndike, A.S. (1984) An estimate of the mean field of Arctic sea ice motion. *Journal of Geophysical Research* **89**(C6), 10623–10629.

Colony, R. and Thorndike, A.S. (1985) Sea ice motion as a drunkard's walk. *Journal of Geophysical Research* **90**(C1), 965–974.

Coon, M.D. (1974) Mechanical behaviour of compacted Arctic ice floes. *Journal of Petroleum Technology* **25**, 466–479.

Coon, M.D. (1980) A review of AIDJEX modeling. In: R.S. Pritchard (ed.), *Proceedings of ICSI/AIDJEX Symposium on Sea Ice Processes and Models* (pp. 12–23). University of Washington, Seattle.

Coon, M.D. and Pritchard, R.S. (1979) Mechanical energy considerations in sea–ice dynamics. *Journal of Glaciology* **24**(90), 377–389.

Coon, M.D., Maykut, G.A., Pritchard, R.S., Rothrock, D.A., and Thorndike, A.S. (1974) Modeling the pack ice as an elastic–plastic material. *AIDJEX Bulletin* **24**, 1–105.

Coon, M.D., Knoke, G.S., Echert, D.C., and Pritchard, R.S. (1998) The architecture of an anisotropic elastic–plastic sea ice mechanics constitutive law. *Journal of Geophysical Research* **103**(C10), 21915–21925.

Crow, E.L. and Shimizu, K. (eds) (1988) *Logarithmic Normal Distribution: Theory and Application*. Marcel Dekker, New York.

Cushman-Roisin, B. (1994) *Introduction to Geophysical Fluid Dynamics*. Prentice Hall, Englewood Cliffs, NJ.

Dempsey, J.P. and Shen, H.H. (eds) (2001) *IUTAM Symposium on Scaling Laws in Ice Mechanics*. Kluwer Academic, Dordrecht, the Netherlands.

Dierking, W. (1995) Laser profiling of the ice surface topography during the Winter Weddell Gyre Study 1992. *Journal of Geophysical Research* **100**(C3), 4807–4820.

Divine, D.V. (2003) Peculiarities of shore–fast ice formation and destruction in the Kara Sea. PhD thesis, University of Bergen, Norway.

Doronin, Yu.P. (1970) On a method of calculating the compactness and drift of ice floes. *Trudy Arkticheskii i Antarkticheskii Nauchno-issledovatel'skii Institut* **291**, 5–17 [English transl. 1970 in *AIDJEX Bulletin* **3**, 22–39].

Doronin, Yu.P. and Kheysin, D.Ye. (1975) *Morskoi led [Sea Ice]* (318 pp.). Gidrometeoizdat, Leningrad [English transl. 1977 by Amerind, New Delhi].

Eicken, H. and Lange, M. (1989) Development and properties of sea ice in the coastal regime of the southern Weddell Sea. *Journal of Geophysical Research* **94**(C6), 8193–8206.

Ekman, V.W. (1902) Om jordrotationens inverkan på vindströmmar i hafvet. *Nyt Magasin för Naturvidenskap B* **40**, 1.

Ekman, V.W. (1904) On dead water. In: *The Norwegian North Polar Expedition 1893–1896: Scientific Results* (edited by F. Nansen, Vol. V, pp. 1–152). Copp, Clark, Mississauga, Ontario.

Erlingsson, B. (1988) Two-dimensional deformation patterns in sea ice. *Journal of Glaciology* **34**(118), 301–308.

Feller, W. (1968) *An Introduction to Probability Theory and Its Applications* (Vol. 1). John Wiley & Sons, New York.

Fily, M. and Rothrock, D.A. (1987) Sea ice tracking by nested correlations. *IEEE Transactions on Geoscience and Remote Sensing* **GE-24**(6), 570–580.

Fissel, D.B. and Tang, C.L. (1991) Response of sea ice drift to wind forcing on the northeastern Newfoundland Shelf. *Journal of Geophysical Research* **96**(C10): 18397–18409.

Flato, G.M. (1993) A particle-in-cell method sea-ice model. *Atmosphere-Ocean* **31**(3), 339–358.

Flato, G.M. and Hibler, W.D., III (1990) On a simple sea-ice dynamics model for climate studies. *Annals of Glaciology* **14**, 72–77.

Flato, G.M. and Hibler, W.D., III (1995) Ridging and strength in modelling the thickness distribution of Arctic sea ice. *Journal of Geophysical Research* **100**(C9), 18611–18626.

Fukutomi, T. (1952). Study of sea ice (the 18th report): Drift of sea ice due to wind in the Sea of Okhotsk, especially in its southern part. *Low Temperature Science* **9**, 137–144.

Garratt, J.R. (1977) Review of drag coefficients over oceans and continents. *Monthly Weather Review*, **105**, 915–929.

Gill, A.E. (1982) *Atmosphere–Ocean Dynamics* (662 pp.). Academic Press, New York.

Glen, J.W. (1958) The flow law of ice. *International Association of Scientific Hydrology* **47**, 171–183.

Glen, J.W. (1970) Thoughts on a viscous model for sea ice. *AIDJEX Bulletin* **2**, 18–27.

Gloersen, P., Zwally, H.J., Chang, A.T.C., Hall, D.K., Campbell, W.J., and Ramseier, R.O. (1978) Time dependence of sea ice concentration and multi-year ice fraction in the Arctic Basin. *Boundary Layer Meteorology* **13**, 339–359.

Goldstein, R., Osipenko, N. and Leppäranta, M. (2000) Classification of large-scale sea-ice structures based on remote sensing imagery. *Geophysica* **36**(1–2), 95–109.

Goldstein, R., Osipenko, N., and Leppäranta, M. (2001) On the formation of large scale structural features. In: J.P. Dempsey and H.H. Shen (eds), *IUTAM Symposium on Scaling Laws in Ice Mechanics and Ice Dynamics* (pp. 323–334). Kluwer Academic, Dordrecht, The Netherlands.

Gorbunov, Yu.A. and Timokhov, L.A. (1968) Investigation of ice dynamics. *Izvestiya Atmospheric and Oceanic Physics* **4**(10), 623–626.

Gordienko, P. (1958) Arctic ice drift. *Proceedings of Conference on Arctic Sea Ice* (Publication No. 598, pp. 210–220). National Academy of Science, National Research Council, Washington, D.C.

Gordon, A.L. and Lukin, V.V. (1992) Ice Station Weddell #1. *Antarctic Journal of the United States* **27**(5), 97–99.

Gothus, Olaus Magnus (1539) *Carta Marina*. Original published in Venice [available at Geenimap Oy, Vantaa, Finland (http://www.genimap.fi/kuluttajatuotteet/)].

Gradinger, R. (1995) Climate change and the biological oceanography of the Arctic Ocean. *Philosophical Transactions of the Royal Society of London* **A352**, 277–286.

Granberg, H.B. and Leppäranta, M. (1999) Observations of sea ice ridging in the Weddell Sea. *Journal of Geophysical Research*, **104**(C11), 25735–25745.

Gray, J.N.M.T. (1999) Loss of hyperbolicity and ill-posedness of the viscous–plastic sea ice rheology in uniaxial divergent flow. *Journal of Physical Oceanography* **29**, 2920–2929.

Gray, J.N.M.T. and Killworth, P. (1995) Stability of viscous–plastic sea ice rheology. *Journal of Physical Oceanography* **25**, 971–978.

Gray, J.N.M.T. and Morland, L.W. (1994) A two-dimensional model for the dynamics of sea ice. *Philosophical Transactions of the Royal Society of London* **A347**, 219–290.

Gudkovic, Z.M. and Doronin, Yu.P. (2001) *Dreif morskikh l'dov*. Gidrometeoizdat, St Petersburg [in Russian].

Gudkovic, Z.M. and Romanov, M.A. (1976) Method for calculation the distribution of ice thickness in the Arctic seas during the winter period. *Ice Forecasting Techniques for the Arctic Seas* (pp. 1–47). Amerind Publishing, New Delhi [original in Russian by Gidrometeorologicheskoe Publishers, Leningrad, 1970].

Gutfraind, R. and Savage, S.B. (1997) Marginal ice zone rheology: Comparison of results from continuum–plastic models and discrete particle simulations. *Journal of Geophysical Research* **102**(C5), 12647–12661.

Haapala, J. (2000) On the modelling of ice-thickness distribution. *Journal of Glaciology* **46**(154), 427–437.

Haapala, J. and Leppäranta, M. (1996) Simulations of the Baltic Sea ice season with a coupled ice–ocean model. *Tellus* **48A**, 622–643.

Haapala, J. and Leppäranta, M. (1997a) The Baltic Sea ice season in changing climate. *Boreal Environment Research* **2**, 93–108.

Haapala, J. and Leppäranta, M. (eds) (1997b) *ZIP-97 Data Report* (Report Series in Geophysics No. 37, 123 pp.). University of Helsinki Department of Geophysics, Helsinki.

Haas, C. (1998) Evaluation of ship based electromagnetic–inductive thickness measurements of summer sea ice in the Bellingshausen and Amundsen seas, Antarctica. *Cold Regions Science and Technology* **27**, 1–6.

Häkkinen, S. (1986) Ice banding as a response of the coupled ice-ocean system to temporally varying winds. *Journal of Geophysical Research* **91**(C4), 5047–5053.

Harlow, F.H. (1964) The particle-in-cell computing method for fluid dynamics. *Methods in Computational Physics* **3**, 319–343.

Harr, M.E. (1977) *Mechanics of Particulate Media. A probabilistic approach* (543 pp.). McGraw-Hill, New York.

Heil, P. and Hibler, W.D., III (2002) Modeling the high-frequency component of Arctic sea ice drift and deformation. *Journal of Physical Oceanography* **32**(11), 3039–3057.

Hibler, W.D., III (1979) A dynamic–thermodynamic sea ice model. *Journal of Physical Oceanography* **9**, 815–846.

Hibler, W.D., III (1980a) Modelling a variable thickness sea ice cover. *Monthly Weather Review* **108**(12), 1943–1973.

Hibler, W.D., III (1980b) Sea ice growth, drift and decay. In: S. Colbeck (ed.), *Dynamics of Snow and Ice Masses* (pp. 141–209). Academic Press, New York.

Hibler, W. D., III (1986) Ice dynamics. In: N. Untersteiner (ed.), *Geophysics of Sea Ice* (pp. 577–640). Plenum Press, New York.

Hibler, W.D., III (2001) Sea ice fracturing on the large scale. *Engineering Fracture Mechanics* **68**, 2013–2043.

Hibler, W.D., III and Ackley, S.F. (1983) Numerical simulation of the Weddell Sea pack ice. *Journal of Geophysical Research* **88**(C5), 2873–2887.

Hibler, W.D., III and Bryan, K. (1987) A diagnostic ice–ocean model. *Journal of Physical Oceanography* **17**, 987–1015.

Hibler, W.D., III and Schulson, E.M. (2000) On modeling the anisotropic failure and flow of flawed sea ice. *Journal of Geophysical Research* **105**(C7), 17105–17120.

Hibler, W.D., III and Tucker, W.B., III (1977) Seasonal variations in apparent sea ice viscosity on the geophysical scale. *Geophysical Research Letters* **4**(2), 87–90.

Hibler, W.D., III, Weeks, W.F., and Mock, S.J. (1972) Statistical aspects of sea-ice ridge distributions. *Journal of Geophysical Research* **77**, 5954–5970.

Hibler, W. D., III, Weeks, W.F., Ackley, S.F., Kovacs, A., and Campbell, W.J. (1973) Mesoscale strain measurements on the Beaufort Sea pack ice (AIDJEX 1971). *Journal of Glaciology* **12**(65), 187–206.

Hibler, W.D., III, Ackley, S.F., Crowder, W.K., McKim, H.L., and Anderson, D.M. (1974a) Analysis of shear zone ice deformation in the Beaufort Sea using satellite imagery. In: J.C. Reed and J.E. Sater (eds), *The Coast and Shelf of the Beaufort Sea* (pp. 285–296). The Arctic Institute of North America, Arlington, VA.

Hibler, W.D., III, Mock, S.J., and Tucker, W.B., III (1974b) Classification and variation of sea ice ridging in the Western Arctic basin. *Journal of Geophysical Research* **79**(18), 2735–2743.

Hibler, W.D., III, Weeks, W.F., Kovacs, A., and Ackley, S.F. (1974c) Differential sea-ice drift: I. Spatial and temporal variations in sea ice deformation. *Journal of Glaciology* **13**(69), 437–455.

Hibler, W.D., III, Tucker, W.B., III, and Weeks, W.F. (1975) Measurement of sea ice drift far from shore using Landsat and aerial photograph imagery. *Proceedings 3rd International Symposium on Ice Problems* (pp. 541–554). IAHR Committee on Ice Problems, Hanover, NH.

Holland, D., Mysak, L.A., and Manak, D.K. (1993) Sensitivity study of a dynamic–thermodynamic sea ice model. *Journal of Geophysical Research* **98**(C2), 2561–2586.

Hopkins, M.A. (1994) On the ridging of intact lead ice. *Journal of Geophysical Research* **99**(C8), 16351–16360.

Hopkins, M.A. and Hibler, W.D., III (1991) On the ridging of a thin sheet of lead ice. *Journal of Geophysical Research* **96**(C3): 4809–4820.

Hopkins, M.A. and Tuhkuri, J. (1999) Compression of floating ice fields. *Journal of Geophysical Research* **104**(C7), 15815–15825.

Hunke, E. and Dukewicz, J.K. (1997) An elastic–viscous–plastic model for sea ice dynamics. *Journal of Physical Oceanography* **27**, 1849–1867.

Hunke, E. and Zhang, Y. (1999) A comparison of sea ice dynamics models at high resolution. *Monthly Weather Review* **127**, 396–408.

Hunkins, K. (1967) Inertial oscillations of Fletcher's Ice Island (T-3). *Journal of Geophysical Research* **72**(4), 1165–1173.

Hunkins, K. (1975) The oceanic boundary-layer and stress beneath a drifting ice floe. *Journal of Geophysical Research* **80**(24), 3425–3433.

Hunter, S.C. (1976) *Mechanics of Continuous Media*. Ellis Horwood, Chichester, UK.

Ikeda, M. (1985) A coupled ice–ocean model of a wind-driven coastal flow. *Journal of Geophysical Research* **90**(C5), 9119–9128.

Joffre, S.M. (1982a) Momentum and heat transfers in the surface layer over a frozen sea. *Boundary Layer Meteorology* **24**, 211–229.

Joffre, S.M. (1982b) Assessment of the separate effects of baroclinicity and thermal stability in the atmospheric boundary layer over the sea. *Tellus*, **34**(6), 567–578.

Joffre, S.M. (1983) Determining the form drag contribution to the total stress of the atmospheric flow over ridged sea ice. *Journal of Geophysical Research* **88**(C7), 4524–4530.

Joffre, S.M. (1984) *The Atmospheric Boundary Layer over the Bothnian Bay: A Review of Work on Momentum Transfer and Wind Structure* (Report No. 40, 58 pp.). Winter Navigation Research Board, Helsinki.

Joffre, S.M. (1985) Effects of local accelerations and baroclinity on the mean structure of the atmospheric boundary layer over the sea. *Boundary-Layer Meteorology*, **32**(3), 237–255.

Jurva, R. (1937) Über die Eisverhältnisse des Baltischen Meeres an den Küsten Finnlands nebst einem Atlas. *Merentutkimuslait. Julk./Havsforskningsinst. Skr.* **243**, 62 [in German].

Kankaanpää, P. (1998) Distribution, morphology and structure of sea ice pressure ridges in the Baltic Sea. *Fennia*, **175**(2).

Keliher, T.E. and Venkatesh, S. (1987) Modelling of Labrador Sea pack ice, with an application to estimating geostrophic currents. *Cold Regions Science and Technology* **13**, 161–176.

Ketchum, R.D., Jr. (1971) Airborne laser profiling of the Arctic pack ice. *Remote Sensing of Environment* **2**, 41–52.

Kheysin, D.Ye. (1978) Relationship between mean stresses and local values of internal forces in a drifting ice cover. *Oceanology* **18**(3), 285–286.

Kheysin, D.Ye. and Ivchenko, V.O. (1973) Numerical model of tidal drift of ice allowing for interaction between ice floes. *Izvestiya AN SSSR Ser. Fiz. Atmosfery i Okeana* **9**(4), 420–429.

Kirillov, A.A. (1957) Calculation of hummockness in determining ice volume. *Problems of the Arctic* **2**, 53–58 [in Russian].

Kolmogorov, A.N. (1941). Über das logaritmisch normale Verteilungsgesetz der Dimensionen der Teilchen bei Zerstückelung. *Doklady Academy of Sciences of USSR* **31**, 99.

Kondratyev, K., Johannessen, O.M., and Melentyev, V.V. (1996) *High Latitude Climate and Remote Sensing* (200 pp.). Wiley-Praxis, Chichester, UK.

Korvin, G. (1992) *Fractal Models in the Earth Sciences* (396 pp.). Elsevier, Amsterdam.

Kottmeier, C. and Hartig, R. (1990) Winter observations of the atmosphere over Antarctic sea ice. *Journal of Geophysical Research* **95**, 16551–16560.

Kovacs, A. and Holladay, J.S. (1990) Airborne sea ice thickness sounding. In: S.F. Ackley and W.F. Weeks (eds), *Sea Ice Properties and Processes* (CRREL Monograph No. 90–1, pp. 225–229). Cold Regions Research and Engineering Laboratory, Hanover, NH.

Kovacs, A. and Mellor, M. (1974) Sea ice morphology and ice as a geological agent in the southern Beaufort Sea. In: J.L. Reed and J.E. Sater (eds), *The Coast and Shelf of the Beaufort Sea* (pp. 113–161). Arctic Institute of North America, Arlington, VA.

Kowalik, Z. (1981) A study of the M_2 tide in the ice-covered Arctic Ocean. *Modeling, Identification and Control* **2**(4), 201–223.

Kowalik, Z. and Murty, T.S. (1993) *Numerical Modeling of Ocean Dynamics* (Advanced Series in Ocean Engineering, 481 pp.). World Scientific, Singapore.

Kowalik, Z. and Proshutinsky, A.Yu. (1995) Topographic enhancement of tidal motion in the western Barents Sea. *Journal of Geophysical Research* **100**(C2), 2613–2637.

Kuznetsov, I.M. and Mironov, E.U. (1986) Studying anomalously high ice drift velocities in Arctic seas. *Meteorologiya i Gidrologiya* **1**, 70–75.

Kwok, R. (1998) The Radarsat geophysical processor system. In: G. Tsatsoulis and R. Kwok (eds), *Analysis of SAR Data of the Polar Oceans* (pp. 235–257). Springer-Verlag, Berlin.

Kwok, R. (2001) Deformation of the Arctic Ocean sea ice cover between November 1996 and April 1997: A qualitative survey. In: J.P. Dempsey and H.H. Shen (eds), *IUTAM Symposium on Scaling Laws in Ice Mechanics and Ice Dynamics* (pp. 315–322). Kluwer Academic, Dordrecht, The Netherlands.

Kwok, R., Curlander, J.C., McConnell, R., and Pang, S.S. (1990) An ice-motion tracking system at the Alaska SAR facility. *IEEE Journal of Oceanic Engineering* **15**(1), 44–54.

Kwok, R., Rignot, E., Holt, B., and Onstott, R. (1992) Identification of sea ice types in spaceborne SAR data. *Journal of Geophysical Research* **97**(C2), 2391–2402.

Kwok, R., Schweiger, A., Rothrock, D.A., Pang, S., and Kottmeier, C. (1998) Sea ice motion from satellite passive microwave imagery assessed with ERS SAR and buoy motion. *Journal of Geophysical Research* **103**(C4), 8191–8214.

Laikhtman, D.L. (1958) O vetrovom dreife ledjanykh poley. *Trudy Leningradskiy Gidrometeorologicheskiy Institut* **7**, 129–137.

Landau, L.D. and Lifschitz, E.M. (1976) *Mechanics* (3rd edn, 169 pp.). Pergamon Press, Oxford, UK.

Lange, M. and Pfirman, S. (1998) Arctic sea ice contamination: Major characteristics and consequences. In: M Leppäranta (ed.), *Physics of Ice-Covered Seas* (Vol. 2, pp. 651–682). Helsinki University Press.

Legen'kov, A.P. (1977) On the strains in the drifting sea ice in the Arctic Basin. *Izvestiya, Atmospheric and Oceanic Physics* **13**(10), 728–733.

Legen'kov, A.P. (1978) Ice movements in the Arctic Basin and external factors. *Oceanology* **18**(2), 156–159.

Legen'kov, A.P., Chuguy, I.V., and Appel, I.L. (1974) On ice shifts in the Arctic basin. *Oceanology* **14**(6), 807–813.

Lehmuskoski, P. and Mäkinen, J. (1978) Gravity measurements on the ice of the Bothnian Bay. *Geophysica* **15**(1).

Lensu, M. (2003) *The Evolution of Ridged Ice Fields* (Report M-280, 140 pp.). Helsinki University of Technology, Ship Laboratory, Espoo, Finland.

Leppäranta, M. (1981a) An ice drift model for the Baltic Sea. *Tellus* **33**(6), 583–596.

Leppäranta, M. (1981b) On the structure and mechanics of pack ice in the Bothnian Bay. *Finnish Marine Research* **248**, 3–86.

Leppäranta, M. (1982) A case study of pack ice displacement and deformation field based on Landsat images. *Geophysica* **19**(1), 23–31.

Leppäranta, M. (1987) Ice observations: Sea Ice-85 experiment. *Meri* **15**.

Leppäranta, M. (1990) Observations of free ice drift and currents in the Bay of Bothnia. *Acta Regiae Societatis Scientiarum et Litterarum Gothoburgensis, Geophysica* **3**, 84–98.

Leppäranta, M. (1993). A review of analytical modelling of sea ice growth. *Atmosphere-Ocean* **31**(1), 123–138.

Leppäranta, M. (1998) The dynamics of sea ice. In: M. Leppäranta (ed.), *Physics of Ice-covered Seas* (Vol. 1, pp. 305–342). Helsinki University Press.

Leppäranta, M. and Hakala, R. (1992) The structure and strength of first-year ice ridges in the Baltic Sea. *Cold Regions Science and Technology* **20**, 295–311.

Leppäranta, M. and Hibler, W.D., III (1984) A mechanism for floe clustering in the marginal ice zone. *MIZEX Bulletin* **3**, 73–76.

Leppäranta, M. and Hibler, W.D., III (1985) The role of plastic ice interaction in marginal ice zone dynamics. *Journal of Geophysical Research* **90**(C6), 11899–11909.

Leppäranta, M. and Hibler, W.D., III (1987) Mesoscale sea ice deformation in the East Greenland marginal ice zone. *Journal of Geophysical Research* **92**(C7), 7060–7070.

Leppäranta, M. and Omstedt, A. (1990) Dynamic coupling of sea ice and water for an ice field with free boundaries. *Tellus* **42A**, 482–495.

Leppäranta, M. and Palosuo, E. (1983) Use of ship's radar to observe two-dimensional ridging characteristics. *Proceedings of 7th International Conference on Port and Ocean Engineering under Arctic Conditions (POAC'83)*, Helsinki, (Vol. I, pp. 138–147).

Leppäranta, M. and Wang, K. (2002) Sea ice dynamics in the Baltic Sea basins. *Proceedings of the 15th IAHR Ice Symposium*, Dunedin, New Zealand.

Leppäranta, M. and Zhang, Z. (1992a) Use of ERS-1 SAR data in numerical sea ice modeling. *Proceedings of the Central Symposium of the "International Space Year" Conference*, Munich, 30 March–4 April 1992 (pp. 123–128, ESA SP-341, July 1992).

Leppäranta, M. and Zhang, Z. (1992b) *A viscous–plastic test model for Baltic Sea ice dynamics* (Internal Report 1992(3)). Finnish Institute of Marine Research, Helsinki.

Leppäranta, M., Lensu, M., and Lu, Q.-m. (1990) Shear flow of sea ice in the marginal ice zone with collision rheology. *Geophysica* **25**(1–2), 57–74.

Leppäranta, M., Lensu, M., Kosloff, P., and Veitch, B. (1995) The life story of a first-year sea ice ridge. *Cold Regions Science and Technology* **23**, 279–290.

Leppäranta, M., Sun, Y., and Haapala, J. (1998) Comparisons of sea-ice velocity fields from ERS-1 SAR and a dynamic model. *Journal of Glaciology*, **44**(147), 248–262.

Leppäranta, M., Zhang, Z., Haapala, J., and Stipa, T. (2001) Sea ice mechanics in a coastal zone of the Baltic Sea. *Annals of Glaciology* **33**, 151–156.

LeSchack, L.A., Hibler, W.D., III, and Morse, F.H. (1971) Automatic processing of Arctic pack ice obtained by means of submarine sonar and other remotes sensing techniques. In: J.B. Lomax (ed.), *Propagation Limitations in Remote Sensing* (AGARD Conf. Proc. 90, pp. 5–1 to 5–19). AGARD, NATO, Neuilly-sur-Seine, France.

Lewis, J.E., Leppäranta, M., and Granberg, H.B. (1993) Statistical properties of sea ice surface topography in the Baltic Sea. *Tellus* **45A**, 127–142.

Li, S., Cheng, W., and Weeks, W.F. (1995) A grid based algorithm for the extraction of intermediate-scale sea ice deformation descriptors from SAR ice motion products. *International Journal of Remote Sensing* **16**(17), 3267–3286.

Li, W.H. and Lam, S.H., 1964. *Principles of Fluid Mechanics*. Addison-Wesley, Reading, MA.

Lichey, C. and Hellmer, H.H. (2001) Modeling giant-iceberg drift under the influence of sea ice in the Weddell Sea, Antarctica. *Journal of Glaciology* **47**(158), 452–460.

Lisitzin, E. (1957) On the reducing influence of sea ice on the piling-up of water due to wind stress. *Commentationes Physico-Mathematicae, Societas Scientiarum Fennica* **20**(7), 1–12.

Liu, A.K. and Cavalieri, D.J. (1998) On sea ice drift from the wavelet analysis of the Defense Meteorological Satellite Program (DMSP) special sensor microwave imager (SSM/I) data. *International Journal of Remote Sensing* **19**(7), 1415–1423.

Løset, S. (1993) Some aspects of floating ice related sea surface operations in the Barents Sea. Ph.D. thesis, University of Trondheim, Norway.

Lu, Q.-m., Larsen, J., and Tryde, P. (1989) On the role of ice interaction due to floe collisions in marginal ice zone dynamics. *Journal of Geophysical Research* **94**(C10), 14525–14537.

Lu, Q.-m., Larsen, J., and Tryde, P. (1990) A dynamic and thermodynamic sea ice model for subpolar regions. *Journal of Geophysical Research* **95**(C8), 13433–13457.

Lurton, X. (2002) *An Introduction to Underwater Acoustics* (347 pp.). Springer-Praxis, Chichester, UK.

Lyon, W.K. (1961) Ocean and sea-ice research in the Arctic Ocean via submarine. *Transactions of the New York Academy of Sciences, Series A* **23**(8), 662–674.

Lytle, V.I. and Ackley, S.F. (1991) Sea ice ridging in the eastern Weddell Sea. *Journal of Geophysical Research* **96**(C10), 18411–18416.

Makshtas, A.P. (1984) *The Heat Budget of Arctic Ice in the Winter*. Gidrometeoizdat, Leningrad [English language version by International Glaciological Society, Cambidge, UK, 1991].

Martinson, D.G. and Wamser, C. (1990) Ice drift and momentum exchange in winter Antarctic pack ice. *Journal of Geophysical Research* **95**(C2), 1741–1755.

Mase, G.E. (1970) *Continuum Mechanics* (Schaum's outline series). McGraw-Hill, New York.

Massom, R., Scambos, T., Lytle, V., Hill, K., Lubin, D., and Giles, B. (2003) Fast ice distribution and its interannual variability in East Antarctica, determined from satellite data analysis. *Proceedings of the 16th International Symposium on Ice* (Vol. 3, p. 271). IAHR and University of Otago, New Zealand.

Maykut, G.A. and Untersteiner, N. (1971) Some results from a time-dependent, thermodynamic model of sea ice. *Journal of Geophysical Research*, **76**, 1550–1575.

McPhee, M.G. (1978) A simulation of the inertial oscillation in drifting pack ice. *Dynamics of Atmospheres and Oceans* **2**, 107–122.

McPhee, M.G. (1980) An analysis of pack ice drift in summer. In: R.S. Pritchard (ed.), *Proceedings of ICSI/AIDJEX Symp. on Sea Ice Processes and Models* (pp. 62–75). University of Washington, Seattle.

McPhee, M.G. (1982) *Sea Ice Drag Laws and Simple Boundary Layer Concepts, Including Application to Rapid Melting* (Report 82-4). US Army Cold Regions Research and Engineering Laboratory, Hanover, NH.

McPhee, M.G. (1986) The upper ocean. In: N. Untersteiner (ed.), *Geophysics of Sea Ice* (pp. 339–394). Plenum Press, New York.

Mellor, M. (1986) The mechanical behavior of sea ice. In: N. Untersteiner (ed.), *Geophysics of Sea Ice* (pp. 165–281). Plenum Press, New York.

Mironov, Ye.U. (1987) Estimation of the volume of sea ice in the Arctic Ocean taking pressure ridges into account. *Polar Geography and Geology* **11**(1), 69–75.

Mock, S.J., Hartwell, A., and Hibler, W.D., III (1972) Spatial aspects of pressure ridge statistics. *Journal of Geophysical Research* **77**, 5945–5953.

Muench, R.D. and Ahlnäs, K. (1976) Ice movement and distribution in the Bering Sea from March to June 1974. *Journal of Geophysical Research* **81**(24), 4467–4476.

Muench, R., Martin, S., and Overland, J.E. (eds) (1987) Marginal ice zone research. *Journal of Geophysical Research* **92**(C7), 7225 pp. (special MIZEX issue).

Multala, J., Hautaniemi, H., Oksama, M., Leppäranta, M., Haapala, J., Herlevi, A., Riska, K. and Lensu, M. (1996) Airborne electromagnetic mapping of sea ice thickness in the Baltic Sea. *Cold Regions Science and Technology* **24**(4), 355–373.

Nansen, F. (1897) *Fram över Polarhavet*. Aschehoug, Kristiania, Norway.

Nansen, F. (1902) The oceanography of the North Polar Basin. *Norwegian North Polar Expedition 1893–1896: Scientific Results* (Vol. III, No. 9, 427 pp.). Longman Green & Co., Kristiania, Norway.

Neralla, V.R., Jessup R.G., and Venkatesh, S. (1988) The Atmospheric Environment Service regional ice model (RIM) for operational applications. *Marine Geodesy* **12**, 135–153.

Nikiforov, E.G. (1957) Changes in the compactness of the ice cover in relation to its dynamics. *Problemy Arktiki* (Vyp. 2). Morskoi Transport Press, Leningrad [in Russian].

Nikiforov, E.G., Gudkovich, Z.M., Yefimov, Yu.N., and Romanov, M.A. (1967) Principles of a method for calculating the ice redistribution under the influence of wind during the navigation period in Arctic seas. *Trudy Arkticheskii i Antarkticheskii Nauchno-issledovatel'skii Institut* **257**, 5–25 [English translation in *AIDJEX Bulletin* **3**, 40–64].

Nye, J.F. (1973) The physical meaning of two-dimensional stresses in a floating ice cover. *AIDJEX Bulletin* **21**, 1–8.

Okubo, A. and Ozmidov, R.V. (1970) Empirical dependence of the coefficient of horizontal turbulent diffusion in the ocean on the scale of the phenomenon in question. *Izvestiya Atmospheric and Oceanic Physics* **6**(5), 534–536.

Omstedt, A. (1998) Freezing estuaries and semi-enclosed basins. In: M. Lepparanta (ed.), *Physics of Ice-covered Seas* (Vol. 2, pp. 483–516). Helsinki University Press.

Omstedt, A., Nyberg, L., and Lepparanta, M. (1994) *A Coupled Ice–Ocean Model Supporting Winter Navigation in the Baltic Sea: Part 1. Ice dynamics and water levels* (Styrelsen för Vintersjöfartsforskning [Winter Navigation Research Board] Rep. No. 47, 17 pp.]. Board of Navigation, Norrköping, Sweden.

Omstedt, A., Nyberg, L., and Lepparanta, M. (1996) The role of ice inertia in ice–ocean dynamics. *Tellus* **48A**, 593–606.

Ono, N. (1978) Inertial period motions of drifting ice. *Low Temperature Science* **A37**, 107–113.

Overgaard, S., Wadhams, P. and Lepparanta, M. (1983) Ice properties in the Greenland and Barents Seas during summer. *Journal of Glaciology* **29**(101), 142–164.

Overland, J.E. (1985) Atmospheric boundary layer structure and drag coefficients over sea ice. *Journal of Geophysical Research*, **90**(C5), 9029–9049.

Overland, J.E. and Pease, C.H. (1988) Modeling ice dynamics of coastal seas. *Journal of Geophysical Research* **93**(C12), 15619–15637.

Overland, J.E., Walter, B.A., Curtin, T.B., and Turet, P. (1995) Hierarchy in sea ice mechanics: A case study from the Beaufort Sea. *Journal of Geophysical Research* **100**(C3), 4559–4571.

Ovsienko, S. (1976) Numerical modeling of the drift of ice. *Izvestiya, Atmospheric and Oceanic Physics* **12**(11), 1201–1206.

Ovsienko, S., Zatsepa, S., and Ivchenko, A. (1999a) Study and modelling of behaviour and spreading of oil in cold water and in ice conditions. *Proceedings 15th Conference on Port and Ocean Engineering under Arctic Conditions, Espoo, Finland*.

Ovsienko, S., Lepparanta, M., Zatsepa, S., and Ivchenko, A. (1999b) The use of ice tank experiment results for mesoscale sea ice modelling. *Proc. 15th Conf. Port and Ocean Eng. Arctic Conditions, Espoo, Finland* (Vol. 2, pp. 480–487).

Palosuo, E. (1953) A treatise on severe ice conditions in the Central Baltic. *Merentutkimuslaitoksen Julkaisu/Havsforskningsinstitutets Skrift* **156**, 130.

Palosuo, E. (1963) The Gulf of Bothnia in winter: II. Freezing and ice forms. *Merentutkimuslartoksen Julkaisu/Havsforskningsinstitutets Skrift* **209**, 64.
Parmerter, R.R. (1975) A model of simple rafting in sea ice. *Journal of Geophysical Research* **80**(15), 1948–1952.
Parmerter, R.R. and Coon, M.D. (1972) Model for pressure ridge formation in sea ice. *Journal of Geophysical Research* **77**, 6565–6575.
Paterson, W.S.B. (1995) *The Physics of Glaciers* (3rd edn). Pergamon Press, Oxford, UK.
Pfirman, S.L., Eicken, H., Bauch, D., and Weeks, W.F. (1995) The potential transport of pollutants by Arctic sea ice. *Science of the Total Environment*, **159**, 129–146.
Pond, S. and Pickard, G.L. (1983) *Introductory Dynamical Oceanography* (2nd edn). Pergamon Press, Oxford, UK.
Pritchard, R.S. (1975) An elastic–plastic constitutive law for sea ice. *Journal of Applied Mechanics* **42E**, 379–384.
Pritchard, R.S. (1977) The effect of strength on simulations of sea ice dynamics. *Proceedings of the 4th International Conference on Port and Ocean Engineering under Arctic Conditions* (12 pp.). Memorial University of Newfoundland, St John's, Canada.
Pritchard, R.S. (ed.) (1980a) *Proceedings of ICSI/AIDJEX Symposium on Sea Ice Processes and Models*. University of Washington, Seattle.
Pritchard, R.S. (1980b) Simulations of nearshore winter ice dynamics in the Beaufort Sea. In: R.S. Pritchard (ed.), *Proceedings of ICSI/AIDJEX Symposium on Sea Ice Processes and Models* (pp. 49–61). University of Washington, Seattle.
Pritchard, R.S. (1981) Mechanical behaviour of pack ice. In: A.P.S. Selvadurai (ed.), *Mechanics of Structured Media* (Part A, pp. 371–405). Elsevier, North-Holland.
Ramsay, W. (1949) *Jääsaarron murtajat [The Breakers of the Ice Barrier]*. K.F. Puromiehen kirjapaino, Helsinki.
Rheem, C.K., Yamaguchi, H., and Hiroharu, K. (1997) Distributed mass/discrete floe model for pack ice rheology computation. *Journal of Marine Science and Technology* **2**, 101–121.
Richter-Menge, J.A. and Elder, B.C. (1998) Characteristics of ice stress in the Alaskan Beaufort Sea. *Journal of Geophysical Research* **103**(C10), 21817–21829.
Richter-Menge, J., McNutt, S.L., Overland, J.E. and Kwok, R. (2002) Relating arctic pack ice stress and deformation under winter conditions. *Journal of Geophysical Research* **107**(C10), SHE 15 (doi:10.1029/2000JC000477).
Rodahl, K. (1954) *T-3* (179 pp.). Gyldendal, Oslo.
Rossby, C.G. and Montgomery, R.G. (1935) *Frictional Influence in Wind and Ocean Currents* (Papers in Physical Oceanography and Meteorology Vol. 3, No. 3, 101 pp.). Massachusetts Institute of Technology and Woods Hole Oceanographic Institute, Cambridge, MA.
Rossiter, J.R. and Holladay, J.S. (1994) Ice thickness measurement. In: S. Haykin, E.O. Lewis, R.K. Ramey, and J.R. Rossiter (eds), *Remote Sensing of Sea Ice and Icebergs* (pp. 141–176) John Wiley & Sons, New York.
Rossiter, J.R., Butt, K.A., Gamberg, J.B., and Ridings, T.F. (1980) Airborne impulse radar sounding of sea ice. *Proceedings of Sixth Canadian Symposium on Remote Sensing* (pp. 187–194).
Rothrock, D.A. (1975a) The energetics of the plastic deformation of pack ice by ridging. *Journal of Geophysical Research* **80**(33), 4514–4519.
Rothrock, D.A. (1975b) The mechanical behavior of pack ice. *Annual Review of Earth and Planetary Sciences* **3**, 317–342.
Rothrock., D.A. (1975c) The steady drift of an incompressible Arctic ice cover. *Journal of Geophysical Research* **80**(3), 387–397.

Rothrock, D.A. (1986). Ice thickness distribution – measurement and theory. In: N. Untersteiner (ed.), *The Geophysics of Sea Ice* (pp. 551–575). Plenum Press, New York.

Rothrock, D.A. and Hall, R. (1975) Testing the redistribution of sea ice thickness from ERTS photographs. *AIDJEX Bulletin* **19**, 1–19.

Rothrock, D.A. and Thorndike, A.S. (1980) Geometric properties of the underside of sea ice. *Journal of Geophysical Research* **85**(C7), 3955–3963.

Rothrock, D.A. and Thorndike, A.S. (1984) Measuring the sea ice floe size distribution. *Journal of Geophysical Research* **89**(C4), 6477–6486.

Rudels, B. (1998) Aspects of Arctic oceanography. In: M. Leppäranta (ed.), *Physics of Ice-covered Seas* (Vol. 2, pp. 517–568). Helsinki University Press.

Sanderson, T.J.O. (1988) Ice Mechanics: Risks to offshore structures (253 pp.). Graham & Trotman, Boston.

Savage, S.B. (1995) Marginal ice zone dynamics modelled by computer simulations involving floe collisions. In: E. Guazelli and L. Oger (eds), *Mobile, Particulate Systems* (pp. 305–330). Kluwer Academic, Norwell, MA.

Schaefer, J.A. and Ettema, R. (1986) Experiments on freeze-bonding between ice blocks in floating ice rubble. *Proceedings of 8th IAHR Ice Symposium, Iowa City, IA* (Vol. 1, pp. 401–413).

Schwarz, J. and Weeks, W.F. (1977) Engineering properties of sea ice. *Journal of Glaciology* **19**(81), 499–531.

Semtner, A. (1976) A model for the thermodynamic growth of sea ice in numerical investigations of climate. *Journal of Physical Oceanography* **6**(3), 379–389.

Shen, H.H., Hibler, W.D., III, and Leppäranta, M. (1986) On applying granular flow theory to a deforming broken ice field. *Acta Mechanica* **63**, 143–160.

Shen, H.H., Hibler, W.D., III, and Leppäranta, M. (1987) The role of ice floe collisions in sea ice rheology. *Journal of Geophysical Research* **92**(C7), 7085–7096.

Shen, H.T., Shen, H H., and Tsai, S.M. (1991) Dynamic transport of river ice. *Journal of Hydraulic Research* **28**(6), 659–671.

Shen, H.T., Chen, Y.C., Wake, A., and Crissman, R.D. (1993) Lagrangian discrete parcel simulation of two-dimensional river ice dynamics. *International Journal of Offshore and Polar Engineering* **3**(4), 328–332.

Shirasawa, K. (1986) Water stress and ocean current measurements under first-year sea ice in the Canadian Arctic. *Journal of Geophysical Research* **91**(C12), 14305–14316.

Shirasawa, K. and Ingram, R.G. (1991) Characteristics of the turbulent oceanic boundary layer under sea ice: Part 1. A review of the ice–ocean boundary layer. *Journal of Marine Systems*, **2**, 153–160.

Shirasawa, K. and Ingram, R. G. (1997) Currents and turbulent fluxes under the first-year sea ice in Resolute Passage, Northwest Territories, Canada. *Journal of Marine Systems*, **11**, 21–32.

Shirokov, K.P. (1977) Vliyanie splochennosti na vetrovoj dreif l'dov. *Sbornik Rabot Leningradskoye Gidrometeorologicheskoye Obsestvo* **9**, 46–53 [in Russian].

Shuleikin, V.V. (1938) Drift of ice fields. *Doklady Academy of Sciences USSR* **19**(8), 589–594.

Smith, S.D. (1972) Wind stress and turbulence over a flat ice floe. *Journal of Geophysical Research* **77**, 3885–3901.

Smith, S.D. (1980) Wind stress and heat flux over the ocean in gale force winds. *Journal of Physical Oceanography* **10**, 709–726.

Squire, V. (1998) The marginal ice zone. In: M. Leppäranta (ed.), *Physics of Ice-covered Seas* (Vol. 1, pp. 381–446). Helsinki University Press.

Squire, V.A., Dugan, J., Wadhams, P., Rottier, P.J., and Liu, A.K. (1995) Of ocean waves and sea ice. *Annual Review of Fluid Mechanics* **27**, 115–168.

Steffen, K. and Lewis, J.E. (1988) Surface temperatures and sea ice typing for northern Baffin Bay. *International Journal of Remote Sensing* **9**(3), 409–422.

Stössel, A., Lemke, P., and Owens, W.P. (1990) Coupled sea ice–mixed layer simulations for the Southern Ocean. *Journal of Geophysical Research* **95**(C6), 9539–9555.

Strass, V.H. and Fahrbach. E. (1998) Temporal and regional variation of sea ice drift and coverage in the Weddell Sea obtained from upward-looking sonars. In: M.O. Jeffries, (ed.), *Antarctic Sea Ice: Physical Processes, Interactions and Variability* (Antarctic Research Series No. 74, pp. 123–139). American Geophysical Union, Washington, DC.

Stull, R.B. (1988) *An Introduction to Boundary-layer Meteorology* (666 pp.). Kluwer Academic, Dordrecht, The Netherlands.

Sun, Y. (1996) Automatic ice motion retrieval from ERS-1 SAR images using the optical flow method. *International Journal of Remote Sensing* **17**(11), 2059–2087.

Sverdrup, H.U. (1928). The wind–drift of the ice on the North Siberian Shelf: The Norwegian north polar expedition with the "Maud" 1918–1925. *Scientific Results* **4**(1), 1–46.

Tabata, T. (1971) Measurements of strain of ice field off the Okhotsk Sea Coast of Hokkaido. *Low Temperature Science* **A29**, 199–211 [with English summary].

Tabata, T. (1972) Radar network for drift ice observations in Hokkaido. In: T. Karlsson (ed.), *Sea Ice: Proceedings of International Conference* (pp. 67–71). National Research Council, Reykjavik.

Tabata, T., Kawamura, T., and Aota, M. (1980) Divergence and rotation of an ice field off Okhotsk Sea coast off Hokkaido. In: R.S. Pritchard (ed.), *Sea Ice Processes and Models* (pp. 273–282). University of Washington Press, Seattle.

Tennekes, H. and Lumley, J.L. (1972) *A First Course in Turbulence* (300 pp.). MIT Press, Cambridge, MA.

Thomas, D.N. and Dieckmann, G.S. (2003) *Sea Ice: An introduction to its physics, chemistry, biology and geology* (402 pp.). Blackwell Publishing, Oxford, UK.

Thorndike, A.S. (1986) Kinematics of sea ice. In: N. Untersteiner (ed.), *Geophysics of Sea Ice* (pp. 489–549). Plenum Press, New York.

Thorndike, A.S. and Colony, R. (1980) Large-scale ice motion in the Beaufort Sea. In: R.S. Pritchard (ed.), *Proceedings of ICSI/AIDJEX Symposium on Sea Ice Processes and Models* (pp. 249–260). University of Washington, Seattle.

Thorndike, A.S. and Colony, R. (1982) Sea ice response to geostrophic winds. *Journal of Geophysical Research* **87**(C8), 5845–5852.

Thorndike, A.S., Rothrock, D.A., Maykut, G.A., and Colony, R. (1975) The thickness distribution of sea ice. *Journal of Geophysical Research* **80**, 4501–4513.

Timco, G.W. and Burden, R.P. (1997) An analysis of the shapes of sea ice ridges. *Cold Regions Science and Technology* **25**, 65–77.

Timmermann, R., Beckmann, A. and Hellmer, H.H. (2002) Simulations of ice–ocean dynamics in the Weddell Sea: 1. Model configuration and validation. *Journal of Geophysical Research* **107**(C3), 10.

Timokhov, L.A. (1998) Ice dynamics models. In: M. Leppäranta (ed.), *Physics of Ice-covered Seas* (Vol. 1, pp. 343–380). Helsinki University Press.

Timokhov, L.A. and Kheysin, D.Ye. (1987) *Dynamika Morskikh L'dov*. Gidrometeoizdat, Leningrad [in Russian].

Tinz, B. (1996) On the relation between annual maximum ice extent of ice cover in the Baltic Sea and sea level pressure as well as air temperature field. *Geophysica*, **32**, 319–341.

Tuhkuri, J. and Lensu, M. (1997) *Ice Tank Tests on Rafting of a Broken Ice Field* (Report M-218). Helsinki University of Technology, Ship Laboratory.

Udin, I. and Ullerstig, A. (1976) *A Numerical Model for the Forecasting of Ice Motion in the Bay and Sea of Bothnia* (Styrelsen för Vintersjöfartsforskning [Winter Navigation Research Board] Report No. 18). Board of Navigation, Norrköping, Sweden.

Ukita, J. and Moritz, R. (1995) Yield curves and flow rules of pack ice. *Journal of Geophysical Research* **100**(C3), 4545–4557.

Untersteiner, N. (1984) The cryosphere. In: J.T. Houghton (ed.), *The Global Climate* (pp. 121–140). Cambridge University Press, New York.

Van Hejst, G.J.F. (1984) An analytical model for ice-edge upwelling. *Geophysical and Astrophysical Fluid Dynamics* **29**, 155–177.

Vasiliev, K.P. (1985) *Sea-ice Forecasting. Marine Meteorology and Related Oceanographic Activities* (Report No. 14, pp. 50–59, Scientific Lectures at CMM–IX). WMO, Geneva.

Venkatesh, S., El-Tahan, H., Comfort, G., and Abdelnour, R. (1990) Modelling the behaviour of oil spills in ice-infested waters. *Atmosphere–Ocean* **28**(3), 303–329.

Vesecky, J.F., Samadani, R., Smith, M.P., Daida, J.M., and Bracewell, R.N. (1988) Observation of sea-ice dynamics using synthetic aperture radar images: Automated analysis. *IEEE Transactions on Geoscience and Remote Sensing* **GE-26**(1), 38–48.

Vinje, T. and Berge, T. (1989) *Upward Looking Sonar Recordings at $75°N–12°W$ from 22 June 1987 to 20 June 1988* (Data report. Nr. 51). Norsk Polarinstitutt, Oslo.

Volkov, V.A., Johannessen, O.M., Borodachev, V.E., Voinov, G.N., Pettersson, L.H., Bobylev, L.P. and Kouraev, A.V. (2002) *Polar Seas Oceanography. An Integrated Case Study of the Kara Sea* (450 pp.). Springer-Praxis, Chichester, UK.

Wadhams, P. (1978) Wave decay in the marginal ice zone measured from a submarine. *Deep-sea Research* **25**, 23–40.

Wadhams, P. (1980a). A comparison of sonar and laser profiles along corresponding tracks in the Arctic Ocean. In: R.S. Pritchard (ed.), *Sea Ice Processes and Models* (pp. 283–299). University of Washington Press, Seattle.

Wadhams, P. (1980b) Ice characteristics in the seasonal sea ice zone. *Cold Regions Science and Technology* **2**, 37–87.

Wadhams, P. (1981) Sea-ice topography of the Arctic Ocean in the region $70°W$ to $25°E$. *Philosophical Transactions of the Royal Society of London* **A302**, 45–85.

Wadhams, P. (1998) Sea ice morphology. In: M. Leppäranta (ed.), *Physics of Ice-covered Seas* (Vol. 1, pp. 231–287). Helsinki University Press.

Wadhams, P. (2000) *Ice in the Ocean*. Gordon & Breach Science, Amsterdam.

Wadhams, P. and Davy, T. (1986) On the spacing and draft distributions for pressure ridge keels. *Journal of Geophysical Research*, **91**(C9), 10697–10708.

Wadhams, P., Martin, S., Johannessen, O.M., Hibler, W.D., III, and Campbell, W.J. (eds) (1981) *MIZEX. A program for mesoscale air–ice–ocean experiments in the Arctic marginal ice zones. I. Research strategy* (CRREL Special Report 81-19, 20 pp.). Cold Regions Research and Engineering Laboratory, Hanover, NH.

Wadhams, P., Lange, M.A., and Ackley, S.F. (1987) The ice thickness distribution across the Atlantic sector of the Antarctic Ocean in midwinter. *Journal of Geophysical Research* **92**(C13), 14535–14552.

Wake, A. and Rumer, R.R. (1983) Great Lakes ice dynamics simulation. *Journal of Waterway, Port, Coastal and Ocean Engineering* **109**, 86–102.

Wang, J., Mysak, L.A., and Ingram, R.G. (1994) A numerical simulation of sea ice cover in Hudson Bay. *Journal of Physical Oceanography* **24**, 2515–2533.

Wang, K., Leppäranta, M., and Kouts, T. (2003) A model for sea ice dynamics in the Gulf of Riga. *Proceedings of Estonian Academy of Sciences, Engineering* **9**(2), 107–125.

Waters, J.K. and Bruno, M.S. (1995) Internal wave generation by ice floes moving in stratified water: Results from a laboratory study. *Journal of Geophysical Research* **100**(C7), 13635–13639.

Weeks, W.F. (1980) Overview. *Cold Regions Science and Technology* **2**, 1–35.

Weeks, W.F. (1998a) Growth conditions and structure and properties of sea ice. In: M. Leppäranta (ed.), *Physics of Ice-covered Seas* (Vol. 1, pp. 25–104). Helsinki University Press.

Weeks, W.F. (1998b) The history of sea ice research. In: M. Leppäranta (ed.), *Physics of Ice-covered Seas* (Vol. 1, pp. 1–24). Helsinki University Press.

Weeks, W.F. and Ackley, S.F. (1986) The growth, structure and properties of sea ice. In: N. Untersteiner (ed.), *Geophysics of Sea Ice* (pp. 9–164). Plenum Press, New York.

Weeks, W.F., Tucker, W.B., III, Frank, M., and Fungcharon, S. (1980) Characterization of surface roughness and floe geometry of sea ice over the continental shelves of the Beaufort and Chukchi Seas. In: R.S. Pritchard (ed.), *Proceedings of ICSI/AIDJEX Symposium on Sea Ice Processes and Models, University of Washington, Seattle* (pp. 300–312) [cited in Rothrock and Thorndike (1984).

Weeks, W.F., Ackley, S.F. and Govoni, J. (1989) Sea ice ridging in the Ross Sea, Antarctica, as compared with sites in the Arctic. *Journal of Geophysical Research* **94**, 4984–4988.

Williams, E., Swithinbank, C.W.M., and Robin, G. de Q. (1975) A submarine sonar study of Arctic pack ice. *Journal of Glaciology* **15**(73), 349–362.

Wittman, W.J. and Schule, J.J., Jr (1966) Comments on the mass budget of Arctic pack ice. *Symposium on Arctic Heat Budget and Atmospheric Circulation* (pp. 215–246). RAND Corporation, Santa Monica, CA.

WMO (1970) *The WMO Sea-Ice Nomenclature* (WMO No. 259, TP-145, with later supplements). WMO, Geneva.

WMO (2000) *Sea-Ice Information Services in the World* (WMO No. 574). WMO, Geneva.

Worby, A.P., Massom, R.A., Allison, I., Lytle, V.I., and Heil, P. (1998) East Antarctic sea ice: A review of structure, properties and drift. In: M.O. Jeffries (ed.), *Antarctic Sea Ice: Physical properties, interactions and variability* (Antarctic Research Series 74, pp. 41–67). American Geophysical Union, Washington, DC.)

Wright, B., Hnatiuk, J., and Kovacs, A. (1978) Sea ice pressure ridges in the Beaufort Sea. *Proceedings of IAHR Ice Symposium*, Luleå, Sweden.

Wu, H.-D. and Leppäranta, M. (1990) Experiments in numerical sea ice forecasting in the Bohai Sea. *Proceedings of IAHR Ice Symposium 1990, Espoo, Finland* (Vol. III, pp. 173–186).

Wu, H., Bai, S., and Li, G. (1997) Sea ice conditions and forecasts in China. *Proceedings of 12th International Symposium on Okhotsk Sea and Sea Ice* (pp. 164–174). Okhotsk Sea & Cold Ocean Research Association, Mombetsu, Hokkaido, Japan.

Wu, H., Bai, S., and Zhang, Z. (1998) Numerical sea ice prediction in China. *Acta Oceanologica Sinica*, **17**(2), 167–185.

Zhang, J., Thomas, D.B., Rothrock, D.A., Lindsay, R.W., and Yu, Y. (2003) Assimilation of ice motion observations and comparisons with submarine ice thickness data. *Journal of Geophysical Research* **108**(C6), 3170, doi: 10.1029/2001JC001041.

Zhang, J., Hibler, W.D., III, Steele, M., and Rothrock, D.A. (1998) Arctic ice–ocean modeling with and without climate restoring. *Journal of Physical Oceanography* **28**, 1745–1748.

Zhang, Z. (2000) Comparisons between observed and simulated ice motion in the northern Baltic Sea. *Geophysica* **36**(1–2), 111–126.

Zhang, Z. and Leppäranta, M. (1995) Modeling the influence of ice on sea level variations in the Baltic Sea. *Geophysica* **31**(2), 31–46.

Zubov, N.N. (1945) *L'dy Arktiki [Arctic Ice]*. Izdatel'stvo Glavsermorputi, Moscow [English translation 1963 by US Naval Oceanographical Office and American Meteorological Society, San Diego].

Zwally, H.J. and Walsh, J.E. (1987) Comparison of observed and modelled ice motion in the Arctic Ocean. *Annals of Glaciology* **9**, 1–9.

Index

Advanced Very High Resolution Radiometer (AVHRR) 18
Acoustic Doppler current profiler (ADCP) 55
advection of sea ice 163
airborne electromagnetic method (AEM) 32–33
airborne laser profilometer 33, 35
airborne observation 13, 17
air–sea interface 5
Alaska x
albedo 5
Antarctica
    drift 189, 191
    drift ice thickness distribution 35
Antarctic regions, long-term modelling applications 224
Archimedes' law 19, 31, 33, 35, 115
Arctic and Antarctic Research Institute (AARI) 3, 221, 232
Arctic Ice Dynamics Joint Experiment (AIDJEX) 3, 61, 91, 120, 209
    elastic–plastic rheology 95–97
    ice dynamics model 211–212
Arctic Ocean 1, 9, 23, 230
    drift ice thickness 30–31
    mean drift field 69–70
Arctic region, long-term modelling applications 222–224
Arctic seas 219–221

Arktica 214
Argos buoy 52–54
Atlantic Conveyor 231

Baltic Sea x, 1–3, 6, 9, 16, 19, 28, 217–219
    ice conditions 226
    ice dynamics model 212–214
    long-term modelling applications 224–227
    sea ice ridge 36–37
    sea surface temperature 226
Barents Sea x, 4, 20
basins *see* sea ice basins
Bay of Bothnia 66, 131
    sea ice structures 65–66, 69
Beaufort Sea 3–4, 16
    drift ice thickness 30
    Gyre 55, 61, 207
    roughness length 120
bergy bits 15
biological productivity ix, 232–233
blini ice *see* pancake ice
Bo Hai Sea 1, 9, 219
    Liaodong Bay 5-day ice forecast 220
boundary conditions 115–116, 204
brash ice 15
buoy *see* drift buoy; Argos buoy; IABP

calving 15
Campbell, William J. 200
Canadian Archipelago, drift ice thickness 30

Index

Canadian Ice Service 232
Cauchy equation of motion 110
central pack 15
channel flow 173–177
   creep 174–175
   geometry of 174
   plastic flow 175–177
Chezy formula 147
circular ice drift 187–191
conservation law *see* ice conservation law
conservation of kinetic energy/divergence/
   vorticity 116–118
continuum deformation
   strain and rotation 47–49
   strain rate and vorticity 49–52, 59–67
corner reflector mast 6
Coriolis effect 1, 5–6, 110, 112, 114, 118,
   134–135, 229
crack 15, 66
creep 174–175

dead water 127
deformation structures 65–67
deformed ice 15
   total thickness 42
Deutschland 1, 52
dimension analysis *see* scale and dimension
   analysis
divergence 163–164
   conservation of 116–118
draft aspect ratio 134
drag force formulae 125–130
   linear drag law 129–130
   neutral stratification 126–128
   non-neutral stratification 128–129
drifter array 63
drift bouy 52–54, 61
   Southern Ocean 61
drift
   see also kinematics
   observations of 52–59, 111
   regimes 116
   zonal sea ice drift 177–191
     circular ice drift 187–191
     configuration of 178
     MIZ 185–187
     steady-state ice thickness and
       compactness profiles 182–184

     steady-state velocity: wind-driven case
       178–182
     steady state with ocean currents 182
     viscous models 184–185
drifting stations 55, 60, 136
drilling 30

East Greenland Current 230
ecology 5
Ekman angle 125
Ekman layer *see* planetary boundary layer
Ekman pumping 118
Ekman spiral 122–123
Ekman, Vagn Walfrid 142
equation of motion 109–139
   derivation of 109–118
   drag force formulae 125–130
   dynamics of a single ice floe 138–139
   planetary boundary layer 118–125
   scale and dimension analysis 130–139
ERS-1 4
European Space Agency (ESA) 4, 18

fast ice 11, 16, 171–173
finger rafting 15
first-year ice 11
Fletcher's ice island 3, 152
forces in drifting sea ice 115
fracture 15
Fram 1, 46, 141
Fram strait 230
   drift ice thickness distribution 35
Franz Joseph's Land 20
frazil ice 11, 176
free drift 141–164
   advection 163
   linear coupled ice–ocean model 154–161
     free drift velocity spectrum 159–161
     general solution 155–156
     inertial oscillations 156–158
     periodic forcing 158–159
   non-steady-state solution 149–154
     drift of a single floe 152–154
     linear model 150–151
     one-dimensional flow with quadratic
       surface stresses 149–150
   steady-state solution 141–149
     classical case, general solution 143–146

linear model 148–149
  one-dimensional channel flow 146–148
  shallow waters 148
freshwater 5
Froude number 134–135
friction *see* internal friction
friction number 134

geostrophic flow 125
geostrophic sea ice drift 142
global warming ix
Gothus, Olaus Magnus 2
granular flow models 103–107
gravity 110
Great Lakes 6
Greenland Sea 5, 9, 230
  drift ice thickness 30, 32
grounded ice 14
ground-penetrating radar (GPR) 33
growlers 15
Gulf of Bothnia 38
  simulated sea level elevation 215
Gulf of Riga 9, 560057
Gulf of St Lawrence 55

Hooke's medium 83
hummocked ice x, 10, 15, 30, 41–42

icebergs 15, 230
ice blocks 19
ice bridges 1–2
ice concentration/compactness 10, 14–15
  Antarctic 17
  compactness vs. drift speed 166–167
ice conservation law 72–80
ice edge 15, 186
  upwelling 187
  zone 16
ice engineering 232, 234, 236
ice field 14
ice floe 10, 15, 19
  clustering 164
  collision 85–86
  dynamics 138
  free drift 152–154
  rotation rate 57–59

scale 19–22
shape 26
size distribution 22–26
ice islands 15
Iceland 17
ice thickness 9
ice state 4, 73–75
ice types 10
internal friction 3, 87–88, 165–197
  channel flow 173–177
  frequency spectrum 170–171
  fast ice 171–173
  modelling of ice tank experiments 191–197
  non-linear 169
  zonal sea ice drift 177–191
internal stress of drift ice 84–87, 110
International Arctic Buoy Programme (IABP) 3, 57
isobaric drift law 3

Jermak 52

Kara Sea 4
Kelvin–Voigt medium 84
kinematics 45–80, 145
  continuum deformation
    strain and rotation 47–49
    strain rate and vorticity 49–52, 59–67
  deformation structures 65–67
  ice conservation law 72–80
  observations of sea ice 52–59
  single floe motion 46–47
  stochastic modelling 67–72
  velocity field 45–52
kinetic energy
  budget 118
  conservation of 116–118
Kirillov's formula 42

Labrador Current 230
Laikhtman model 185
Lake Ladoga 6
Lake Vänern 6
landfast ice *see* fast ice
laser geodimeter 6
latent heat 5
leads x, 10, 15, 231
level ice
lotus ice *see* pancake ice

Manning's formula 147
marginal ice zone (MIZ) 16, 64, 177,
    185–187, 216
Marginal Ice Zone Experiment (MIZEX)
    4–5, 59, 63–64, 98, 167–168
Marine sediments 6
Markov process model 3
Maud 52
mechanical growth of drift ice 29–30
Mellor, Malcolm 166
modelling
    numerical modelling 199–227
        calibration/validation 205–206
        Eulerian/Langrangian frames 202–203
        geophysical parameters 201
        grids 203–204
        ice tank experiments 191–197
        initial and boundary conditions 204
        long-term modeling applications
            222–227
        numerical design parameters 201
        numerical integration 205
        numerical technology 202–205
        short-term modelling applications
            214–222
        system of equations 199–202
    plastic models 91–103
        AIDJEX elastic–plastic rheology 95–97
        Hibler's viscous–plastic rheology 97–100
        Mohr–Coulomb rheology 94–95
        plastic drift ice 91–94, 175–177
        scaling of ice strength 100–103
    sea ice dynamic models
        AIDJEX 211–212
        Baltic Sea 212
        Campbell and Cronin 206–208
        Hibler 208–211, 223
    stochastic modelling 67–72
        2-D motion using complex variables
            67–69
        diffusion 71
        random walk 71
Monin–Obukhov similarity theory 128
Mt Oyama 58
multi-year ice 11

Nansen–Ekman drift law 3, 199
Nansen, Fridtjof 46

Nansen number 134
National Snow and Ice Data Centre
    (NSIDC) 17, 19
Navier–Stokes equation 81
new ice 11
Newton's medium 83
Newton's second law 109
Niagara River 6
nilas 11, 15
Nimbus 54
North Atlantic Oscillation 230
Northern Research and Trade Expedition 3
Northern Sea Routes x, 3, 232
North Pole drifting station 3
    programme 4
    Soviet Union North Pole drifting station
        130
Nova Zemlya 20
numerical modelling 199–227
    calibration/validation 205–206
    Eulerian/Langrangian frames 202–203
    geophysical parameters 201
    grids 203–204
    ice tank experiments 191–197
    initial and boundary conditions 204
    long-term modelling applications 222–227
        Antarctic regions 224
        Arctic regions 222–224
        Baltic Sea 224–227
    numerical design parameters 201
    numerical integration 205
    numerical technology 202–205
    short-term modelling applications 214–222
        research work 214–217
        sea ice forecasting 217–222
    system of equations 199–202

oil spill ix, 221–222
Okhotsk Sea x, 3, 19, 58

pack ice *see* drift ice
paleoclimatology 5
paleoceanography 5
pancake ice 13, 15
Papanin, Ivan 130
passive microwave imagery 4, 17, 33
perennial sea ice zone 9
piling 30

planetary boundary layer 118–125
   Ekman layer 121–123
   stratification 123–125
   surface layer 120–121
plastic models 91–103
   AIDJEX elastic–plastic rheology 95–97
   Hibler's viscous–plastic rheology 97–100
   Mohr–Coulomb rheology 94–95
   plastic drift ice 91–94, 175–177
   scaling of ice strength 100–103
plate ice *see* pancake ice 13
pollution ix
polynya 15, 176, 231
Prandtl's mixing length hypothesis 120
productivity *see* biological productivity

Radarsat 18
rafted ice 15, 29
random breakage model 25–26
Reiner–Rivlin fluid model 88
remote-imaging stations 54, 58
remote-sensing satellites 3
Reynolds number 125
rheology 81–107, 229
   granular flow models 103–107
   internal friction 87–88
   internal stress of drift ice 84–87
   models 83–84
   plastic laws 91–103
      AIDJEX elastic–plastic rheology 95–97
      Hibler's viscous–plastic rheology 97–100
      Mohr–Coulomb rheology 94–95
      plastic drift ice 91–94
      scaling of ice strength 100–103
   viscous laws 88–91
      linear viscous models 88–90
      non-linear viscous models 90–91
ridges x, 10, 15–16, 35–42, 95
   size 39
   spacing 39–40
   structure 35–38
   ridging measures 40–41
Rossby number 124, 134–135
rotation 47–49
roughness length 120
R/V Aranda 154
R/V Polarbjørn 5, 63

Sakhalin Island x
Sanderson's curve 102
scale and dimension analysis 130–139
   basin scales 137–138
   dimensionless form 133–137
      stationary ice 136–137
   magnitude 130–133
scanning multichannel microwave
      radiometer (SMMR) 54
Sea Ice Mechanics Initiative (SIMI) 4
sea ice 1, 11, 14, 19
   modelling in a tank 191–197
sea ice banding 65, 68
sea ice basins 11
sea ice charting 17–19, 33
   Arctic Ocean 18
sea ice cover 9–19
sea ice cracks 15, 66
sea ice distribution 59
sea ice dynamic models
   AIDJEX 211–212
   Baltic Sea 212
   Campbell and Dronin 206–208
   Hibler 208–211, 223
sea ice field 19–20
sea ice information services 232
sea ice landscape 10–15
sea ice particles 21–22
sea ice state 42–44
sea ice thickness 27–35
   distribution 34–35, 75–80, 229
   measurement method 30–34
   mechanical growth 29–30
   observations of 30
   thermal growth 27–28
sea ice velocity 5, 45–52
   Arctic Ocean 62, 69–70
   Baltic Sea 61, 62
sea ice zones 12, 15–17
sea surface slope 167–168
seal hunting 1–2
seasonal sea ice zone (SISZ) 9
sequential remote-sensing imagery 54, 56–57
shear zone 16
shipping x
Shuleikin, Vasilii Vladimirovich 110
Siberian Shelf 230
   drift ice thickness 30
sonar, upward looking 30

Southern Ocean 9
Soviet Union North Pole drifting station 130
special sensor microwave imager (SSM/I) 17, 54
strain 47–49, 83
strain rate 49–52, 59–67, 83, 168
stratification
    neutral stratification 126–128
    non-neutral stratification 128–129
    planetary boundary layer 123–125
stress 5, 81–83
    see also internal stress of drift ice
stochastic modelling 67–72
    2-D motion using complex variables 67–69
    diffusion 71
    random walk 71
Strouhal number 134–135
St Venant's medium 83
surface layer see planetary boundary layer
Svalbard 20
synthetic aperture radar (SAR) 4, 18, 33, 35, 54

T-3 3, 152
Tabata, Tadashi 82
Terra 57
thermal growth of drift ice 27–28
thermal infrared mapping 33
Transpolar Drift Stream 55, 61, 206
turning angle see Ekman angle

undeformed ice see level ice
upward-looking sonar 30, 35
US Defense Meteorological Satellite Programme (DMSP) 17

viscous laws 88–91
    linear viscous models 88–90
    non-linear viscous models 90–91
von Karman's constant 120
vorticity 49–52, 163–164
    conservation of 116–118

Weddell-1 (US–Russian Ice Station) 4
Weddell Sea 1, 4, 9, 28
    sea ice thickness 225
    sea ice velocity 225

young ice 11

zonal sea ice drift 177–191
    circular ice drift 187–191
    configuration of 178
    MIZ 185–187
    steady-state ice thickness and compactness profiles 182–184
    steady-state velocity: wind-driven case 178–182
    steady state with ocean currents 182
    viscous models 184–185
Zubov, Nikolaevich 10
Zubov's isobaric drift rule 143

Printing: Mercedes-Druck, Berlin
Binding: Stein+Lehmann, Berlin